Lecture Notes in Computer Science 14987

Founding Editors

Gerhard Goos
Juris Hartmanis

Editorial Board Members

Elisa Bertino, *Purdue University, West Lafayette, IN, USA*
Wen Gao, *Peking University, Beijing, China*
Bernhard Steffen ⓘ, *TU Dortmund University, Dortmund, Germany*
Moti Yung ⓘ, *Columbia University, New York, NY, USA*

The series Lecture Notes in Computer Science (LNCS), including its subseries Lecture Notes in Artificial Intelligence (LNAI) and Lecture Notes in Bioinformatics (LNBI), has established itself as a medium for the publication of new developments in computer science and information technology research, teaching, and education.

LNCS enjoys close cooperation with the computer science R & D community, the series counts many renowned academics among its volume editors and paper authors, and collaborates with prestigious societies. Its mission is to serve this international community by providing an invaluable service, mainly focused on the publication of conference and workshop proceedings and postproceedings. LNCS commenced publication in 1973.

Heiko Hamann · Marco Dorigo ·
Leslie Pérez Cáceres · Andreagiovanni Reina ·
Jonas Kuckling · Tanja Katharina Kaiser ·
Mohammad Soorati · Ken Hasselmann ·
Eduard Buss
Editors

Swarm Intelligence

14th International Conference, ANTS 2024
Konstanz, Germany, October 9–11, 2024
Proceedings

Editors
Heiko Hamann
University of Konstanz
Konstanz, Germany

Marco Dorigo
IRIDIA, Université Libre de Bruxelles
Brussels, Belgium

Leslie Pérez Cáceres
Pontificia Universidad Católica de Valparaíso
Valparaíso, Chile

Andreagiovanni Reina
University of Konstanz
Konstanz, Germany

Jonas Kuckling
University of Konstanz
Konstanz, Germany

Tanja Katharina Kaiser
University of Technology Nuremberg
Nuremberg, Germany

Mohammad Soorati
University of Southampton
Southampton, UK

Ken Hasselmann
Royal Military Academy
Brussels, Belgium

Eduard Buss
University of Konstanz
Konstanz, Germany

ISSN 0302-9743   ISSN 1611-3349 (electronic)
Lecture Notes in Computer Science
ISBN 978-3-031-70931-9   ISBN 978-3-031-70932-6 (eBook)
https://doi.org/10.1007/978-3-031-70932-6

© The Editor(s) (if applicable) and The Author(s), under exclusive license
to Springer Nature Switzerland AG 2024

This work is subject to copyright. All rights are solely and exclusively licensed by the Publisher, whether the whole or part of the material is concerned, specifically the rights of translation, reprinting, reuse of illustrations, recitation, broadcasting, reproduction on microfilms or in any other physical way, and transmission or information storage and retrieval, electronic adaptation, computer software, or by similar or dissimilar methodology now known or hereafter developed.
The use of general descriptive names, registered names, trademarks, service marks, etc. in this publication does not imply, even in the absence of a specific statement, that such names are exempt from the relevant protective laws and regulations and therefore free for general use.
The publisher, the authors and the editors are safe to assume that the advice and information in this book are believed to be true and accurate at the date of publication. Neither the publisher nor the authors or the editors give a warranty, expressed or implied, with respect to the material contained herein or for any errors or omissions that may have been made. The publisher remains neutral with regard to jurisdictional claims in published maps and institutional affiliations.

This Springer imprint is published by the registered company Springer Nature Switzerland AG
The registered company address is: Gewerbestrasse 11, 6330 Cham, Switzerland

If disposing of this product, please recycle the paper.

Preface

These proceedings contain the papers presented at ANTS 2024, the 14th International Conference on Swarm Intelligence, which took place on October 9–11, 2024. The conference was held at the University of Konstanz, Germany.

The ANTS series of conferences started in 1998 with the First International Workshop on Ant Colony Optimization (ANTS'98), held in Brussels in 1998 and organized by Marco Dorigo. Since then ANTS, which is held biennially, has gradually become an international forum for researchers in the wider field of swarm intelligence. In 2004, this development was acknowledged by the inclusion of the term "Swarm Intelligence" (next to "Ant Colony Optimization") in the conference title. Starting in 2010, the ANTS conference has been officially devoted to the field of swarm intelligence as a whole, without any bias towards specific research directions. This is reflected in the current title of the conference: "International Conference on Swarm Intelligence."

This volume contains 19 papers selected from 33 initial submissions. Of these, 14 were accepted as full-length papers, and five were accepted as short papers. This corresponds to an overall acceptance rate of about 58%. Also included in this volume are six extended abstracts. Submissions received an average of four single-blind reviews.

All papers were presented orally in a plenary session and in poster sessions. Extended versions of the full-length papers presented at the conference will be published in a special issue of the journal *Swarm Intelligence*. The conference also invited a number of researchers from biology who gave short presentations and posters on collective animal behaviour, fostering a fruitful exchange between technological and natural sciences.

We take this opportunity to thank the large number of people that were involved in making this conference a success. We express our gratitude to the authors who contributed their work and to the members of the international Program Committee, and to the additional referees for their qualified and detailed reviews.

We hope that this collection of papers is a valuable read both as a reference disseminating current research in swarm intelligence and as a starting point for future work in swarm robotics and swarm intelligence.

July 2024

Heiko Hamann
Marco Dorigo
Leslie Pérez Cáceres
Andreagiovanni Reina
Jonas Kuckling
Tanja Katharina Kaiser
Mohammad Soorati
Ken Hasselmann
Eduard Buss

Organization

Organizing Committee

General Chair

Heiko Hamann   University of Konstanz, Germany

Honorary Chair

Marco Dorigo   Université Libre de Bruxelles, Belgium

Local Organisation and Publicity Chairs

Heiko Hamann   University of Konstanz, Germany
Jonas Kuckling   University of Konstanz, Germany
Tanja Katharina Kaiser   University of Technology Nuremberg, Germany

Technical Program Chairs

Leslie Pérez Cáceres   Pontificia Universidad Católica de Valparaíso, Chile
Andreagiovanni Reina   University of Konstanz, Germany

Sponsorship Chair

Mohammad Soorati   University of Southampton, UK

Publication Chair

Ken Hasselmann   Royal Military Academy, Belgium

Paper Submission Chair

Eduard Buss   University of Konstanz, Germany

Program Committee

Ashraf Abdelbar	Brandon University, Canada
Dario Albani	Technology Innovation Institute, United Arab Emirates
Merihan Alhafnawi	Princeton University, USA
Francesco Amigoni	Politecnico di Milano, Italy
Martyn Amos	Northumbria University, UK
Farshad Arvin	Durham University, UK
Palina Bartashevich	Humboldt University of Berlin, Germany
Jacob Beal	BBN Technologies, USA
Giovanni Beltrame	École Polytechnique Montréal, Canada
Spring Berman	Arizona State University, USA
Mauro Birattari	Université Libre de Bruxelles, Belgium
Tim Blackwell	Goldsmiths, University of London, UK
Roland Bouffanais	University of Geneva, Switzerland
Darko Bozhinoski	Université Libre de Bruxelles, Belgium
Nicolas Bredèche	Université Pierre et Marie Curie, France
Christian Camacho	Université Libre de Bruxelles, Belgium
Timoteo Carletti	University of Namur, Belgium
Marco Castellani	University of Birmingham, UK
Stephen Chen	York University, Canada
Anders Lyhne Christensen	University of Southern Denmark, Denmark
Maurice Clerc	Independent Consultant, France
Leandro Coelho	Pontifícia Universidade Católica do Paraná, Brazil
Carlos Coello Coello	CINVESTAV-IPN, Mexico
Óscar Cordón	Universidad de Granada, Spain
Nicolas Coucke	Ghent University, Belgium
Michael Crosscombe	University of Tokyo, Japan
Sanjoy Das	Kansas State University, USA
Guido de Croon	Delft University of Technology, The Netherlands
Gonzalo De Polavieja	Champalimaud Foundation, Portugal
Karl Doerner	University of Vienna, Austria
Mohammed El-Abd	American University of Kuwait, Kuwait
Andries Engelbrecht	University of Stellenbosch, South Africa
Eliseo Ferrante	Vrije Universiteit Amsterdam, The Netherlands
George Fricke	University of New Mexico, USA
Hector Garcia de Marina	Universidad de Granada, Spain
José García-Nieto	University of Málaga, Spain
Simon Garnier	New Jersey Institute of Technology, USA
David Garzón Ramos	University of Bristol, UK
Ebi George	University of Lausanne, Switzerland

Carlos Gershenson	SUNY Binghamton, USA
Roderich Gross	Technical University of Darmstadt, Germany
Bahar Haghighat	University of Groningen, The Netherlands
Julia Handl	University of Manchester, UK
Kiyohiko Hattori	University of Electro-Communications, Japan
Sabine Hauert	University of Bristol, UK
Mary Katherine Heinrich	Université Libre de Bruxelles, Belgium
Mardé Helbig	Griffith University, Australia
Tim Hendtlass	Swinburne University, Australia
Edmund Hunt	University of Bristol, UK
Takashi Ikegami	University of Tokyo, Japan
Simon Jones	University of Bristol, UK
Tanja Katharina Kaiser	University of Technology Nuremberg, Germany
Andrew J. King	Swansea University, UK
Liang Li	University of Konstanz, Germany
Simone Ludwig	North Dakota State University, USA
Vittorio Maniezzo	University of Bologna, Italy
Richard Mann	University of Leeds, UK
Alcherio Martinoli	École Polytechnique Fédérale de Lausanne, Switzerland
Yi Mei	Victoria University of Wellington, New Zealand
Bernd Meyer	Monash University, Australia
Alan Millard	University of York, UK
Genki Miyauchi	University of Sheffield, UK
Nicolas Monmarché	Université de Tours, France
Sanaz Mostaghim	University of Magdeburg, Germany
Johannes Nauta	University of Padua, Italy
Frank Neumann	University of Adelaide, Australia
Kazuhiro Ohkura	Hiroshima University, Japan
Michael Otte	University of Maryland, USA
Jacopo Panerati	École Polytechnique Montréal, Canada
Konstantinos Parsopoulos	University of Ioannina, Greece
Sujit P. B.	IISER Bhopal, India
Paola Pellegrini	IFSTTAR, France
Gilbert Peterson	US Air Force Institute of Technology, USA
Tatjana Petrov	University of Konstanz, Germany
Carlo Pinciroli	Worcester Polytechnic Institute, USA
Michal Pluhacek	Tomas Bata University in Zlín, Czech Republic
Günther Raidl	Vienna University of Technology, Austria
Nicolás Rojas	UTFSM, Chile
Andrea Roli	University of Bologna, Italy
Lorenzo Sabattini	University of Modena and Reggio Emilia, Italy

Erol SahinMiddle East Technical University, Turkey
Mohammad SalahshourMax Planck Institute of Animal Behavior, Germany
Albin SalazarUniversity of Konstanz, Germany
Thomas SchmicklUniversity of Graz, Austria
Roman SenkerikTomas Bata University in Zlín, Czech Republic
Pieter SimoensGhent University, Belgium
Christine SolnonINSA Lyon, France
Mohammad SooratiUniversity of Southampton, UK
Volker StrobelUniversité Libre de Bruxelles, Belgium
Daniel StroembomLafayette College, USA
Thomas StützleUniversité Libre de Bruxelles, Belgium
Dirk SudholtUniversity of Passau, Germany
Mohamed Salah TalamaliUniversity of Sheffield, UK
Danesh TaraporeUniversity of Southampton, UK
Guy TheraulazCNRS CRCA, France
Vito TrianniISTC-CNR, Italy
Elio TuciUniversité de Namur, Belgium
Ali Emre TurgutMiddle East Technical University, Turkey
Vivek Shankar VaradharajanPolytechnique Montreal, Canada
Andrew VardyMemorial University of Newfoundland, Canada
Rolf WankaFriedrich-Alexander-Universität Erlangen-Nürnberg, Germany
Tom WenseleersKatholieke Universiteit Leuven, Belgium
Carsten WittTechnical University of Denmark, Denmark
Cheng XuUniversity of Science and Technology Beijing, China

Additional Reviewers

Cyrill BaumannÉcole Polytechnique Fédérale de Lausanne, Switzerland
Davis CathermanWorcester Polytechnic Institute, USA
Matthew FrickeUniversity of New Mexico, USA
Junyan HuDurham University, UK
Karthik SomaÉcole Polytechnique de Montréal, Canada
Kefan WuDurham University, UK
Raina ZakirUniversité Libre de Bruxelles, Belgium

Contents

Full Papers

A Comparative Study of Energy Replenishment Strategies for Robot Swarms 3
 Genki Miyauchi, Mohamed S. Talamali, and Roderich Groß

A Data-Driven Method to Identify Fault Mitigation Strategies in Robot Swarms 16
 Suet Lee and Sabine Hauert

Achieving Human-Inspired Drift Diffusion Consensus in Swarm Robotics 29
 Gal Sajko and Jan Babič

Byzantine Fault Detection in Swarm-SLAM Using Blockchain and Geometric Constraints 42
 Angelo Moroncelli, Alexandre Pacheco, Volker Strobel, Pierre-Yves Lajoie, Marco Dorigo, and Andreagiovanni Reina

Collective Bayesian Decision-Making in a Swarm of Miniaturized Robots for Surface Inspection 57
 Thiemen Siemensma, Darren Chiu, Sneha Ramshanker, Radhika Nagpal, and Bahar Haghighat

Extinguishing Wildfires in Large Scale Scenarios Using Swarms of UAVs 71
 Georgios Tzoumas, Lucio Salina, Alex McConville, Tom Richardson, and Sabine Hauert

Grasshopper Optimization Algorithm (GOA): A Novel Algorithm or A Variant of PSO? 84
 Negin Harandi, Arnout Van Messem, Wesley De Neve, and Joris Vankerschaver

Group-Level Behavioral Switch in a Robot Swarm Using Blockchain 98
 Himank Gupta, Volker Strobel, Alexandre Pacheco, Eliseo Ferrante, Enrico Natalizio, and Marco Dorigo

Heterogeneity Can Enhance the Adaptivity of Robot Swarms to Dynamic Environments 112
 Raina Zakir, Mohammad Salahshour, Marco Dorigo, and Andreagiovanni Reina

Impact of Individual Defection on Collective Motion 127
    Swadhin Agrawal, Jitesh Jhawar, Andreagiovanni Reina,
    Sujit P. Baliyarasimhuni, Heiko Hamann, and Liang Li

Minimalist Protocols for Quorum Sensing in Robot Swarms 141
    Fabio Oddi, Andreagiovanni Reina, and Vito Trianni

Self-organized Flocking in Three Dimensions 155
    Tugay Alperen Karagüzel, Fuda van Diggelen, Andres Garcia Rincon,
    and Eliseo Ferrante

Swarm-Inspired Controller: An Inference-Free Approach to Distributed
Manipulation 168
    Nicolas Bessone, Kasper Stoy, and Payam Zahadat

Swarming Out of the Lab: Comparing Relative Localization Methods
for Collective Behavior 181
    Rafael Gomes Braga, Vivek Shankar Varadharajan,
    Giovanni Beltrame, and David St-Onge

Short Papers

BittyBuzz: A Swarm Robotics Runtime for Tiny Systems 197
    Ulrich Dah-Achinanon, Emir Khaled Belhaddad, Guillaume Ricard,
    and Giovanni Beltrame

Collective Random Walks of Flocking Agents Through Emergent Implicit
Leadership 206
    Andres Garcia Rincon, Tugay Alperen Karagüzel, Fuda van Diggelen,
    and Eliseo Ferrante

Decentralized Conflict Resolution for Navigation in Swarm Robotics 215
    Sebastian Mai and Sanaz Mostaghim

On the Design of Control Mechanisms for a Site Selection Task
in a Simulated Swarm of Robots 224
    Ahmed Almansoori, Dari Trendafilov, Muhanad Alkilabi, and Elio Tuci

Development of a Pheromone-Based Aggregation Method for Swarm
Robots 233
    Atakan Botasun, Mehmet Şahin, Ali Emre Turgut, and Erol Şahin

Extended Abstracts

Ant-Search Algorithm for Distributed Knowledge Graphs 243
   Oleksandr Chepizhko, Péter Forgács, and Melanie Schranz

LARS: Light Augmented Reality System for Swarm 246
   Mohsen Raoufi, Pawel Romanczuk, and Heiko Hamann

Moving Depot (MOD): An Efficient Depot Motion Strategy
for Multi-Robot Foraging 248
   Pratik Ingle, Ananya Gandhi, and Sujit Baliyarasimhuni

Statistical Study of Worker Activity Relying on Location in Ant Colonies 251
   Masashi Shiraishi and Hiraku Nishimori

Swarm in the City: Inspirations from Urban Street Networks for Swarm
Robot Aggregation 253
   Dalia S. Ibrahim and Andrew Vardy

The Two-Bridge Ant Experiment as an Interactive NetLogo Library Model 255
   Martina Umlauft and Melanie Schranz

Author Index 257

Full Papers

A Comparative Study of Energy Replenishment Strategies for Robot Swarms

Genki Miyauchi[1(✉)], Mohamed S. Talamali[1], and Roderich Groß[1,2]

[1] Department of Automatic Control and Systems Engineering,
The University of Sheffield, Sheffield, UK
{g.miyauchi,m.s.talamali}@sheffield.ac.uk
[2] Department of Computer Science, Technical University of Darmstadt,
Darmstadt, Germany
roderich.gross@tu-darmstadt.de

Abstract. To enable long-term operations of swarms of energy-constrained robots, they need to manage both their in-flow and out-flow of energy. We consider two strategies for doing so: In the first strategy, all robots work at a remote location but due to their limited storage capacity must return to charge. In the second strategy, dedicated mobile chargers with finite storage capacity deliver energy to the remote location, substantially shortening the worker robots' commute. We compare the work performed and the energy efficiency of these strategies using physics-based simulations and reveal conditions under which their performance is close to theoretically derived upper bounds. We assess several factors, including the number of mobile chargers, their storage capacity, transfer losses, and the ratio of energy expended while working and traveling. Our findings confirm that mobile chargers can help increase the work performed, and even overall energy efficiency provided that their energy storage is larger than that of workers.

1 Introduction

When playing their part in real-world applications, swarms of robots will have to operate autonomously over extended periods of time. Examples include applications in environmental monitoring, surveillance, agriculture, construction, and mining [7,13,21]. In this context, energy is a key consideration, and both its in-flow (i.e. how a swarm replenishes its energy) and out-flow (i.e. how a swarm invests its energy) need careful consideration. At every moment of time, the robots ultimately devote some of their energy towards performing work versus securing energy to replenish.

Several studies consider swarms where the individual robots alternate between performing work and visiting charging points to replenish their finite energy storage [1,2,16,19,22]. To reduce charging times, stations offering robots to hot-swap their batteries have been considered [18]. To reduce travel times, optimized placements of (mobile) charging stations have been considered [5].

Other studies consider swarms where energy is transferred among members of finite storage capacity, a process referred to as energy trophallaxis [4,8,12,20]. Such an approach can promote division of labor, enabling some individuals to focus on certain tasks [3], while also reducing congestion near a shared resource [15] (e.g. at the charging station). To reduce transfer times, prior work [17] proposed robots that exchange energy by swapping batteries. Recent multi-robot platforms such as Freebot [6] have demonstrated low transfer times using super-capacitors, achieving duty cycles of up to 98%. However, several factors including ineffective strategies for sharing energy across the swarm or high energy transfer loss may potentially hinder uptake of this technology.

In this paper, we compare two strategies for regulating the in-flow and out-flow of energy of a swarm of robots that is required to perform work at a remote location for an extended period of time. The first strategy makes exclusive use of fixed charging points. Every robot alternates between performing work at the remote location and recharging back at the base. The second strategy introduces the use of mobile charging units of limited storage capacity, which deliver energy to the working robots and recharge back at the base. The remaining robots alternate between performing work and replenishing via energy transfer from nearby mobile chargers. We formally derive some upper bounds for the amount of work performed and energy efficiency. Through a series of physics-based simulations, we identify conditions in which either strategy becomes favorable, considering among others the cost of being idle, moving, and working, as well as the ratio of storage capacity of mobile chargers versus workers.

The paper is organized as follows. Section 2 presents the problem formulation and strategy for energy replenishing using fixed charging points, and presents a formal analysis. Section 3 presents the problem formulation and strategy for energy replenishing using mobile chargers. Section 4 describes the implementation of the strategies. Section 5 presents the results. Section 6 concludes the paper.

2 Energy Replenishing Using Fixed Charging Stations

2.1 Problem Scenario

Consider the 2D environment illustrated in Fig. 1a. It comprises three regions: (i) a *base* region (green), (ii) a *commuter* region (white), and (iii) a *work* region (red).

The environment contains a population of n_w robots (green circles), known as *workers*. At each time step, each robot can choose whether to move or remain stationary. When within the work region, the robot can choose in addition to perform work. A working robot performs work at a rate of 1 unit per time step.

Each worker has the capacity to store a maximum of $c_{w,max}$ units of energy. Its energy consumption is as follows:

1. A worker that is neither moving nor working consumes $\nu_{w,min}$ units of energy per time step. This is to support its core operations.

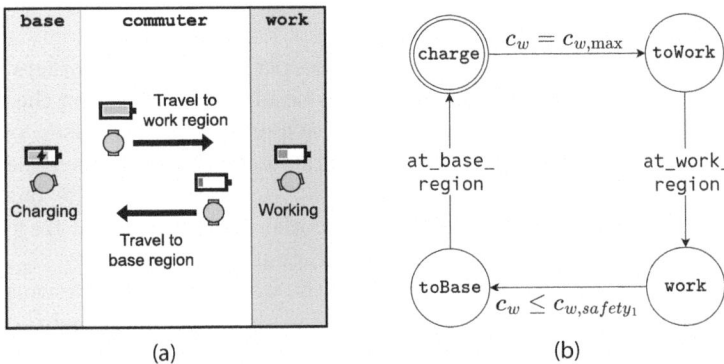

Fig. 1. Energy replenishing using fixed charging stations. (a) The workers, represented by green disks, accumulate energy in the base region (green rectangle) before moving to the work region (red rectangle) to perform work. When a worker's energy storage runs low, it moves back to the base region to recharge. (b) Finite state machine executed by each robot. (Color figure online)

2. A worker that is moving but not working consumes $\nu_{w,min} + \nu_{w,move}$ units of energy per time step.
3. A worker that is working but not moving consumes $\nu_{w,min} + \nu_{w,work}$ units of energy per time step.
4. A worker that is both moving and working consumes $\nu_{w,min} + \nu_{w,move} + \nu_{w,work}$ units of energy per time step.

While residing within the base region, a worker can accumulate (gross) energy at a rate of $\nu_{w,charge}$ units per time step. Its (net) accumulation of energy is $\nu_{w,charge} - \nu_{w,min}$ if stationary, and $\nu_{w,charge} - \nu_{w,min} - \nu_{w,move}$ otherwise. To reach the work region from the base region, the worker has to travel through the commuter region and vice versa.

We consider a mission over a finite duration τ. Initially, each worker's energy storage is assumed to be at full capacity. The workers' objective is to perform as many units of work as possible without any worker depleting their energy storage while outside the base region. As $\tau \to \infty$, theoretically, this allows the workers to keep operating autonomously for an indefinite period of time.

2.2 Strategy

Figure 1b depicts a finite state machine that is executed by each worker. Initially, all workers are assumed to reside within the base region. Hence, at time zero, the worker is charging (state **charge**). Once its level of stored energy reaches full capacity, the worker travels towards the work region (state **toWork**). Once reaching it, the worker performs work (state **work**). Once the level of its stored energy is below some safety threshold, $c_{w,safety_1}$, the worker returns to the base region (state **toBase**).

2.3 Analysis

We formally derive an upper bound for the performance of workers. In this theoretical model, we assume that (i) there is no interference among the workers' bodies (i.e., the workers may move through each other); (ii) the base, commuter, and work regions are closed sets; (iii) all workers commute between the subsets of the boundary of the commuter region that are shared with the base region and work region, respectively, and (iv) the workers follow optimal trajectories.

At time 0, the worker leaves the base region at full capacity $c_{w,max}$. Let $\Delta_{w,commute}$ denote the time that it takes for the worker to reach the work region. At time $\Delta_{w,commute}$, the worker reaches the work region at capacity $c_{w,max} - (\nu_{w,min} + \nu_{w,move})\Delta_{w,commute}$, and chooses to perform work from this moment. Once having only $(\nu_{w,min} + \nu_{w,move})\Delta_{w,commute}$ units of energy left, the worker returns to the base region, arriving at the moment its energy storage reaches zero. It then charges its level of energy to full capacity, which takes $\Delta_{w,charge} = \frac{c_{w,max}}{\nu_{w,charge} - \nu_{w,min}}$ units of time. The cycle then repeats.

In every cycle, the time a worker performs work is given by

$$\Delta_{w,work} = \frac{c_{w,max} - 2(\nu_{w,min} + \nu_{w,move})\Delta_{w,commute}}{\nu_{w,min} + \nu_{w,work}}$$

The worker's duty cycle is the proportion of time it is working. It is given by

$$D_w = \frac{\Delta_{w,work}}{\Delta_{w,cycle}}$$

where $\Delta_{w,cycle} = \Delta_{w,work} + 2\Delta_{w,commute} + \Delta_{w,charge}$ is the duration of a worker's cycle.

The system's total amount of work performed is

$$W = n_w D_w \tau \qquad (1)$$

We define the system's energy efficiency as the proportion of energy spent on performing work[1]:

$$E = 1 - \frac{(\nu_{w,min} + \nu_{w,move})2\Delta_{w,commute} + \nu_{w,min}(\Delta_{w,work} + \Delta_{w,charge})}{c_{w,max}} \qquad (2)$$

$$= 1 - \Delta_{w,cycle}\frac{\nu_{w,min}}{c_{w,max}} - 2\Delta_{w,commute}\frac{\nu_{w,move}}{c_{w,max}} \qquad (3)$$

3 Energy Replenishing Using Mobile Charging Stations

3.1 Problem Scenario

Consider the 2D environment illustrated in Fig. 2a. Compared to the scenario with fixed charging stations, we have a fourth region (blue) called *transfer* region, which sits in between the commuter and work regions.

---

[1] Note that devoting energy to core operations while performing work reduces a worker's efficiency.

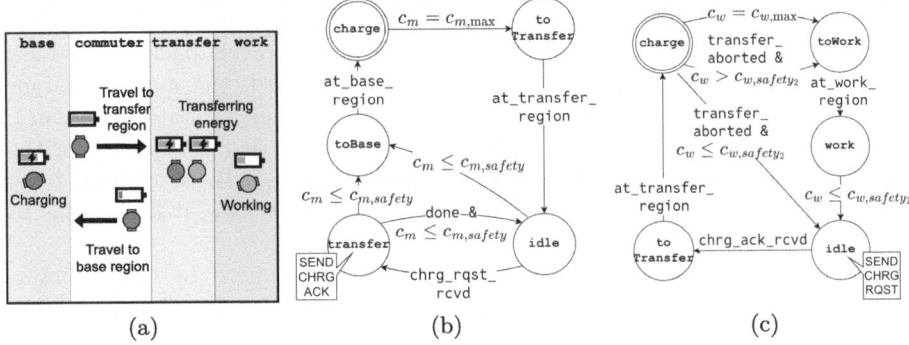

Fig. 2. Energy replenishing with mobile chargers of limited storage capacity. (a) Mobile chargers (represented as blue disks) transport energy from the base region to the transfer region (blue rectangle), where workers with low levels of energy arrive to recharge. (b) Finite state machine executed by each mobile charging unit. (c) Finite state machine executed by each working robot. (Color figure online)

The environment contains n_w workers and n_m mobile chargers. The workers have the same capabilities as described earlier. The mobile chargers can move but are unable to perform any work.

Each mobile charger has the capacity to store a maximum of $c_{m,max}$ units of energy. Its energy consumption is as follows:

1. A mobile charger that is not moving consumes $\nu_{m,min}$ units of energy per time step. This is to support its core operations.
2. A mobile charger that is moving consumes $\nu_{m,min} + \nu_{m,move}$ units of energy per time step.

While residing within the base region, a mobile charger accumulates (gross) energy at a rate of $\nu_{m,charge}$ units per time step. Its (net) accumulation of energy is $\nu_{m,charge} - \nu_{m,min}$ if stationary, and $\nu_{m,charge} - \nu_{m,min} - \nu_{m,move}$ otherwise.

While residing within the transfer region, a mobile charger can agree to donate (gross) energy at a rate of $\nu_{m,transfer}$ units per time step to a worker in that region. The worker accumulates (gross) energy at a rate of $\xi \nu_{m,transfer}$ units per time step, where $\xi \in (0,1]$ denotes the transfer loss. While energy is being transferred, both mobile chargers and workers consume energy for their core operations, and, if applicable, their movement.

3.2 Strategy

Figure 2b depicts a finite state machine that is executed by each mobile charger. Initially, all mobile chargers are assumed to reside within the base region. Hence, a mobile charger is charging at time zero (state **charge**). Once its level of stored energy reaches full capacity, the mobile charger travels towards the transfer region (state **toTransfer**). Once within the transfer region, the mobile charger pauses (state idle), waiting for a transfer request by a worker. If reaching a critical energy threshold, $c_{m,safety}$, the mobile charger travels back to the base region (state toBase). Otherwise, if a request is received, the mobile charger awaits the worker and upon arrival starts transferring energy (state transfer). The transfer is stopped as soon as the worker's storage is at full capacity, or the mobile charger's storage reaches a critical limit, $c_{m,safety}$. In the former case, the mobile charger transitions to state idle; in the latter case, it transitions to state toBase.

Figure 2c depicts a finite state machine that is executed by each worker. Initially, all workers are assumed to reside within the base region. Hence, a worker is charging at time zero (state charge). Once its level of stored energy reaches full capacity, the worker moves towards the work region (state toWork). When within the work region, the worker performs work (state work). If the level of its stored energy gets below some safety threshold, $c_{w,safety_2}$, the worker suspends work (state idle) and (repetitively) requests an energy transfer. When its request gets acknowledged by a mobile charger, the worker approaches this charger (state toTransfer), and the transfer starts (state charge). Once its storage reaches full capacity, the worker moves towards the work region (state toWork). If the transfer is aborted prior to reaching full capacity, the worker probes whether the stored energy exceeds the safety threshold, $c_{w,safety_2}$. If it does, it moves towards the work region (state toWork). Otherwise, it seeks a transfer from a different mobile charger (state idle).

4 Implementation

To evaluate the energy replenishment strategies, we implement the state machines depicted in Figs. 1b, 2b and 2c on the e-puck [11] platform. The latter is a mobile differential-wheeled robot of diameter 7 cm and maximum speed $v_{\max} = 12$ cm/s. We assume the robot is equipped with a range-and-bearing system enabling relative localization and communication within a local neighborhood of radius 0.533 m. All robots update their states using the state machines and use virtual forces to determine their direction of movement. Let $\mathbf{p}_{ij}$ denote robot j's position in the local coordinate system of robot i. The virtual force of robot i is given by

$$\mathbf{u}_i = \alpha \mathbf{u}_i^a + \beta \mathbf{u}_i^{rn} + \gamma \mathbf{u}_i^{ro}$$

where α, β and γ are positive scalars to weigh the influence of the components. Component $\mathbf{u}_i^a$ represents the attraction towards a goal, $\mathbf{g}_i$. It is defined as

$$\mathbf{u}_i^a = \frac{\mathbf{g}_i}{||\mathbf{g}_i||} \min(||\mathbf{g}_i||, v_{\max})$$

where goal $\mathbf{g}_i$ provides the position vector of a point in the base region, work region, or transfer region, respectively, where all vectors are defined relative to the position of robot i.

Component $\mathbf{u}_i^{rn}$ represents the repulsion from neighboring robots. It is defined as

$$\mathbf{u}_i^{rn} = -\frac{1}{|\mathcal{N}_i|} \sum_{j \in N_i} \frac{\sigma^\lambda}{||\mathbf{p}_{ij}||^\lambda} \frac{\mathbf{p}_{ij}}{||\mathbf{p}_{ij}||}$$

where $\mathcal{N}_i$ denotes the set of robots in the neighbouring of robot i, σ is the desired separation between robots and λ is the exponent.

To avoid collisions with obstacles such as the walls, component $\mathbf{u}_i^{ro}$ uses the e-puck's eight proximity sensors which are distributed around the robot's circumference [11]. It is defined as

$$\mathbf{u}_i^{ro} = -\frac{1}{8d_{\max}} \sum_{j=1}^{8} (d_{\max} - d_j)\hat{\mathbf{v}}_j$$

where $d_{\max} = 10\,\text{cm}$ is the range of the proximity sensors, d_j is the distance extracted from the jth sensor and $\hat{\mathbf{v}}_j$ is the unit vector pointing from the robot's center to the jth sensor. Where sensor j detects no object, we set $d_j = d_{\max}$.

5 Results

We use physics-based simulations that consider collisions among robots to quantify the performance of the energy replenishment strategies in terms of both work performed and energy efficiency. Simulations are performed in ARGoS [14] with the physics and state machines updated every 0.1 s. The arena has a dimension (H×W) of 1.6 m×1.6 m and is bounded within $x, y \in [-0.8, 0.8]$. The base, work, and transfer regions are each of dimension 0.3×1.6 m. We use $\alpha = 1$, $\beta = 100$, $\gamma = 1$, $\sigma = 0.1$ m and $\lambda = 24$ for the robot's motion and no transfer loss ($\xi = 1$) per default. Trials are terminated after 10 min. Video recordings of the simulation can be found at [9]. The source code can be found at [10].

5.1 Fixed Vs Mobile Charging Stations

For both strategies, an identical amount of workers is used, while for one strategy, mobile chargers are used as well. This is motivated by the observation that replacing a worker with a mobile charger does under no circumstances increase the amount of work being performed. On the contrary, it typically leads to a reduction[2]. Mobile chargers could nevertheless be useful as they are not required to perform any work and hence do not need potentially expensive work tools. They could be designed to move efficiently and offer increased energy storage.

We conduct trials with $n_w = 6$ workers. For the mobile charger strategy, we vary the number of chargers $n_m = \{1, 2, 4, 6, 8, 10, 12\}$ and their storage capacity $c_{m,max} = \phi c_{w,max}$ where $\phi = \{1, 1.5, 2, 4, 6\}$ to examine their effect on the aforementioned metrics.

---
[2] This is because the transfer of energy from mobile chargers to workers requires time, and during this time all parties consume a base level of energy.

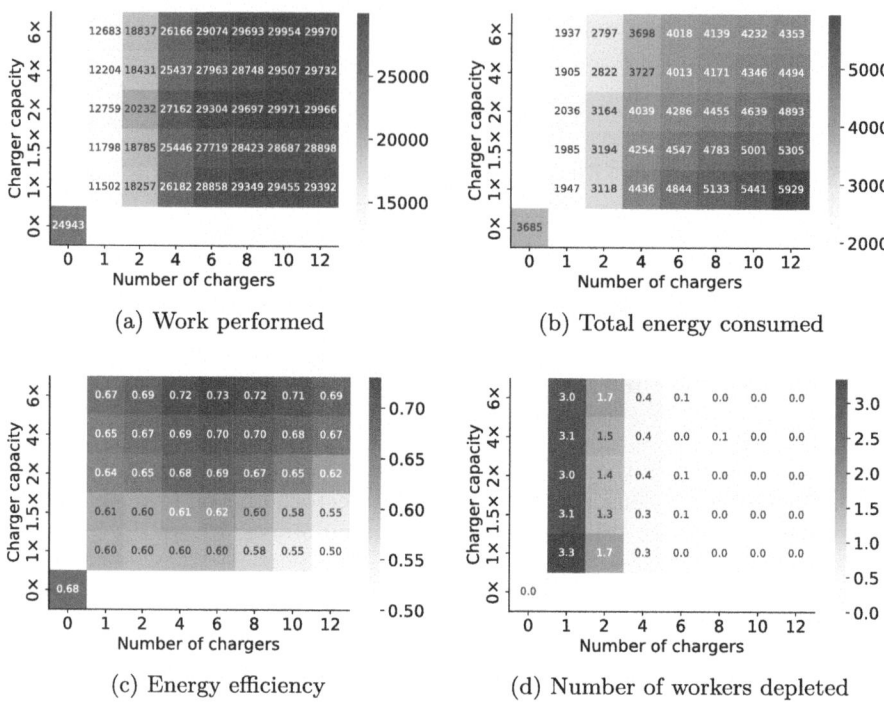

Fig. 3. The effect of the number of mobile chargers and their storage capacity on (a) the amount of work performed, (b) the total energy consumed, (c) the proportion of energy used to perform work, and (d) the number of depleted workers. The results of the static charger strategy are shown in 0 chargers and 0× charger capacity cells. Each configuration was tested for 50 trials.

Figure 3a shows that 24943 units of work are performed if the workers charge at the base. This is 92.8% of the theoretical upper bound, obtained from Eq. (1), suggesting that the workers perform close to optimal in embodied simulations. When workers obtain their energy from mobile chargers, the amount of work performed increases with the number of mobile chargers, n_m, and plateaus after around $n_m = n_w$. The work performed with 6 and 12 mobile chargers, respectively, is 89.3% and 92.0% of what could optimally be achieved if the transfer region offered a constant supply of energy, which is equivalent to setting $\Delta_{w,commute} = 0$ in Eq. (1). Performing substantially more work would require additional workers.

Figure 3b shows the total energy consumed by all robots, whereas Fig. 3c shows the energy efficiency, that is the proportion of total energy consumed that was devoted to performing work. Workers that charge at the base achieve an energy efficiency of 68.2%. This is only 12.6% less efficient than the upper bound obtained from Eq. (3). For workers obtaining energy from mobile chargers, the highest energy efficiency is consistently observed when $n_m = n_w$. This

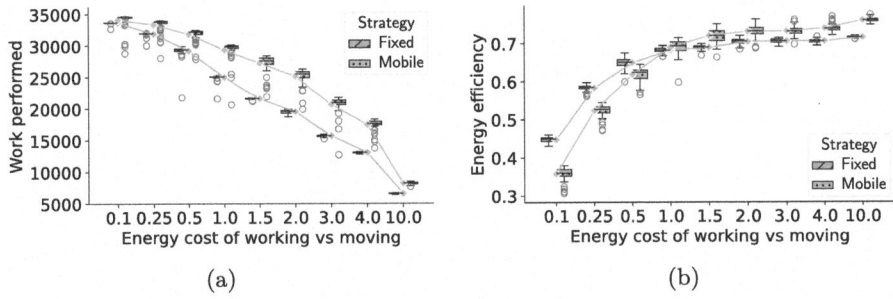

Fig. 4. Comparison of the two strategies across different energy consumption ratios $\nu_{*,work}/\nu_{*,move}$. The solid lines connects the average values indicated by the diamond marker for each strategy. For *mobile* chargers, results are shown for 6 chargers and 2× charger capacity. (a) The amount of work performed and (b) the proportion of energy used to perform work. Each configuration was tested for 50 trials.

suggests that using fewer or additional mobile chargers would result in them or the workers wasting energy due to the time they spend waiting for each other. Moreover, deploying many mobile chargers with small capacity proves inefficient. Nevertheless, mobile chargers were able to help perform more work and even improve energy efficiency when their storage capacity was larger than the workers.

Figure 3d reveals that a small amount of mobile chargers is unable to supply sufficient energy to all workers, even if they are of high capacity. This is because each mobile charger can serve only a single robot at a time, which can cause some robots to deplete all their energy while waiting in the transfer region.

5.2 Impact of Workload

In the previous section, the rates of energy for performing work and for moving were assumed equal (i.e., $\nu_{*,move} = \nu_{*,work}$). Here, we investigate the impact of changing the energy rate required to perform work. We use $n_m = 6$ and $c_{m,max} = 2c_{w,max}$. This setting performed more work while maintaining a similar energy efficiency to the strategy without mobile chargers (see Fig. 3c).

Figure 4a shows that as performing work becomes more energy-intense, less work is being performed. Furthermore, the strategy with mobile chargers outperforms the strategy without them. Figure 4b shows that using mobile chargers results in higher energy efficiency when performing work that costs as much or more energy than moving. Otherwise, having workers return to the base for recharging is more energy-efficient.

5.3 Impact of Charging Rate

We explore the effect of the rate at which the agents charge and transfer energy. For the mobile charger strategy, we use $n_m = 6$ and $c_{m,max} = 2c_{w,max}$.

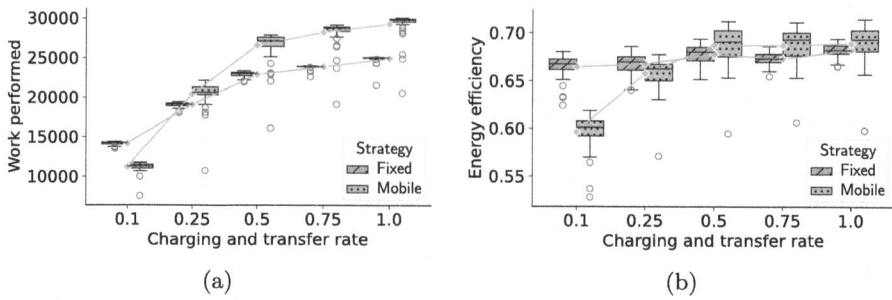

Fig. 5. Comparison of the two strategies across different charging and transfer rates, $\nu_{*,charge} = \nu_{m,transfer}$. The solid lines connect the average values indicated by the diamond marker for each strategy. For *mobile* chargers, results are shown for 6 chargers and 2× charger capacity. (a) The amount of work performed and (b) the proportion of energy used to perform work. Each configuration was tested for 50 trials.

Figure 5a shows that, for both strategies, the amount of work performed decreases as the charging and transfer rates, $\nu_{*,charge}$ and $\nu_{m,transfer}$, decreases. This is because robots need to spend a longer time to charge, reducing the time available for working.

Figure 5b illustrates that the energy efficiency using fixed chargers remains fairly constant, presumably as movement dominates charging both in terms of duration and rate of energy consumption [see Eq. (3)]. In contrast, when using mobile chargers, the energy efficiency, while fairly constant at high values of $\nu_{*,charge}$, drops for $\nu_{*,charge} \leq 0.25$. As the charging times start to dominate, this disproportionately impacts the strategy with mobile chargers, as it requires twice the time for workers to charge.

5.4 Impact of Energy Transfer Loss

In this section, we introduce energy transfer loss, where a proportion of energy dissipates to the environment when mobile chargers transfer energy to workers. We examine $n_m = 6$ and $c_{m,max} = 2c_{w,max}$.

Figure 6a reveals a non-monotonic dependency. The work performed first decreases as the transfer loss increases up to 40%, then increases at 50%, and finally decreases again from 60%. Upon close inspection of the simulation runs, it was observed that this was caused by the particular timings of the chargers and workers' activities. At a transfer loss of 50%, the mobile chargers could only transfer available energy to a single worker and then had to return to the base. At a transfer loss of 40%, they attempted to transfer energy to a second worker but had to abort the process prematurely to return to the base.

As expected, Fig. 6b shows that transfer loss negatively affects energy efficiency when using mobile chargers as some of the energy is lost while transferring energy to workers.

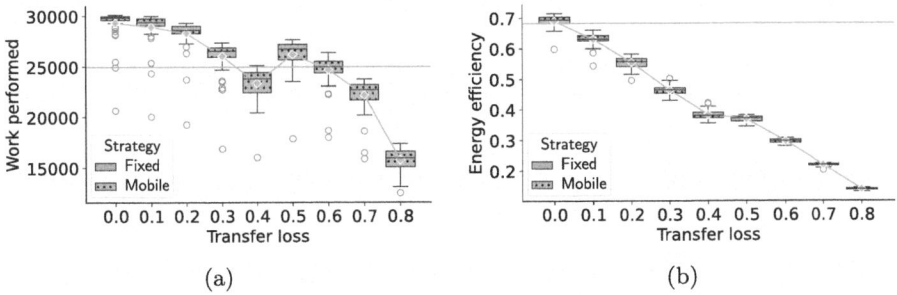

Fig. 6. Comparison of the two strategies for different levels of transfer losses ξ. The solid line connects the average values indicated by the diamond marker for each strategy. For *mobile* chargers, results are shown for 6 chargers and 2× charger capacity. (a) The amount of work performed and (b) the proportion of energy used to perform work. Each configuration was tested for 50 trials.

6 Conclusion

This paper compared two energy replenishment strategies for a swarm of robots. In the first strategy, all robots were responsible for their energy and commuted between charging stations and remote locations of work. In the second strategy, robots capable of performing work at remote locations depended on the delivery and transfer of energy from mobile chargers. We compared the two strategies based on the amount of work performed and energy efficiency using physics-based simulations and revealed conditions under which their performance is close to theoretically derived upper bounds. Results show that mobile chargers are beneficial when there is a sufficient, though not excessive, number of them and when they have a large energy capacity. Moreover, they are beneficial when more energy is required while performing work than while navigating. However, slow recharging rates as well as energy transfer losses negatively affect both the amount of work performed and energy efficiency.

Future work will investigate improved theoretical upper bounds for using mobile chargers, propose a hybrid strategy to recharge the robots, consider faster mobile chargers optimized for navigation, individual differences in battery health and validate our findings on a physical robotic platform.

Acknowledgements. This document is issued within the frame and for the purpose of the OpenSwarm project. This project has received funding from the European Union's Horizon Europe Framework Programme under Grant Agreement No. 101093046. Views and opinions expressed are however those of the author(s) only and the European Commission is not responsible for any use that may be made of the information it contains.

Disclosure of Interests. The authors declare no competing interests.

References

1. Chour, K., Reddinger, J.P., Dotterweich, J., Childers, M., Humann, J., Rathinam, S., Darbha, S.: An agent-based modeling framework for the multi-UAV rendezvous recharging problem. Robot. Auton. Syst. **166**, 104442 (2023)
2. Kannan, B., Marmol, V., Bourne, J., Dias, M.B.: The autonomous recharging problem: formulation and a market-based solution. In: 2013 IEEE International Conference on Robotics and Automation, pp. 3503–3510. IEEE (2013)
3. Kreider, J.J., Janzen, T., Bernadou, A., Elsner, D., Kramer, B.H., Weissing, F.J.: Resource sharing is sufficient for the emergence of division of labour. Nat. Commun. **13**(1), 7232 (2022)
4. Kubo, M., Melhuish, C.: Robot trophallaxis: managing energy autonomy in multiple robots. In: Proceedings of Towards Autonomous Robotic Systems, pp. 77–84 (2004)
5. Li, G., Svogor, I., Beltrame, G.: Long-term pattern formation and maintenance for battery-powered robots. Swarm Intell. **13**(1), 21–57 (2019)
6. Liu, M., et al.: Demo abstract: FreeBot, a battery-free swarm robotics platform. In: The 21st ACM Conference on Embedded Networked Sensor Systems (SenSys 2023), 12–17 November 2023, Istanbul, Turkiye (2023)
7. Marques, J.V.A., Lorente, MT., Groß, R.: Multi-robot systems research: a data-driven trend analysis. In: Bourgeois, J., et al. Distributed Autonomous Robotic Systems, DARS 2022, SPAR, vol. 28, pp. 537–549. Springer, Cham (2024). https://doi.org/10.1007/978-3-031-51497-5_38
8. Melhuish, C., Kubo, M.: Collective energy distribution: maintaining the energy balance in distributed autonomous robots using trophallaxis. In: Alami, R., Chatila, R., Asama, H. (eds.) Distributed Autonomous Robotic Systems, vol. 6, pp. 275–284. Springer, Tokyo (2007). https://doi.org/10.1007/978-4-431-35873-2_27
9. Miyauchi, G., Talamali, M.S., Groß, R.: Online supplementary material (2024), https://doi.org/10.15131/shef.data.25561923
10. Miyauchi, G., Talamali, M.S., Groß, R.: Robot controller source code (2024). https://github.com/genkimiyauchi/swarm-energy-replenishment
11. Mondada, F., et al.: The e-puck, a robot designed for education in engineering. In: Proceedings of the 9th Conference on Autonomous Robot Systems and Competitions, vol. 1, no. 1, pp. 59–65 (2009)
12. Moonjaita, C., Philamore, H., Matsuno, F.: Trophallaxis with predetermined energy threshold for enhanced performance in swarms of scavenger robots. Artif. Life Robot. **23**(4), 609–617 (2018)
13. Pearson, S., et al.: Robotics and autonomous systems for net zero agriculture. Current Robot. Rep. **3**(2), 57–64 (2022)
14. Pinciroli, C., et al.: ARGoS: a modular, parallel, multi-engine simulator for multi-robot systems. Swarm Intell. **6**(4), 271–295 (2012)
15. Pini, G., Brutschy, A., Birattari, M., Dorigo, M.: Interference reduction through task partitioning in a robotic swarm. In: Sixth International Conference on Informatics in Control, Automation and Robotics–ICINCO, pp. 52–59 (2009)
16. Rappaport, M., Bettstetter, C.: Coordinated recharging of mobile robots during exploration. In: 2017 IEEE/RSJ international conference on intelligent robots and systems (IROS), pp. 6809–6816. IEEE (2017)
17. Schioler, H., Ngo, T.D.: Trophallaxis in robotic swarms-beyond energy autonomy. In: 2008 10th International Conference on Control, Automation, Robotics and Vision, pp. 1526–1533. IEEE (2008)

18. Vaussard, F., Rétornaz, P., Roelofsen, S., Bonani, M., Rey, F., Mondada, F.: Towards long-term collective experiments. In: Lee, S., Cho, H., Yoon, K.J., Lee, J. (eds.) Intelligent Autonomous Systems 12, pp. 683–692. Springer, Berlin, Heidelberg (2013). https://doi.org/10.1007/978-3-642-33932-5_64
19. Warsame, Y., Edelkamp, S., Plaku, E.: Energy-aware multi-goal motion planning guided by monte carlo search. In: 2020 IEEE 16th International Conference on Automation Science and Engineering (CASE), pp. 335–342. IEEE (2020)
20. Winfield, A.F., Kernbach, S., Schmickl, T.: Collective foraging: cleaning, Energy harvesting, and trophallaxis. In: Handbook of Collective Robotics. Jenny Stanford Publishing (2013)
21. Xie, L., Shi, Y., Hou, Y.T., Lou, A.: Wireless power transfer and applications to sensor networks. IEEE Wirel. Commun. **20**(4), 140–145 (2013)
22. Yu, K., Budhiraja, A.K., Buebel, S., Tokekar, P.: Algorithms and experiments on routing of unmanned aerial vehicles with mobile recharging stations. J. Field Robot. **36**(3), 602–616 (2019)

A Data-Driven Method to Identify Fault Mitigation Strategies in Robot Swarms

Suet Lee[✉] and Sabine Hauert

Bristol Robotics Laboratory, University of Bristol, Bristol, UK
{suet.lee,sabine.hauert}@bristol.ac.uk

Abstract. As robot swarms are increasingly deployed in the real-world, making them safe will be critical to improving adoption and trust. A robot swarm is composed of many individual robots each susceptible to failure at any given time, which may decrease the performance of the swarm as a whole. The ability to mitigate critical faults is therefore necessary. The difficulty with designing an effective mitigation strategy lies in the complexity of the swarm as a system, where individual interactions give rise to emergent behaviour. In this paper, we present a data-driven method to identify effective local actions available to faulty robots in the swarm. We make the assumption that robots are able to self-detect faults and that pre-coded actions are indeed available. An effective action should mitigate any negative impact of faults on overall swarm performance. We consider two intralogistics scenarios where the swarm must retrieve and deliver boxes. The first concerns single robot transport (one robot per box) and the second, collective transport (four robots per box). Our method is able to identify effective actions for particular fault types. We also consider the impact of actions across ratios of fault in the swarm. Interestingly, faults do not always benefit from mitigations, with mitigations causing overall lower system performance for certain fault types.

1 Introduction

Swarms have potential in the real-world: they are decentralized systems where an individual only has access to knowledge of its local environment [25]. The overall behaviour of the swarm emerges from local interactions. Examples of real-world applications include search and rescue [23], construction [33], and space exploration [6,26,32]. However, in taking these applications from a research environment to real-world implementation, there will need to be a consideration of safety and, in particular, a method to mitigate faults.

Traditionally, swarms have been considered to be robust through redundancy of a large number of individuals [25], although it has now been demonstrated that this assumption can be false with the presence of partial failures [3]. The complexity of swarm dynamics means that it is not obvious which mitigation will work best for a given fault. In particular, a faulty robot may not always require mitigation depending on the mode of failure. In this paper, we present a data-driven method in which, given a set of potential faults and actions available

© The Author(s), under exclusive license to Springer Nature Switzerland AG 2024
H. Hamann et al. (Eds.): ANTS 2024, LNCS 14987, pp. 16–28, 2024.
https://doi.org/10.1007/978-3-031-70932-6_2

to faulty individuals in the swarm, we are able to evaluate the effectiveness of actions against fault type. The method is data-driven in the sense that we learn on sample sizes of 1000 across 144 trial configurations. We select actions which can be implemented by faulty robots in an asynchronous and distributed manner. The individual actions taken by faulty robots give rise to an overall swarm mitigation strategy.

We focus on two variations of an intralogistics scenario, labelled *task A* and *task B*. In task A, the swarm must retrieve and deliver boxes with one robot required to transport each box. Task B adopts the same setup but four robots are required to transport a box and communication between robots is necessary for coordination. In task B, the robots in the swarm have a higher degree of interdependency than in task A. The two tasks provide a rich point of comparison as failure modes will differ in each case and, as we find, so do the actions which prove to be most effective for mitigation. We evaluate, in simulation, the impact of mitigation actions taken by faulty robots on the overall swarm performance, measured as box delivery over time. We assume that robots are aware at all times of the presence of a fault. Such detection can be done using one of our previous data-driven techniques: this work extracts salient metrics for detecting a range of fault types, metrics which can then be applied to find thresholds for faulty states in a fault detection model [16]. Finally, we examine the effectiveness of a selected mitigation when varying the faulty/non-faulty ratio in a swarm. We do so for two fault types, one of which is a partial failure and the other critical.

2 Related Work

Fault mitigation for swarms is a relatively new area of research whereas there is extensive work for single robot systems. At the level of the swarm, work has been done in ensuring that emergent swarm behaviour meets specified requirements which are defined based on the hazards and risks present in the system [1]. Considering the mitigation strategies available to individuals in the swarm, we can identify two approaches in current work: 1) direct mitigation where a fault is identified and a mitigation action taken accordingly, and 2) indirect mitigation where fault tolerance is embedded in the swarm behaviour allowing the swarm to adapt to the presence of faults.

For direct mitigation, work has been done in learning mitigation strategies through reinforcement learning and evolutionary algorithms [21]. Immune system inspired approaches have considered granuloma formation for energy transfer to recover a robot with low power supply [31]. In both cases, non-faulty robots take direct action to mitigate a fault. Mitigation actions are behaviour-based, for example a robot may lead a faulty robot to a repair station. These mitigations require explicit coordination between two or more robots. The identity of the faulty robot is also assumed to be known, and in a real-world mitigation pipeline some method of fault detection will be required. A strong body of work in fault detection and diagnosis for swarms ranges across model-based and data-driven approaches [5, 7, 15, 16, 20, 29, 30].

For indirect mitigation, faulty states do not need to be defined nor detected as there is no explicit mitigation action to apply. Instead, the swarm adapts its behaviour to overcome sub-optimal states for a given performance metric. A diverse set of behaviours, controllers for an individual in the swarm, can be generated or defined *offline* and this allows the swarm to adapt *online* by selecting the best performing behaviours which may be shared in the the collective [4,11,22]. Additionally, work has been done in learning swarm controllers which are robust to the presence of faults and perturbation in the environment [19,24]. In this approach, the ability to tolerate faults is embedded in the designed behaviours of the robots, and so there is no direct mitigation. Further, the system is able to adapt or tolerate faults without explicit fault detection. However the resulting performance may depend heavily on the *quality* and *diversity* of pre-defined behaviours where greater diversity of behaviours increases the likelihood of adapting to unknown faults. To generate such behaviours requires either expert knowledge of the system or extensive computation.

Finally, we cast an eye to fault mitigation approaches in the broader multi-robot literature which include adaptation of the robot controller to faults (indirect mitigation), and isolating a faulty robot through exclusion from the networked system (direct mitigation) [2,8–10,17,27,28]. In this work we consider direct mitigation: we present a data-driven method to select and evaluate mitigation actions targeting faults directly. In comparison to current work, we consider actions which are taken only by self-detected faulty robots, actions which do not require explicit co-ordination between robots nor intervention external to the system. We demonstrate the method can be used in different scenarios, in this work, single transport and collective transport tasks. Given an initial set of actions, our method is able to identify which actions are most effective for mitigating specific fault types for the two scenarios considered.

3 Method

3.1 Scenario

Swarms have the potential to be used out-of-the-box for intralogistics, in messy real-world environments for example [13]. In our scenario, the robots operate in a 500 cm × 500 cm bounded arena: robots must retrieve and deliver boxes to the drop-off zone, a 75 cm-length horizontal strip extending along the width of the upper boundary. Robots are able to detect objects (robot, box or wall) via ArUco tags [12]. Following previous work in our team, the parameters of the scenario have been selected to match as closely as possible the robot platform and the arena we have available, our aim being to close the reality gap with real-world tests [13,14]. Table 1 and Fig. 1 summarize the scenario configuration.

We run experiments on a C++ simulator developed for the purpose of studying the intralogistics scenario[1]. It adheres to real-world physical constraints and

---

[1] https://bitbucket.org/suet_lee/swarm_mitigation_cpp/src/master/.

Table 1. Scenario configuration

	Property	Value
Arena	Dimensions	500 × 500 cm
	Number of boxes	(Task A) 10, (Task B) 2
	Number of robots	10
	Box diameter	(Task A) 25 cm, (Task B) 70 cm
	Deposit zone	75 cm-length horizontal strip along upper boundary
Robot	Diameter	25 cm
	Cameras	4 × 120° FOV video cameras equidistant on
		perimeter, 100 cm range, used to detect objects;
		1 × 120° FOV video camera upward-facing for
		precise positioning under boxes.
	Proximity	16 x IR laser ToF, 3 m range, for collision avoidance
	Communication	Bluetooth, 100 cm range
	Robot max speed	200 cm/s (real-world time)

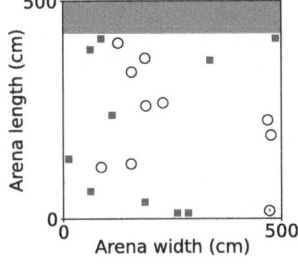 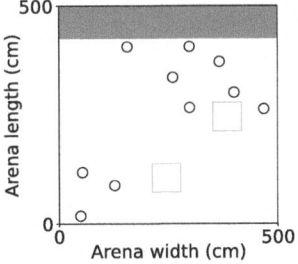

Fig. 1. Task A (left): Single transport; Task B (right): Collective transport; Arena setup: robots are represented by circles, boxes are represented by small blue squares (task A) and large orange squares (task B). The deposit zone is the green-shaded strip at the top of the arena. (Color figure online)

implements a point model with relevant speed. One simulation timestep is equivalent to 0.02 s in real-time. A further consideration made in trials is that robot and box positions are intialized at random in the simulated arena.

3.2 Task Variations

In both tasks, robots employ a random walk until they come into contact with a box. A new heading is chosen every 0.4 s. In single transport, task A, a single robot is able to pick up a detected box and will move stochastically until it enters the deposit zone. In collective transport, task B, robots must position themselves under detected boxes and wait for a complete team (four robots) to be formed. Once complete, robots in the team 'vote' for the next collective action: either *lift box, move,* or *deposit box*. Upon depositing a box, the team is disbanded and robots search for a new box. We assume one robot needs to be positioned at each corner of the box. In collective movement, each robot chooses a heading at random and communicates the chosen heading to its teammates. The final heading chosen is taken as the average of individually chosen headings. The team is assumed to move at the velocity of the slowest robot.

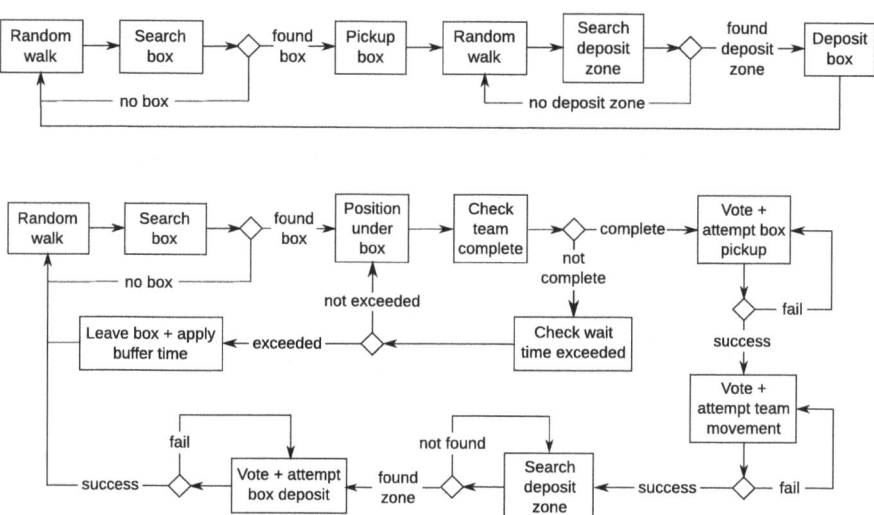

Fig. 2. Robot controllers for: (top) task A, single transport task, (bottom) task B, collective transport task.

One consideration made for the robot controller in task B is the addition of a timeout variable, t_w, which acts as a threshold for the time that a robot will wait for a complete team to form. If the threshold is exceeded, the robot will leave its position under a box. A robot that leaves a team in this manner is not able to rejoin or join a new team for a given buffer time, t_b. We outline the robot controllers for each task in Fig. 2.

Finally, we choose trial duration $t_d = 200$ seconds for task A and $t_d = 2000$ seconds for task B. It is desirable to choose a trial duration which is sufficient for the swarm to complete the tasks, up to a certain degree of faults present. For some critical failures, the swarm never completes the task. We have therefore chosen the trial duration with a view to maximise the potential performance scores of the swarm. In selecting the timeout and buffer variables for task B, we evaluate performance across a parameter sweep and find parameters $t_w = 120$ seconds, $t_b = 4$ seconds to produce a consistently high-performing baseline, with highest mean performance overall. With these selected parameters, the swarm completed the task successfully in delivering all boxes across all 100 trials. The range of parameters considered were $t_w \in \{2, 4, 6, 8, 19, 12, 14, 16, 18, 20, 40, 80, 120, 160, 200\}$ and $t_b \in \{4, 8, 12, 16, 20\}$, with units in seconds. In choosing parameters for a high-performing controller, we create a baseline against which we can compare the impact of faults and mitigations.

3.3 Fault Injection

In previous work where we injected faults in simulation, we considered a realistic sample of faults with respect to the capabilities of the robots and the scenario [16]. We consider the same set of faults in this work with the addition of communication faults in task B (e.g. failure to send or receive messages). The faults considered are based on individual robot capabilities (i.e. motors, sensors and communication modules available). We also account for granularity of faults: varying degrees of reduced speed, for example. We do not consider battery failure, however, as this can be considered to render a robot stationary and inactive, in which case it is unable to apply any action at all [34]. When evaluating mitigation actions, we cover all possible numbers of fault (from one to all faulty robots) across fault types, excluding the case of 0 faults as there are no faulty robots to apply mitigation in this case:

F1 0% maximum speed
F2 10% maximum speed
F3 50% maximum speed
F4 Perimeter camera failure
F5 Can't pick up boxes
F6 Can't deposit boxes
F7 IR laser failure
F8 Can't receive messages (task B only)
F9 Can't send messages (task B only)

It is assumed that all cameras or IR lasers fail in the cases of $F4$, $F7$. The faults are injected at the beginning of each trial and persist for the whole duration. Our aim is to align actions to specific faults, and so we only inject a single type of fault in each trial.

3.4 Mitigation Actions

Similar to our consideration of faults, the mitigation actions we consider are based on an individual's capabilities. We aim to choose distinct action "building blocks", from which strategies might emerge on the level of the swarm. We only consider actions that individual robots may take, which do not rely on interference external to the swarm:

A1 Decrease speed 50%
A2 Stop moving
A3 Don't join team (task B only)
A4 Leave team (task B only)
A5 Drop box (task A only)
A6 Bias towards nearest robot
A7 Bias towards nearest box
A8 Bias towards nearest wall
A9 Bias left
A10 Bias away from nearest robot
A11 Bias away from nearest box
A12 Bias away from nearest wall
A13 Attract robots in range
A14 Repel robots in range

A15 Stop receiving messages (task B only)   *A16* Stop sending messages (task B only)

We focus on actions which can be performed asynchronously and in a distributed manner by individual robots. In particular, movement-related actions impact spatial organization of the swarm: for example, bias movement may prevent faulty robots from becoming an obstacle to non-faulty robots. Pre-emptive actions *don't join* or *leave team* (*A3*, *A4*) directly impact team formation. Communication-related actions to *stop receiving* or *stop sending messages* (*A15*, *A16*) directly impact team action. Actions to attract or repel robots rely on hard-coded behaviours in the robot controller where a robot will respond as appropriate to an *attract* or *repel* signal from a neighbour. Choosing actions at the behavioural level may also provide a degree of human-readability as we are able to reason directly about why a mitigation action is effective or not.

In each trial, the same mitigation action is activated by all faulty robots which we assume to be known. Indeed, our previous work showed robots could self-detect the faults proposed in this paper [16]. Actions are active for the duration of the trial for the purpose of data generation, allowing for comparison against trials with no mitigation. We also consider the effect of removing faulty robots from the swarm entirely. This is not included in the action set above as it is mitigation in the form of external intervention, however, it provides a baseline for comparison. We label the action A^*.

3.5 Evaluation Measures

Performance: For each trial we evaluate performance P, defined as the integral under a step function $f(t)$, the number of boxes delivered at time $t \in [0, T]$. T is the trial duration in simulation timesteps, B is the number of boxes in the arena. We have $f(t) \in \{0, 1, ..., B\}$.

Mitigation Power: We evaluate each action independently against each type of fault using a method based on the Mann-Whitney U test and effect size analysis. We applied this test in previous work towards a measure of discriminatory power [16]. Here, we adapt the method to compare performance score samples for mitigation versus no mitigation, D_{mit} and D_{none} respectively. Each data sample is representative of performance across all numbers of possible fault. The Mann-Whitney U test evaluates the group difference of the two samples, whereas effect size analysis accounts for significance and sample size. The test makes no assumptions about the underlying distribution of samples which is a key consideration in its application in this work. Let $S(A, F)$ be the *mitigation power* of an action A for fault F:

$$S(A, F) = 2\,h \left| \frac{U}{n_1 n_2} - 0.5 \right| \tag{1}$$

$$h = \begin{cases} 1, & \text{if } R_{mit} > R_{none} \\ 0, & \text{if } R_{mit} = R_{none} \\ -1, & \text{otherwise} \end{cases} \qquad (2)$$

U is the Mann-Whitney U test statistic and n_1, n_2 are the size of the data samples for comparison. Variable h takes into account the *directionality* of group difference by comparing the rank totals for the two samples. Rank totals for each sample, R_{mit} and R_{none}, are produced by ranking the sample data in aggregate [18]. Intuitively, the formulation of $S(A, F)$ normalizes the test statistic, U, to sit in the range of values $[-1.0, 1.0]$, where a negative score indicates that a mitigation action has negative impact on performance, and a positive score indicates an improvement. This provides an easily interpretable value for our purposes of evaluating mitigation actions.

4 Results

For each simulation configuration (selecting for fault type, number of faults, and mitigation action), we run 100 trials. Figures 3 and 4 show the mitigation power of each action against fault types for task A and B.

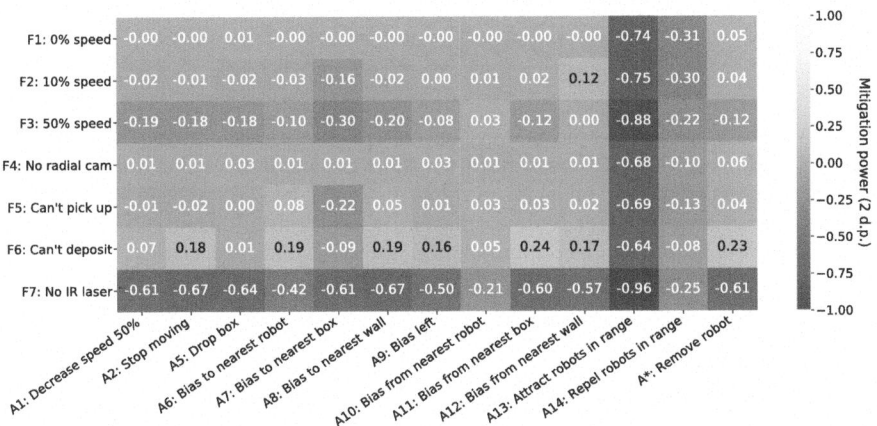

Fig. 3. Evaluation of mitigation power for task A, single transport, for actions against all fault types. For partial fault *no IR laser*, we see that all actions have negative power: it is better to take no action for this fault type. For fault *can't deposit box*, movement-related actions have positive power and are thus effective mitigation strategies.

4.1 Task A: Single Transport

Figure 3 shows mitigation power for the single transport task. Our method finds that movement-related actions have positive power for the critical failure *can't deposit box* - these may be effective in preventing faulty robots from picking up boxes which they are unable to deliver. For the partial failure *no IR laser*, all actions have negative power and in this case, no action is better than taking any action at all. We also notice that actions to attract and repel other robots have negative power across all fault types - this makes sense intuitively as these actions may disrupt the behaviour of the non-faulty individuals in the swarm.

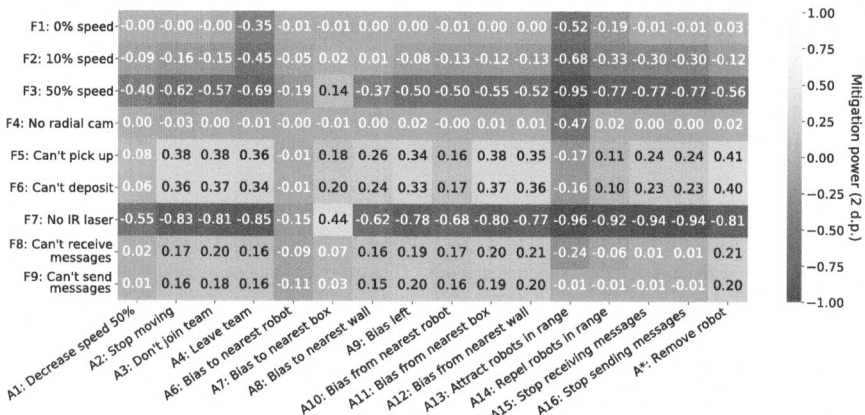

Fig. 4. Evaluation of mitigation power for task B, collective transport, for actions against all fault types. For partial faults *no IR laser* and *50% speed*, we see that all actions have negative power apart from *bias to nearest box*. We reason that the bias action aids the swarm in faster team formation, whereas in all other cases it is better to take no action. Lifter and communication faults, which are critical failures, are effectively mitigated by a wide range of actions relating to team formation and bias movements.

4.2 Task B: Collective Transport

Figure 4 shows mitigation power for the collective transport task. Critical lifter faults are effectively mitigated by faulty robots taking movement-related actions or by not engaging with a team e.g. leaving a team or not joining a team. Communication-related faults are mitigated by a similar subset of actions to a lesser degree. Additionally, we see that if faulty robots take action to stop receiving or sending messages, this is effective mitigation for lifter-related faults. Stopping communication prevents a faulty robot from participating in, and negatively impacting a team. Finally we notice that, similar to task A, for partial failure *50% speed*, all actions besides one (*bias to nearest box*) have negative power. Reduced robot speed may slow down team movement resulting in longer time to deliver boxes. However, the formation of a complete team may be the bottleneck for performance in this task - hence even a slow-moving team member may contribute positively to overall swarm performance and it is largely better to take no action. Indeed, the action *bias to nearest box* may aid in faster team formation which explains the improved performance for even partial failures.

4.3 Fault Ratios

Previously we evaluated the effectiveness of actions across the aggregate of possible number of faults that could occur in the swarm. However, we can examine more closely the performance of the swarm for specific ratios of fault. We focus on task B and the impact of mitigation action *don't join the team* (*A3*) for both critical lifter failure *can't deposit box* (*F6*) and partial failure *movement at 50% speed* (*F3*) across 1, 3, 5 and 7 faults. Figure 5 shows swarm performance scores in each case. We choose these configurations for examination as the mitigation power of the selected action has large magnitude for both fault types and it is interesting to compare a critical and partial failure.

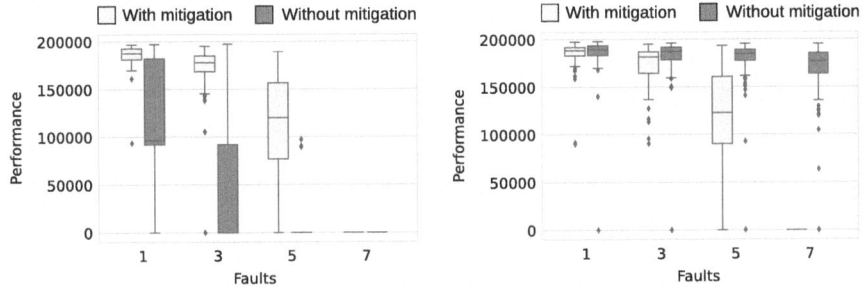

Fig. 5. Performance with and without mitigation in task B for: (left) a critical failure *can't deposit box*, (right) a partial failure *50% speed*; Performance is evaluated across ratios of fault. The mitigation action applied in both cases is *don't join team*. For the critical lifter failure, mitigating even a single robot improves performance. For the partial speed failure, mitigation starts to have large negative impact on performance for 5 faults and above. The impact of mitigation thus varies for different numbers of fault for each fault type.

Fault *can't deposit box* is a critical failure and we see in Fig. 5 (left) that mitigation of even a single faulty robot in the swarm has a large positive impact on the performance: the positive impact of mitigation increases with the number of faults, up to a limit of 7 faults at which point the swarm is unable to form any complete teams. Mitigation in the case of 7 or more faulty robots is entirely ineffective. For the partial failure, *movement at 50% speed*, we see in Fig. 5 (right) that the impact of mitigation becomes significant when we reach 5 faults. For 7 faults, mitigation results in a zero performance score since the swarm is unable to form complete teams in this case. The results in this section highlight how the impact of a mitigation action may vary when considering either a critical or partial failure, and for different ratios of fault.

4.4 Summary

We make three key observations: first, the type of mitigation which is effective will widely depend on the fault type and the task. Given an initial set of local actions, our data-driven method is able to automatically differentiate between actions which have positive mitigation power (effective mitigation), neutral (little effect) or negative power (detrimental effect to swarm performance). A faulty robot may then apply the action with highest positive power corresponding to fault type. Second, choosing the right type of mitigation may improve performance relative to baseline *no mitigation*, comparable with removing the faulty robots entirely. In the case where no effective mitigation was found, even the external action of removing the faulty robots has little or negative impact on performance compared to baseline. This suggests that self-mitigation may not be sufficient in these cases, requiring a more complex, co-ordinated mitigation strategy. Non-faulty robots may also be able to take effective action. Thirdly, some behaviours do not require mitigation or mitigation even lowers the performance relative to baseline: the swarm is thus robust to these types of fault.

5 Conclusion

We have demonstrated a data-driven method for identifying effective mitigations across fault types, for two variations of the intralogistics task: a single transport and a collective transport task. We consider actions that an individual in a robot swarm may take to mitigate self-detected faults, with the sum of individual actions optimising for overall swarm performance.

We show that where removal of a fault (an external action) improves swarm performance, a local action with corresponding mitigation power can be found. We also see that mitigation in the case of partial failures may have negative impact on swarm performance for both the single and collective transport tasks. In these cases no mitigation may generally be better than any mitigation. This is because robots with partial failures can still contribute positively to swarm performance. Performance here is defined as box delivery over time but safety considerations extend beyond reduced speed of delivery - we could also consider the number of collisions as a safety performance metric in which case, an IR laser fault might be classed as a critical failure [1].

We have also demonstrated how the impact of mitigation depends on the ratio of faults in the swarm. The analysis of fault ratios combined with mitigation power points the way to more complex fault mitigation strategies, where faulty (and non-faulty) robots may select appropriate mitigation actions based on local metrics and individual states. Whist we have selected a representative set of fault types, this may be extended by considering dynamic faults or combinations of fault types in the swarm - these examples could be considered as "fault types" themselves in which case our method is still applicable. The action set may also be extended: for example, an action where a robot *copies* the mitigation action of its neighbour. In the future, we hope to learn strategies which can adapt to new or dynamic faults, both in simulation and real-world robots.

Acknowledgements. This work was supported by the US Air Force Office of Scientific Research, European Office of Aerospace Research and Development, the UKRI Trustworthy Autonomous Systems Node in Functionality (EP/V026518/1), and the European Union under Grant Agreement 101070918 and UKRI grant number 10038942.

References

1. Abeywickrama, D.B., et al.: Aeros: assurance of emergent behaviour in autonomous robotic swarms. arXiv preprint arXiv:2302.10292 (2023)
2. Bai, Y., Wang, J.: Fault detection and isolation using relative information for multi-agent systems. ISA Trans. **116**, 182–190 (2021)
3. Bjerknes, J., Winfield, A.: On fault tolerance and scalability of swarm robotic systems **83**, 431–444 (2013). https://doi.org/10.1007/978-3-642-32723-0_31
4. Bossens, D.M., Tarapore, D.: Rapidly adapting robot swarms with swarm map-based bayesian optimisation. In: 2021 IEEE International Conference on Robotics and Automation (ICRA), pp. 9848–9854 (2021). https://doi.org/10.1109/ICRA48506.2021.9560958
5. Carrasco, R.A., Núñez, F., Cipriano, A.: Fault detection and isolation in cooperative mobile robots using multilayer architecture and dynamic observers. Robotica **29**(4), 555–562 (2011)
6. Carrillo-Zapata, D., et al.: Mutual shaping in swarm robotics: user studies in fire and rescue, storage organization, and bridge inspection. Front. Robot. AI **7**, 53 (2020)
7. Christensen, A.L., OGrady, R., Dorigo, M.: From fireflies to fault-tolerant swarms of robots. IEEE Trans. Evol. Comput. **13**(4), 754–766 (2009)
8. Gallehdari, Z., Meskin, N., Khorasani, K.: An h^∞ cooperative fault recovery control of multi-agent systems. Automatica **84**, 101–108 (2017)
9. Ghedini, C., Ribeiro, C., Sabattini, L.: Toward fault-tolerant multi-robot networks. Networks **70**(4), 388–400 (2017)
10. Guo, M., Dimarogonas, D.V., Johansson, K.H.: Distributed real-time fault detection and isolation for cooperative multi-agent systems. In: 2012 American Control Conference (ACC), pp. 5270–5275. IEEE (2012)
11. Hogg, E., Harvey, D., Hauert, S., Richards, A.: Social exploration in robot swarms. In: Bourgeois, J., et al. (ed.) Distributed Autonomous Robotic Systems, DARS 2022, SPAR, vol. 28, pp. 69–82. Springer, Cham (2024). https://doi.org/10.1007/978-3-031-51497-5_6
12. Jones, S., Hauert, S.: Frappe: fast fiducial detection on low cost hardware. J. Real-Time Image Proc. **20**(6), 119 (2023)
13. Jones, S., Milner, E., Sooriyabandara, M., Hauert, S.: Distributed situational awareness in robot swarms. Adv. Intell. Syst. **2**(11), 2000110 (2020)
14. Jones, S., Milner, E., Sooriyabandara, M., Hauert, S.: Dots: an open testbed for industrial swarm robotic solutions. arXiv preprint arXiv:2203.13809 (2022)
15. Lau, H., Bate, I., Cairns, P., Timmis, J.: Adaptive data-driven error detection in swarm robotics with statistical classifiers. Robot. Auton. Syst. **59**(12), 1021–1035 (2011)
16. Lee, S., Milner, E., Hauert, S.: A data-driven method for metric extraction to detect faults in robot swarms. IEEE Robot. Autom. Lett. **7**(4), 10746–10753 (2022). https://doi.org/10.1109/LRA.2022.3189789

17. Minelli, M., Panerati, J., Kaufmann, M., Ghedini, C., Beltrame, G., Sabattini, L.: Self-optimization of resilient topologies for fallible multi-robots. Robot. Auton. Syst. **124**, 103384 (2020)
18. Nachar, N.: The mann-whitney u: a test for assessing whether two independent samples come from the same distribution. Tutorials Quant. Methods Psychol. **4** (2008). https://doi.org/10.20982/tqmp.04.1.p013
19. Neupane, A., Goodrich, M.A.: Learning resilient swarm behaviors via ongoing evolution. In: Dorigo, M., et al. (ed.) Swarm Intelligence, ANTS 2022, LNCS, vol. 13491, pp. 155–170. Springer, Cham (2022). https://doi.org/10.1007/978-3-031-20176-9_13
20. O'Keeffe, J., Tarapore, D., Millard, A.G., Timmis, J.: Fault diagnosis in robot swarms: an adaptive online behaviour characterisation approach. In: 2017 IEEE Symposium Series on Computational Intelligence (SSCI), pp. 1–8. IEEE (2017)
21. Oladiran, O.: Fault recovery in swarm robotics systems using learning algorithms learning algorithms (2019)
22. Parker, L.: Alliance: an architecture for fault tolerant multirobot cooperation. IEEE Trans. Robot. Autom. **14**(2), 220–240 (1998). https://doi.org/10.1109/70.681242
23. Penders, J., et al.: A robot swarm assisting a human fire-fighter. Adv. Robot. **25**, 93–117 (2011). https://doi.org/10.1163/016918610X538507
24. Putter, R., Nitschke, G.: Evolving morphological robustness for collective robotics. In: 2017 IEEE Symposium Series on Computational Intelligence (SSCI), pp. 1–8. IEEE (2017)
25. Şahin, E.: Swarm robotics: From sources of inspiration to domains of application. In: Şahin, E., Spears, W.M. (eds.) Swarm Robotics, pp. 10–20. Springer, Berlin, Heidelberg (2005). https://doi.org/10.1007/978-3-540-30552-1_2
26. Schranz, M., Umlauft, M., Sende, M., Elmenreich, W.: Swarm robotic behaviors and current applications. Front. Robot. AI **7**, 36 (2020)
27. Shiliang, S., Ting, W., Chen, Y., Xiaofan, L., Hai, Z.: Distributed fault detection and isolation for flocking in a multi-robot system with imperfect communication. Int. J. Adv. Rob. Syst. **11**(6), 86 (2014)
28. Subha, N.A.M., Mahyuddin, M.N.: Distributed adaptive cooperative control with fault compensation mechanism for heterogeneous multi-robot system. IEEE Access **9**, 128550–128563 (2021)
29. Tarapore, D., Lima, P.U., Carneiro, J., Christensen, A.L.: To err is robotic, to tolerate immunological: fault detection in multirobot systems. Bioinspiration Biomimetics **10**(1), 016014 (2015)
30. Tarapore, D., Timmis, J., Christensen, A.L.: Fault detection in a swarm of physical robots based on behavioral outlier detection. IEEE Trans. Rob. **35**(6), 1516–1522 (2019)
31. Timmis, J., Ismail, A.R., Bjerknes, J., Winfield, A.: An immune-inspired swarm aggregation algorithm for self-healing swarm robotic system. Biosystems **146** (2016). https://doi.org/10.1016/j.biosystems.2016.04.001
32. Vassev, E., Sterritt, R., Rouff, C., Hinchey, M.: Swarm technology at NASA: building resilient systems. IT Professional **14**(2), 36–42 (2012)
33. Werfel, J., Petersen, K., Nagpal, R.: Designing collective behavior in a termite-inspired robot construction team. Science (New York, N.Y.) **343**, 754–758 (2014)
34. Winfield, A.F., Nembrini, J.: Safety in numbers: fault-tolerance in robot swarms. Int. J. Model. Ident. Control **1**(1), 30–37 (2006)

Achieving Human-Inspired Drift Diffusion Consensus in Swarm Robotics

Gal Sajko[1,2] and Jan Babič[1]

[1] Laboratory of Neuromechanics and Biorobotics, Department of Automation, Biocybernetics and Robotics, Jožef Stefan Institute, Ljubljana, Slovenia
`{gal.sajko,jan.babic}@ijs.si`
[2] Jožef Stefan International Postgraduate School, Ljubljana, Slovenia

Abstract. We present a human inspired approach to collective decision-making in swarm robotics, leveraging a social drift diffusion model that models the decision making process of group of humans. We adapt its principles to robotic swarms to address collective perception tasks. Our method introduces the *social factor* parameter that allows direct control over the trade-off between decision speed and accuracy. It enables robotic swarms to reach consensus on environmental characteristics more efficiently, with the possibility to prioritize either speed or accuracy, depending on the task requirements. Experimental simulations across various environmental complexities demonstrate our method's superior performance compared to traditional algorithms like the voter model and majority rule. The results highlight the effectiveness of human-inspired decision-making mechanisms in enhancing the capabilities of swarm robotics.

1 Introduction

Swarm robotics is a branch of robotics that draws inspiration from social organisms in nature, such as ants, bees, flocks of birds, etc. [5]. The aim of imitating such organisms is to increase the system's robustness and leverage the advantages that come with the numerosity of agents in such a swarm. The main characteristic of such organisms is the absence of a central control element. The operation of the swarm as a whole is a result of the behavior of each individual element of the swarm [9]. Individuals follow their own local rules and their decisions lead to the desired behavior of the swarm as a whole [13]. One of the main problems in designing a robotic swarm is how to define local rules that will cause the desired global operation of the swarm [1,3]. This paper is particularly focused in the decision-making scenario for the task of collective perception [8,17,18,22]. Collective perception is a task that requires the swarm to collectively sense the environment and make a group decision (consensus) about the characteristics of the environment they are exploring. The advantage of using a decentralized system for this task is that each agent can explore only a small part of the environment and then, through ways of communicating with other agents and making decisions based on collected information, influence the global decision of the swarm. Specifically we were interested in whether the group decision-making mechanism used by humans can be transferred to a robotic swarm and if it can improve consensus efficiency in the task of collective perception, compared to the traditional methods. The efficiency of consensus is measured by two values: the time that the swarm needs to reach a consensus and the percentage of attempts in which the decision about the environment is correct. Our method uses the principles of the social drift diffusion model [20], which describes the way humans make group decisions. The results show that by choosing different parameters of the method, we can directly influence the speed of consensus and its accuracy. Among the most common and widely used decision making algorithms are the voter model (VM) [23] and majority rule (MR) [24]. Consequently, we used them as baseline methods and compared the results with the results of our method.

The article is organized as follows. In Sect. 2, we describe all the methods used. The following Sect. 3 presents the obtained results, their analysis, and comparison with baseline methods. The paper is concluded with Sect. 4 where we discuss some main limitations of our method and provide suggestions for future work.

2 Methods

2.1 Human Decision Making Model

The drift diffusion model is frequently employed to characterize human decision-making process when choosing between two alternatives [11,16]. During decision-making process, a person collects (accumulates) evidence in support of one option or the other [15]. The evidence is gathered by observing the environment and is translated through a complex process in the brain into a numerical value called decision variable [12]. The decision variable expresses the value of all the evidence accumulated. Decision boundaries determine the threshold values that the decision variable must exceed to make a decision for one option or the other. If a person has no initial bias at the beginning of the decision-making process (does not lean towards one decision or another due to previous experiences, knowledge, preferences, etc.), the initial value of the decision variable is equal to the average value of both decision boundaries. During the decision making process, a person accumulates evidence in favor of one option or the other, which causes the drift of the decision variable towards one or the other boundary [19]. When a person is placed in a social environment, external components also start to influence his or her decisions. In addition to the information obtained based on his or her own observation (personal information), information obtained through passive or active interaction with other people in the vicinity (social information) becomes important. These additional pieces of information must be considered in the accumulation of evidence. The social drift diffusion model [20] extends the described drift diffusion model by adding a social component of information that a person gathers during the accumulation of evidence and updating of the decision variable.

Fig. 1. Experimental environment with agents in different states: small green circles - receiving state, big yellow circles - dissemination state with communication range, small black (possibly white) circles inside the agents - opinions. (Color figure online)

2.2 Experimental Setup

The experiment is conducted in a simulation, which we implemented using the Python programming language and the Pygame library. The problem is summarized from [10, 22] and is often used to simulate problems of collective perception. The task of the swarm is to collectively decide which color of tiles on the 2D surface with the dimensions of 2 m × 2 m predominates. The swarm is homogeneous and consists of 20 agents, as can be seen in Fig. 1. During the operation, each agent transitions between four possible states: (I.) exploration state, (II.) receiving state, (III.) dissemination state, and (IV.) decision-making state. The order of transitioning between states depends on the used decision-making method.

The ratio of the number of black to white tiles represents the complexity of the problem. It is given as $c = \frac{N_{black}}{N_{white}}$. We conducted the experiment for four different environmental complexities $c \in (0.25, 0.52, 0.67, 0.82)$ (values are taken directly from [22]), both for cases where black prevails and for cases where white prevails. This showed that our method is not subject to any bias in favor of one color of tiles over the other.

The described probabilistic finite state machine is adapted from [22] where the agent moves randomly through the space at all time. Movement is implemented as sequences of linear movement in a random direction for a randomly determined time (exponential distribution, mean value $\overline{t_{mov}} = 40$ s) and stationary states that simulate turning in place, for a randomly determined time (uniform distribution, $t_{rot} \in [0, 4.5]$ s). If the agent reaches the edge of the environment or collides with another agent, the movement stops, followed by a simulated turn in place and then continuation of movement in a new random direction.

In the exploration state, the agent collects data about the surroundings, meaning it records information on how much time it has spent over each color (information is only recorded while moving, not while turning). The time in exploration state is denoted with t_{exp} and it has an exponential distribution with mean value $\overline{t_{exp}} = 10$ s. The agents spend $t_{rec} = 3$ s in each receiving state. In this state, it receives information from agents that are in the dissemination state during the communication time. Communication is possible if those agents are within the communication radius, which is set to 70 cm. An agent in the dissemination state communicates its current opinion to an agent in the receiving state. The time that the agent spends in dissemination state depends on the quality of the agent's opinion or the degree of belief, with the maximum time being $T_{diss} = 10$ s. This models the positive feedback effect, causing agents that are more convinced of their opinion to broadcast it for a longer time [7,21]. Consequently, there is a greater chance that more agents in the receiving state will receive this opinion and consider it in their own decision-making process. The last in the cycle of states is the decision-making state. In this state, the agent updates its opinion based on the received and collected information. The methods used for this purpose will be described in more detail in the following two subsections.

Experiments were conducted by calculating the percentage of correct decisions out of 100 repetitions, i.e., accuracy α and the average time $\overline{T_c}$, the swarm needed to reach consensus. Time for each repetition was limited to 400 s.

2.3 Voter Model and Majority Rule Methods

Both the VM and MR models are based on the use of a probability finite state machine for modulating positive feedback. Both models employ the same state machine, where agents transition between states in the following sequence: (I.) exploration state, (II.) dissemination state, (III.) receiving state, and (IV.) decision-making state. The models differ only in the decision-making state, with the main difference being in how they consider information obtained from other agents during the receiving state.

Decision-making using VM is implemented in such a way that an agent randomly selects one of all the opinions gathered and adjusts its own opinion accordingly. If the randomly selected opinion is the same as the one it currently holds, the agent's opinion remains unchanged. Using MR, the agent determines the majority opinion in the set of gathered opinions, to which it adds its own current opinion. If the majority opinion obtained in this way is different from its current one, the agent adjusts its opinion accordingly. If the result is tied, or if the majority opinion is the same as the current one, the agent does not change its opinion. VM is considered to be more accurate and robust because, by considering only one randomly selected opinion, potential incorrect opinions spread slower. Consequently, the time that the swarm needs to reach consensus is on average longer than when using the MR method. A common feature of both methods is that they are not capable of adapting to a dynamic environment after the swarm has already reached a consensus. The reason is that an agent, in determining its opinion, considers only the opinions of other agents. Once the swarm reaches a consensus, consequently, a change in the opinion of an individual agent, which would then trigger potential changes in the opinions of other agents, is no longer possible.

2.4 Swarm Opinion Drift Model

In our Swarm Opinion Drift Model (SODM) method, we used the same probability finite state machine as in the VM and MR method, but we modified the order of transitioning between states. We also implemented the modulation of positive feedback differently. Following the example of the social drift diffusion model, during the decision-making process an agent collects evidence in support of one of the two options. Evidence is collected in the form of personal information, which represents the agent's own measurements, and social information, which represents the current opinions of other agents. The opinion is represented as the current value of the decision variable (d_i). d_i can take values on the interval $[-1\frac{2}{3} \cdot D, 1\frac{2}{3} \cdot D]$, where D is the decision boundary that d_i must exceed for the agent to make a decision for one option or the other. The boundary $+D$ represents the decision for black, and $-D$ for white. To prevent runaway towards infinity the values of d_i are limited to $1\frac{2}{3} \cdot D$.

All agents start in the exploration state, and their initial value of d_i is equal to 0. By that we avoid any uneducated guesses which, if not correct, could cause drift of decision variable towards the incorrect decision and by that increasing the possibility of reaching the wrong consensus or at least prolonging the consensus time. At the end of the exploration state, the agent determines personal information based on the measurements gathered, which is defined as

$$p_i = log(\frac{q_b}{1 - q_b}), \quad (1)$$

where the i represents the index of the agent, and q_b is defined as

$$q_b = \begin{cases} \frac{t_b}{t_{exp}}, & \text{if } t_{exp} \neq 0 \\ 0.5, & \text{else} \end{cases} \qquad (2)$$

where t_b represents the time that the agent spends over black tiles. If $t_b = \frac{1}{2} \cdot t_{exp}$ than in the exploration state the agent spends an equal amount of time over black and white tiles. In this case, it cannot gain any information about which color is predominant, so $p_i = 0$. For cases where $t_b = 0$ or $t_b = t_{exp}$, p_i tends towards $-\infty$ or $+\infty$ respectively, so we limited q_b to $q_b \in [\epsilon, 1-\epsilon]$, where $\epsilon = 0.001$. Consequently, p_i is limited to values $p_i \in [-6.907, 6.907]$. Since the problem is formulated in such a way that the number of black tiles directly determines also the number of white tiles, we can write

$$P(x_i = \text{black}) = 1 - P(x_i = \text{white}), \qquad (3)$$

where x_i is an individual measurement of the agent, and $P(x_i)$ is the probability that the measurement takes one value or another, which the agent determines directly by measuring the time spent over a specific color of tiles. Consequently, we can write

$$\frac{P(x_i = \text{black})}{P(x_i = \text{white})} = \frac{\frac{t_b}{t_{exp}}}{t_{exp} - \frac{t_b}{t_{exp}}} = \frac{q_b}{1-q_b}. \qquad (4)$$

In Eq. (1), the term (4) is used in a logarithmic function [14], which gives us two advantages: (i) the function is symmetric around the value of 0.5, consequently measurements where $q_b < 0.5$ cause a negative p_i and a drift of d_i towards the lower decision boundary and vice versa; and (ii) p_i will have relatively much larger absolute values at the end of the exploration state for those observations that are more unambiguous.

The exploration state is followed by the receiving state where the agent receives the current value of d_i exactly once from each agent in the dissemination state that comes into its communication range during this time. At the end of the receiving state, the agent determines social information as the average value of all received d_i, which we write as

$$s_i = (\frac{1}{M} \cdot \sum d_j)^\eta, \qquad (5)$$

where M is the number of agents from which it receives data, d_j is the value of the decision variable from the agent with index j and η is defined as

$$\eta = 1 + \frac{M}{N}, \qquad (6)$$

where N is the total number of agents in the swarm, which in our case equals 20. The parameter η additionally weights the social information that is obtained from a larger number of agents.

When an agent enters the decision-making state, it has both personal information, gathered during the exploration state, and social information, calculated at the end of the receiving state, available. Based on these two pieces of information, it can update its current opinion about the environment, that is the value of d_i. The update of d_i is written as

$$d_{i_k} = d_{i_{k-1}} + p_i + \gamma \cdot s_i, \qquad (7)$$

where γ is a parameter that amplifies social information and directly affects the drift of d_i, referred to as the *social factor*, and the index k represents the current time step. The moment the value of d_i exceeds one of the decision boundaries $\pm D$, the agent makes a decision about the majority color of the tiles. Once all agents make the same decision, the process concludes.

At the end of each cycle of transitioning between states of the probability finite state machine is the dissemination state. Within this state, the agent broadcasts its current value of d_i for a duration of t_{diss}. Positive feedback modulation is implemented in such a way that agents closer to making a decision (i.e., agents with a larger absolute value of d_i) broadcast their opinion for a longer period. The time during which an agent broadcasts its opinion is calculated as

$$t_{diss} = \begin{cases} \frac{|d_i|}{D} \cdot T_{diss}, & \text{if } |d_i| < D \\ T_{diss}, & \text{else} \end{cases} \qquad (8)$$

After completing the dissemination state, the agent returns to the exploration state. The entire described process repeats until the agents reach a consensus.

3 Results

In the following subsections we analyze the effect of personal and social information and the *social factor* on accuracy and speed of decision making process. We also compare the results of our method with the results of the baseline methods.

3.1 Effect of Personal and Social Information

If agents disregard personal and consider only social information, the swarm is incapable of reaching consensus, since social information depends only on the current values of the decision variable of other agents, as stated in Eq. (5). That is why social information alone is not capable of initiating a drift of d_i from the initial value towards one of the decision boundaries, if the initial value is 0. Consequently, the values of d_i for all agents remain at the initial value, and none of the agents make a decision.

Conversely, if agents in the swarm use only personal information and the *social factor* γ is equal to 0, the agent updates its d_i based solely on its own perceptions of the environment during the exploration state. At higher environmental complexities, the mere use of the personal information does not allow for consensus during the limited time, as the measurements captured by individual agents are too random to enable a sufficiently fast and steady drift of d_i towards one of the decision boundaries. This shows that an individual agent in the swarm

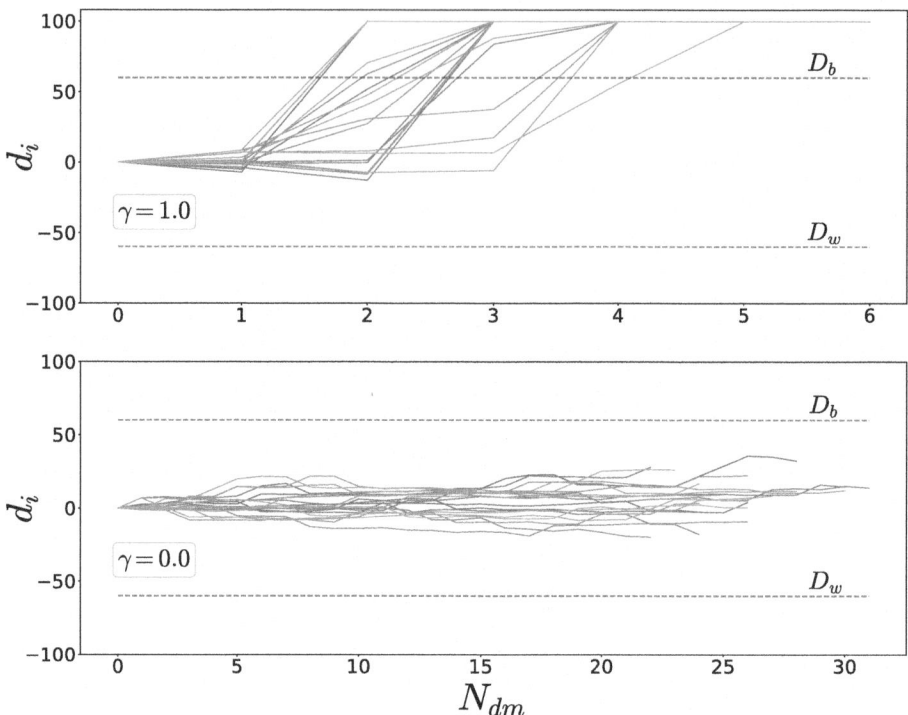

Fig. 2. Above: the trajectory of d_i for each agent at a parameter value of $\gamma = 1.0$. Below: the trajectory of d_i for each agent, at $\gamma = 0.0$. Decision boundaries for black and white color are denoted with D_b and D_w respectively. N_{dm} represents the number of decision making states.

is not capable of performing the task alone and that cooperation with others is necessary. The trajectory of d_i values for all agents for the case when the γ is equal to 0 and 1 is shown with graphs in Fig. 2 (both cases for complexity $c = 0.82$). For values of $\gamma = 1.0$, the drift of d_i of all agents is very directed and moves towards the decision boundary for the black color D_b which all agents also exceed, meaning the swarm reaches consensus. At $\gamma = 0.0$ the drift of d_i is much weaker and less directed. Therefore, d_i of no agent exceeds any of the decision boundaries within the given time, meaning the swarm does not reach a consensus. We see, then, that the γ parameter affects both the direction and the speed of the drift of d_i. The sufficient speed of the drift is necessary for d_i to exceed the decision boundaries within a limited time, while the direction causes the drift of d_i of all agents to proceed towards the same decision boundary, allowing the swarm to reach a consensus.

3.2 Optimal Values of *Social Factor* Parameter and Speed vs. Accuracy Trade-off

We empirically determined the optimal values of the *social factor* parameter γ. The experiment was run for a set of γ values from 0.1 to 2, across all four environmental complexities c. The value of $\gamma = 0$ was omitted for reasons described in Subsect. 3.1. The graphs in Fig. 3 show the progression of accuracy α and average consensus times $\overline{T_c}$ depending on the γ, for each of the four values of environmental complexity c. As we can see, both α and the $\overline{T_c}$ decrease with increasing values of the γ parameter, which very clearly demonstrates the speed vs. accuracy trade-off. Such results are expected, as a higher parameter value gives greater weight to social information. Consequently, an agent adjusts its opinion to others in the swarm faster, and consensus is reached more quickly. On the other hand, higher γ values also lead to faster spread of incorrect opinions, thereby negatively affecting accuracy. Specifically at $c = 0.25$, the highest achieved accuracy is $\alpha = 1.0$. This accuracy is achieved at parameter values $\gamma \in (0.1, 0.2, 0.3, 0.4, 0.6, 0.8)$. The shortest consensus time to reach maximum accuracy is at $\gamma = 0.8$, with $\overline{T_c} = 63.4$ s. At $c = 0.52$, the swarm achieves accuracy $\alpha = 1.0$ at parameters $\gamma \in (0.1, 0.2, 0.3, 0.4, 0.6)$. The shortest consensus time at this accuracy is $\overline{T_c} = 86.3$ s, achieved at $\gamma = 0.6$. At $c = 0.67$, the highest achieved accuracy is $\alpha = 0.99$ at parameters $\gamma \in (0.2, 0.3)$, with a shorter consensus time at $\gamma = 0.3$ being $\overline{T_c} = 147.82$ s. At $c = 0.82$, the swarm achieves the highest accuracy $\alpha = 0.93$ at $\gamma = 0.2$, with an average decision time of $\overline{T_c} = 240.49$ s. At complexities $c = 0.67$ and $c = 0.82$ with a *social factor* value of $\gamma = 0.1$, the swarm achieves lower accuracies than at some higher γ values. The reason is that the swarm did not reach consensus within the prescribed time in 2% and 30% (for $c = 0.67$ and $c = 0.82$ respectively), which is treated the same as a wrong decision.

3.3 Comparison to the Baseline Methods

The graphs in Fig. 4 show a comparison of the results of our method with the results of the VM and MR methods across all four environmental complexity values. The first graph compares the best accuracies achieved by all methods. On the second graph, we compared the consensus times at the highest accuracies of our method with the consensus times of the baseline methods. The third graph shows a comparison of the consensus times of both baseline methods with the consensus times of our method at comparable accuracies. As we can see, our method achieves or exceeds the accuracies of both baseline methods at all environmental complexities. The difference in accuracies increases with the complexity of the environment. The largest difference is at $c = 0.82$, where VM achieves an accuracy of $\alpha_{VM} = 0.71$, MR $\alpha_{MR} = 0.61$, and our method $\alpha_{SODM} = 0.93$. Accuracy decreases most significantly with increasing complexity in the MR method, where the difference between the highest and the lowest accuracy is $\Delta_{MR} = 0.39$, in VM this difference is $\Delta_{VM} = 0.29$ and in our method only $\Delta_{SODM} = 0.07$. When comparing consensus times at maximum accuracies, we

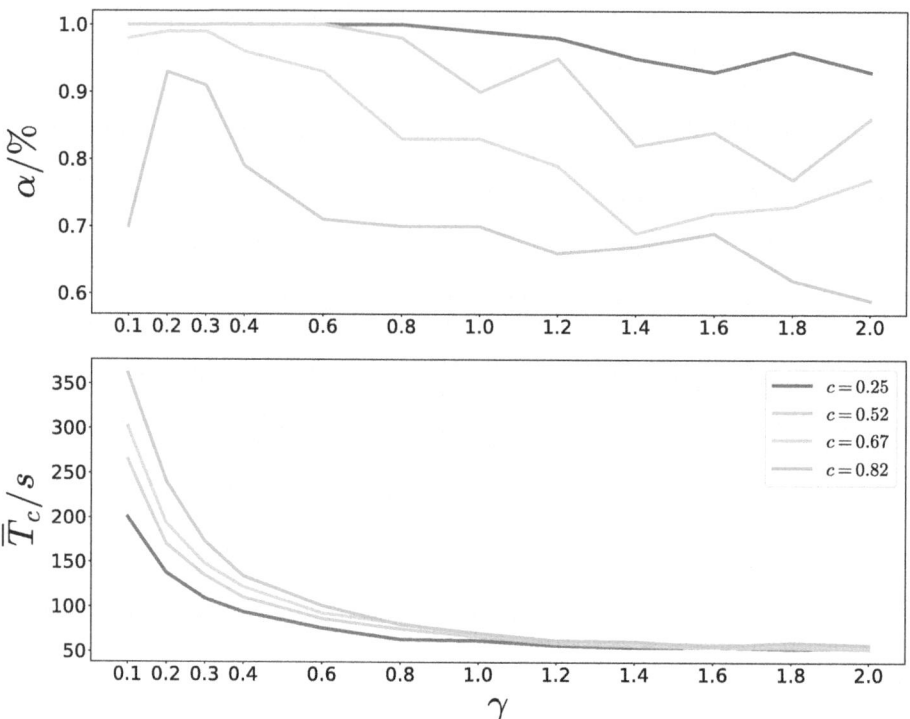

Fig. 3. Above: Decision accuracy α depending on the *social factor* parameter γ, for each of the four complexities. Below: Average consensus times depending on γ, for each of the four complexities.

see that for complexities $c \in (0.25, 0.52, 0.67)$ the consensus times of our method are somewhere between the times of the VM and MR methods but closer to the time of the VM method. At complexity $c = 0.82$ the average consensus time of our method is longer than both times of the baseline methods. When comparing the standard deviations of consensus times, we see that the consensus times of our method are the least dispersed (lowest standard deviations). Comparing the consensus times achieved by methods at comparable accuracies, we see that the consensus times of our method are always shorter (except when compared with the MR method for complexity $c = 0.25$) and with a smaller standard deviation.

4 Discussion

In this paper, we introduced a decision-making method inspired by the group decision-making process used by humans. This process was adapted for use on robotic swarms for the task of collective perception. We demonstrated that our method can achieve higher accuracies and shorter consensus times compared to the VM and MR methods. The advantage of our method is primarily in that

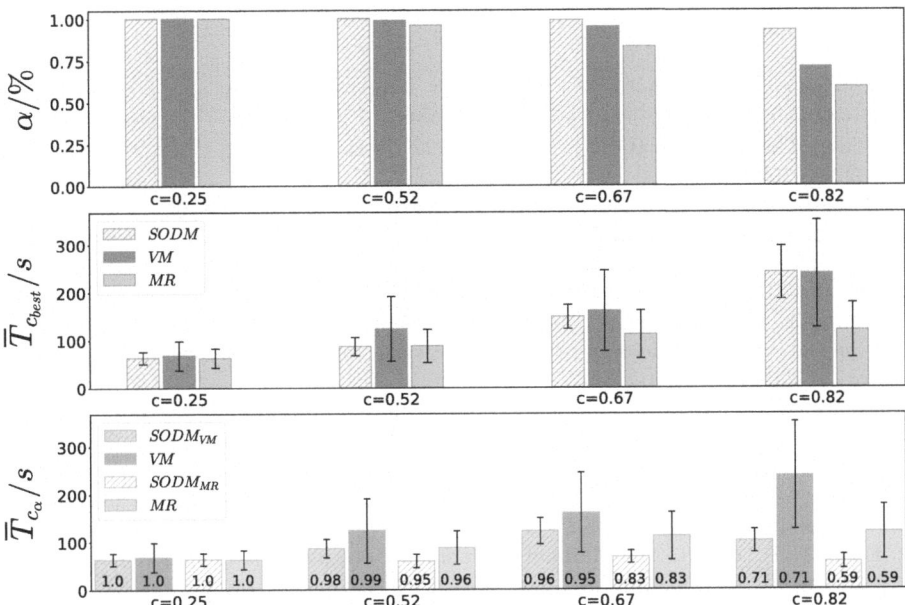

Fig. 4. Above: comparison of the accuracies of the methods across all four complexities. Middle: comparison of consensus times at the best accuracies. Below: comparison of the consensus times at comparable accuracies (the numbers within individual column represent accuracies). Error bars represent the standard deviation.

it uses the *social factor* parameter, whose value directly defines the speed vs. accuracy trade-off. Higher parameter values allow for a faster consensus time of the swarm but negatively affect the accuracy. At the same time, lower values enable high accuracies with longer consensus times, which are still shorter or nearly equal to the times achieved by the VM method. An important aspect of our future work will be to compare our method to more similar algorithms that also consider evidence accumulation and social impact on the decision making process (i.e. [2,6]). In described work the consensus detection is still done in a centralized way. We acknowledge that in real-world application also the consensus detection needs to be decentralized. Besides, our method does not allow for automatic adjustment of the *social factor* parameter value, meaning it must be determined manually before starting the task. In our future work, we will focus on this limitation and try to find a way for agents to determine the optimal *social factor* value autonomously while collecting information about the environment. This process will likely lead to a transition from initially homogeneous swarm to a heterogeneous one [4], as the information about the environment collected by each agent is slightly different, resulting in different *social factor* parameters.

References

1. Almansoori, A., Alkilabi, M., Tuci, E.: A comparative study on decision making mechanisms in a simulated swarm of robots. In: 2022 IEEE Congress on Evolutionary Computation (CEC), pp. 1–8 (2022). https://doi.org/10.1109/CEC55065.2022.9870208
2. Bartashevich, P., Mostaghim, S.: Ising model as a switch voting mechanism in collective perception. In: Moura Oliveira, P., Novais, P., Reis, L.P. (eds.) EPIA 2019. LNCS (LNAI), vol. 11805, pp. 617–629. Springer, Cham (2019). https://doi.org/10.1007/978-3-030-30244-3_51
3. Brambilla, M., Ferrante, E., Birattari, M., Dorigo, M.: Swarm robotics: a review from the swarm engineering perspective. Swarm Intell. **7**, 1–41 (2013). https://doi.org/10.1007/s11721-012-0075-2
4. Dorigo, M., et al.: Swarmanoid: a novel concept for the study of heterogeneous robotic swarms. IEEE Robot. Autom. Mag. **20**, 60–71 (2013). https://doi.org/10.1109/MRA.2013.2252996
5. Dorigo, M., Theraulaz, G., Trianni, V.: Reflections on the future of swarm robotics. Sci. Robot. **5**(49), eabe4385 (2020). https://doi.org/10.1126/scirobotics.abe4385
6. Ebert, J.T., Gauci, M., Mallmann-Trenn, F., Nagpal, R.: Bayes bots: collective bayesian decision-making in decentralized robot swarms. In: 2020 IEEE International Conference on Robotics and Automation (ICRA), pp. 7186–7192 (2020). https://doi.org/10.1109/ICRA40945.2020.9196584
7. Golman, R., Hagmann, D., Miller, J.H.: Polya's bees: a model of decentralized decision-making. Sci. Adv. **1**(8), e1500253 (2015). https://doi.org/10.1126/sciadv.1500253
8. Hamann, H.: Evolution of collective behaviors by minimizing surprise. In: Artificial Life Conference Proceedings, vol. ALIFE 14: The Fourteenth International Conference on the Synthesis and Simulation of Living Systems, pp. 344–351 (2014). https://doi.org/10.1162/978-0-262-32621-6-ch055
9. Hamann, H.: Swarm Robotics: A Formal Approach. Springer, Cham (2018). https://doi.org/10.1007/978-3-319-74528-2
10. Kaiser, T.K., Potten, T., Hamann, H.: Evolution of collective decision-making mechanisms for collective perception. In: 2023 IEEE Congress on Evolutionary Computation (CEC), pp. 1–8 (2023). https://doi.org/10.1109/CEC53210.2023.10253996
11. Kimura, M., Moehlis, J.: Group decision-making models for sequential tasks. SIAM Rev. **54**(1), 121–138 (2012), http://www.jstor.org/stable/41642574
12. Kira, S., Yang, T., Shadlen, M.N.: A neural implementation of wald's sequential probability ratio test. Neuron **85**, 861–873 (2015). https://doi.org/10.1016/j.neuron.2015.01.007
13. Kuckling, J.: Recent trends in robot learning and evolution for swarm robotics. Front. Robot. AI **10** (2023). https://doi.org/10.3389/frobt.2023.1134841
14. Marshall, J.A., Brown, G., Radford, A.N.: Individual confidence-weighting and group decision-making. Trends Ecol. Evol. **32**(9), 636–645 (2017). https://doi.org/10.1016/j.tree.2017.06.004
15. Myers, C., Interian, A., Moustafa, A.: A practical introduction to using the drift diffusion model of decision-making in cognitive psychology, neuroscience, and health sciences. Front. Psychol. **13**, 1039172 (2022). https://doi.org/10.3389/fpsyg.2022.1039172

16. Ratcliff, R., McKoon, G.: The diffusion decision model: theory and data for two-choice decision tasks. Neural Comput. **20**, 873–922 (2008). https://doi.org/10.1162/neco.2008.12-06-420
17. Schranz, M., Umlauft, M., Sende, M., Elmenreich, W.: Swarm robotic behaviors and current applications. Front. Robot. AI **7** (2020). https://doi.org/10.3389/frobt.2020.00036
18. Shan, Q., Mostaghim, S.: Collective decision making in swarm robotics with distributed bayesian hypothesis testing. vol. 12421 LNCS, pp. 55–67. Springer Science and Business Media Deutschland GmbH (2020).https://doi.org/10.1007/978-3-030-60376-2_5
19. Thieu, T., Melnik, R.: Social human collective decision-making and its applications with brain network models, pp. 103–141 (2023). https://doi.org/10.1007/978-3-031-46359-4_5
20. Tump, A.N., Pleskac, T.J., Kurvers, R.H.J.M.: Wise or mad crowds? the cognitive mechanisms underlying information cascades. Sci. Adv. **6**, eabb0266 (2020). https://doi.org/10.1126/sciadv.abb0266
21. Valentini, G.: Achieving Consensus in Robot Swarms, vol. 706. Springer, Cham (2017). https://doi.org/10.1007/978-3-319-53609-5
22. Valentini, G., Brambilla, D., Hamann, H., Dorigo, M.: Collective perception of environmental features in a robot swarm, pp. 65–76, September 2016. https://doi.org/10.1007/978-3-319-44427-7_6
23. Valentini, G., Hamann, H., Dorigo, M.: Self-organized collective decision making: the weighted voter model. In: Proceedings of the 2014 International Conference on Autonomous Agents and Multi-Agent Systems, AAMAS 2014, pp. 45–52. International Foundation for Autonomous Agents and Multiagent Systems, Richland, SC (2014)
24. Valentini, G., Hamann, H., Dorigo, M.: Efficient decision-making in a self-organizing robot swarm: On the speed versus accuracy trade-off. In: Proceedings of the 2015 International Conference on Autonomous Agents and Multiagent Systems, AAMAS 2015, pp. 1305–1314. International Foundation for Autonomous Agents and Multiagent Systems, Richland, SC (2015)

Byzantine Fault Detection in Swarm-SLAM Using Blockchain and Geometric Constraints

Angelo Moroncelli[1,2,3](✉), Alexandre Pacheco[1], Volker Strobel[1], Pierre-Yves Lajoie[4], Marco Dorigo[1], and Andreagiovanni Reina[1,5,6](✉)

[1] IRIDIA, Université Libre de Bruxelles, Brussels, Belgium
alexandre.melo.pacheco@gmail.com, volker.strobel@ulb.be,
mdorigo@ulb.ac.be
[2] DEIB, Politecnico di Milano, Milan, Italy
[3] IDSIA, USI-SUPSI, Lugano, Switzerland
angelo.moroncelli@idsia.ch
[4] Department of Software and Computer Engineering, Polytechnique Montréal, Montreal, Canada
pierre-yves.lajoie@polymtl.ca
[5] CASCB, Universität Konstanz, Konstanz, Germany
andreagiovanni.reina@uni-konstanz.de
[6] Department of Collective Behaviour, Max Planck Institute of Animal Behavior, Konstanz, Germany

Abstract. Effective methods for Simultaneous Localisation And Mapping (SLAM) are key to enabling autonomous robots to navigate unknown environments. Multi-robot collaborative SLAM (C-SLAM) offers the opportunity for higher performance thanks to parallel execution of mapping and localisation by a distributed team of robots but it also introduces challenges in system scalability and consistent data aggregation, exposing the system to potential security risks. In particular, we show that the state-of-the-art decentralised C-SLAM framework for swarm robotics is vulnerable to Byzantine robots, which are robots that behave incorrectly, possibly due to malfunctioning or hacking. We propose a solution that uses a blockchain to achieve data consistency and a smart contract that manages robots' reputations to identify and neutralise Byzantine robots. Each robot's contribution to collaborative mapping is peer-reviewed by other robots by verifying its correctness through geometric constraints. Our multi-robot simulation results show the existence of a trade-off between fault tolerance and efficiency in terms of map generation speed. With this work, we also release open-source research software that interfaces a custom blockchain with the ROS 2 framework.

1 Introduction

Autonomous robotic systems face the challenge of navigating unknown environments without relying on external localisation systems. To address this challenge, robots employ Simultaneous Localisation And Mapping (SLAM) algorithms [12,47]. Through SLAM, robots build a map of the environment and determine their positions within this map [40]. Several efficient solutions to single-robot SLAM have been proposed thanks to decades of research that focused on this crucial topic [3,9,14,30,33,36,54]. More recently, research has started to investigate multi-robot collaborative SLAM (C-SLAM) [28], where groups of robots collaborate to build maps. Thanks to parallelisation, C-SLAM offers opportunities for increased efficiency, localisation accuracy, and robustness to errors.

A particularly promising type of multi-robot system is a robot swarm, which comprises typically a large number of autonomous robots. A key characteristic of robot swarms is decentralisation as robots only interact with their near neighbours and lack a centralised controller that orchestrates the actions of every robot [11]. To allow robot swarms to perform their operations, they must be able to navigate their environment, therefore implementing C-SLAM algorithms for robot swarms can be particularly useful [22,23].

However, robot swarms introduce new challenges to C-SLAM due to the large number of robots participating in the process, and the lack of a central server that aggregates potentially conflicting data generated by different robots. Additionally, although robustness is often indicated as an intrinsic characteristic of swarm robotics, recent research [49,50] has shown that robot redundancy and parallelisation of operations are not sufficient to achieve system robustness against misbehaving robots. In fact, a small proportion of misbehaving robots—called Byzantine robots—is often sufficient to disrupt the entire swarm system [2,48,52,55]. Because it is reasonable to assume that a subset of robots may misbehave—for example, due to internal errors or external malicious tampering—implementing robust and secure algorithms is of utmost importance [13,19,21,29,41,43].

This paper studies the potential security vulnerabilities of Swarm-SLAM [25], the state-of-the-art framework for C-SLAM with decentralised robot swarms (Sect. 2). We first discuss and characterise security issues that Swarm-SLAM, in particular, and C-SLAM, in general, face (Sect. 3). We show that Swarm-SLAM is highly vulnerable to the presence of different types of Byzantine robots. Inspired by recent research successes on protecting robot swarms via blockchain technology [4,10,17,38,42,48–50,50,55], we build a security layer for Swarm-SLAM through a blockchain-based smart contract, which is a distributed tamper-proof algorithm running on data stored in the blockchain (Sect. 4). We test our solution with physics-based simulations of groups of eight robots using ROS 2 and a custom blockchain framework (Sect. 5). The results show that the proposed blockchain-based solution makes Swarm-SLAM tolerant to a relatively large number of Byzantine robots. However, this comes at the cost of a decrease in the map construction speed (i.e., lower system efficiency). In Sect. 6, we conclude the paper by discussing such a robustness-efficiency trade-off and suggesting potential future research in blockchain-based swarm robotics.

2 Background and Related Work

Multi-robot Collaborative SLAM. Individual robots use SLAM to autonomously explore and map unknown environments, but through collaboration, multiple robots can more effectively navigate large spaces to create comprehensive maps. This multi-robot collaborative SLAM (C-SLAM) approach may mitigate exploration costs, map error, computational load, and single-point failure risks; but achieving this coordination is a complex task [6,23]. Initially, single-robot SLAM algorithms were adapted for multi-robot use (e.g., by employing Kalman filters [44], or cooperative localisation algorithms [35]). Methods that formulate C-SLAM as a mathematical optimisation problem have become prevalent due to their higher performance than traditional C-SLAM methods that use filters to estimate the robots' poses and the map [45]. While recent research has shown great progress in C-SLAM methods (e.g., through advanced multi-source data fusion and deep learning to enhance adaptability and reduce the likelihood of failures [6]), most applications remain limited to small robot teams, and addressing the problems of perceptual aliasing [51], heterogeneous robot teams [5], and real-time distributed multi-robot coordination remain open challenges. Existing open-source frameworks for C-SLAM [5,7,8,15,20,25,27,46,51,56] produce accurate results in the tested configurations but still have limitations in efficient data management, scalability to larger robot teams, and robustness against single points of failure, either because they use a centralised component to aggregate the maps and coordinate robots' movements or because the decentralised approach requires onboard computation by robots with computation and communication limits.

When we consider swarm robotics systems, Swarm-SLAM [25] stands out as a unique framework (based on the ROS 2 libraries [31]) to perform C-SLAM with a decentralised swarm of resource-limited robots. Swarm-SLAM outperforms other methods by allowing robots to use diverse sensors and operate with sporadic connectivity and significantly reduced communication demands. While Swarm-SLAM is a promising framework, there are still pending research questions on how to improve system scalability, achieve consistent data aggregation, and mitigate security risks. In this paper, we address the relatively unexplored problem of security in C-SLAM in general and Swarm-SLAM in particular. Indeed, when authors refer to Swarm-SLAM's robustness, they indicate the problem of perceptual aliasing [26]. However, in robotic systems operating in the real world, robots that exhibit non-ideal behaviour—e.g., due to faults or malicious intentions— may compromise the reliability of the entire system.

Securing Robot Swarms Using Blockchains. Swarm robotics, originally inspired by natural collectives, aims to create decentralised, robust, and scalable behaviour for groups of robots [11,18]. However, recent research has shown that protecting the swarm against Byzantine robots can be difficult and requires dedicated strategies [21,50]. A new and promising line of research suggests that blockchain-based smart contracts [53] can increase the Byzantine fault tolerance of robot swarms [10]. This research has shown that a solution to prevent the spreading of erroneous information is implementing a reputation management system where only information from high-reputation robots is used [48,50,52,55]. Reputation management is implemented using a blockchain to record all robots' information exchanges and a smart contract to implement outlier detection algorithms and assign reputation to robots whose actions align with the majority. We build on these promising results to design a smart contract that exploits geometric constraints in the collectively constructed map to identify and neutralise Byzantine robots.

3 Vulnerability of Swarm-SLAM to Byzantine Robots

Swarm-SLAM can be compromised by Byzantine robots in three ways: two involve the corruption of loop closures and one consists of tampering with the creation of the pose graph, an optimisation process that creates a collective map based on existing loop closures. In single-robot SLAM, loop closures enhance map and self-localisation accuracy by detecting when a robot revisits the same location. In multi-robot C-SLAM, this process can become more accurate and faster when two different robots that visit the same location generate inter-robot loop closures. However, this exchanging of information makes the system vulnerable to incorrect data sent by Byzantine robots. Current C-SLAM systems are not resilient to incorrect data points that significantly deviate from the overall map being constructed. Although some systems incorporate techniques to negate the effects of perceptual outliers [5, 8, 15, 20, 25, 27, 51], these systems remain vulnerable to Byzantine robots that conspire to overcome the rejection of these outliers, or to Sybil attacks, in which a single Byzantine robot forges many identities to gain control of the system. Protecting against these attacks is crucial, yet research on Byzantine fault tolerance in C-SLAM is lacking.

In C-SLAM, an inter-robot loop closure happens when robot A recognises a scene that another robot B saw previously. When they meet, B shares an image of the scene, which is then used by A to calculate a geometric loop closure. We found that a loop closure can have two main sources of error: (1) incorrect calculation by A (using correct information sent by B), or (2) incorrect information sent by B (while A makes the correct computation based on the received information).

The third security issue concerns the pose graph optimisation (PGO) management, which is a crucial step in merging the partial maps acquired by each robot. The pose graph is the mathematical representation of the relationships between robot poses during their trajectories and the scenes in the map. The configuration of this graph that best satisfies all the constraints can subsequently be used for building a map and simultaneously correcting the location of the robot poses inside the map. Always assigning the task of performing PGO to the same robot creates a single point of failure where a Byzantine robot in this role could compromise the entire SLAM operation, while on the opposite end, having all the robots perform the PGO could be unnecessarily costly. Figure 1 illustrates a comparison between a correct Swarm-SLAM process and the three scenarios where the actions of a Byzantine robot can lead to a wrong outcome.

4 Securing Swarm-SLAM Through Blockchain Technology

We employ blockchain technology to enhance the Byzantine fault tolerance and reliability of Swarm-SLAM. Blockchain is a shared, tamper-proof ledger

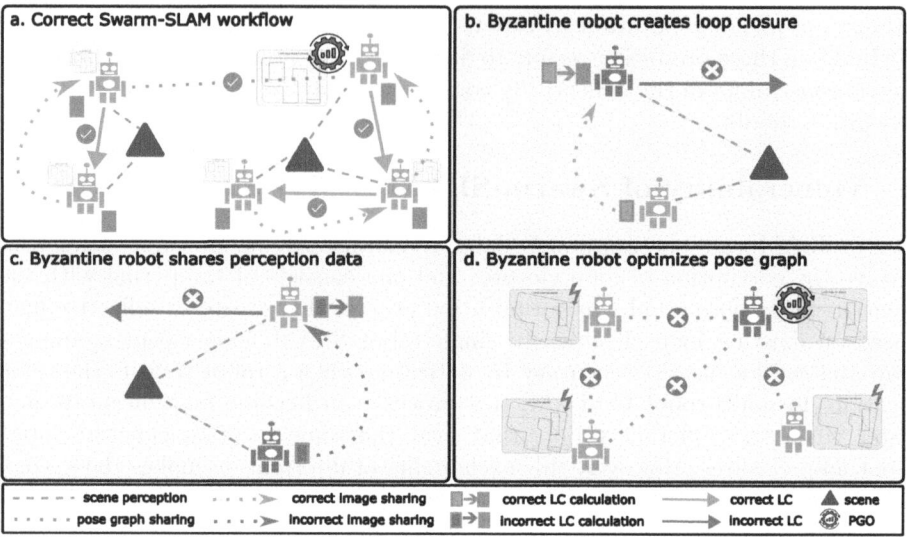

Fig. 1. Panel a represents the ideal workflow in a Swarm-SLAM application with reliable (green, non-Byzantine) robots, where robots acquire information about the same scene (purple triangle). When two of these robots meet, they have two matching descriptors for the scene and thus one of the robots (the receiver robot) shares its image (blue square) that has the matching descriptor with the sender robot, which calculates an inter-robot loop closure towards the receiver. These loop closures are then used by one robot (top right) to perform pose graph optimisation (PGO). Panels b-d depict three possible scenarios where the action of a Byzantine robot leads to incorrect global pose graph results. In panel b the sender is a Byzantine (red robot) which generates an incorrect loop closure (pointing to a wrong position) from itself to the receiver (green robot), while the image is shared correctly from the receiver. Panel c represents the construction of an incorrect loop closure due to false information shared by the Byzantine receiver robot (red robot): it shares an incorrect image with the sender, which calculates a wrong loop closure although being a reliable robot. Panel d depicts the action of PGO, which, in Swarm-SLAM, is performed by one single elected robot (in this case the optimiser is the red Byzantine robot). A Byzantine robot performing the PGO can introduce significant errors in the pose graph which is then shared throughout the robot swarm. Once an incorrect loop closure is introduced in the PGO, this error cannot be recovered nor verified by other robots. (Colour figure online)

that enables secure information storage and decentralised transaction validation [34]. This removes the need for a central authority, thus offering the opportunity to manage robots' permissions and reputation in swarm robotics [10]. Blockchain technology offers several benefits that are key to improving Swarm-SLAM's reliability and Byzantine fault tolerance: decentralisation eliminates single points of failure, tamper-resistance protects data integrity, and smart contracts enable the execution of algorithms among untrusting agents. Blockchain technology can have a limited computational overhead, making it also suitable for resource-constrained devices such as robots, as demonstrated in previous studies [38,39,50].

The integration of blockchain with Swarm-SLAM consists of a smart contract that validates loop closures and manages robots' reputations based on the correctness of the contributed loop closures. This smart contract uses geometric relationships to identify and reject incorrect loop closures, preventing the injection of faulty data in the construction of the collaborative map, whether they are intentionally wrong or by mistake. While this step introduces some latency, validating loop closures before they are utilised in the PGO can dramatically improve map accuracy in the presence of Byzantine robots.

An important aspect of Swarm-SLAM is that it minimises communication costs by requiring only one robot to share its (usually large) raw data [25]. The receiving robot then uses this data to compute the loop closure, which is directly used for PGO, making it difficult to assess the correctness of both the data and the loop closure. With our solution, loop closures are first stored in the blockchain, and only used in PGO once they are validated by other robots (see below). As the blockchain stores loop closures, which are lightweight geometric transformations, the system's scalability is improved by avoiding redundant sharing of raw image data. Storage requirements can also be reduced by letting robots delete the raw data once a corresponding loop closure is stored in the blockchain.

Our main contribution is a blockchain-based smart contract that leverages geometric constraints to validate loop closures in Swarm-SLAM and address the threat of Byzantine robots generating incorrect loop closures (Figs. 1b-c). To validate a loop closure, a second and a third loop closure are required to establish a triangle, as shown in Fig. 2. The triangle identity can then be used as a geometric constraint to validate the loop closures. The presence of more inter-robot loop closures from different robots leads to the formation of multiple triangles, thus increasing the confidence in the correctness of the loop closures.

By creating blockchain transactions, robots can store their loop closures in the blockchain, triggering the smart contract to evaluate whether the new loop closure formed any new validation triangle. The triangle identity constraint is validated when: $l_x^i + l_x^j + l_x^k < \epsilon$ and $l_y^i + l_y^j + l_y^k < \epsilon$, where l_x^i is robot i's loop closure on the x coordinate and ϵ is a sensitivity threshold. The threshold ϵ is a parameter to set the maximum Euclidean distance between the transformation head and tail—which should ideally match at each vertex—controlling the smart contract's sensitivity to errors in loop closure calculations. This threshold defines what the smart contract labels as Byzantine (above ϵ) and what it labels as noise (below ϵ and accepted as triangle identity). The best threshold largely depends on the application in which the algorithm is used. The three loop closures involved in a validation triangle must be submitted by three different robots (possibly at different times) and follow a cyclic orientation of transformations where each robot is only once a receiver and once a sender (see Fig. 2c). In this way, we promote peer-to-peer validation and prevent a single Byzantine robot from inputting multiple incorrect (but consistent) loop closures that pass

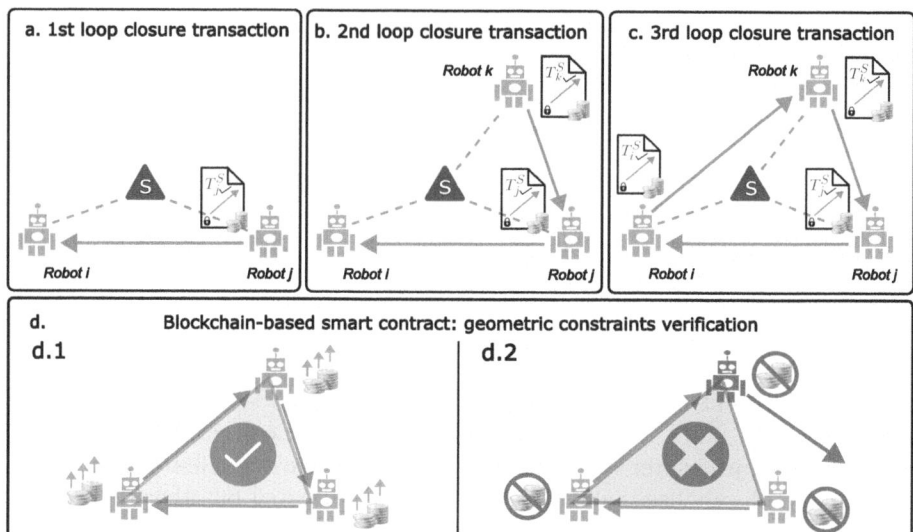

Fig. 2. The figure depicts the process by which our smart contract validates loop closures (LC). The blue arrows correspond to correct LCs proposed by reliable robots (green robots), while the red arrow illustrates an incorrect inter-robot LC created by a Byzantine robot (red robot). (a-c) The robots j, k, and i make blockchain transactions (T_j^S, T_k^S, and T_i^S, respectively) which contain the proposed LC (blue arrow) for the purple scene S and some authorisation tokens that the robots must deposit along with every transaction. (d) There are two possible outcomes of a LC validation. (d.1) The triangle identity is validated and the robots receive back their authorisation tokens and one reputation token. (d.2) The triangle is rejected and none of the robots receives back their reputation tokens, since it is not possible to infer which robot is Byzantine. (Colour figure online)

the geometric test. However, two colluding Byzantine robots could bypass this protection mechanism. For this reason, we introduce a parameter called *security level* (discussed below) to increase security against collusion. During PGO, only the individual robots' maps linked by validated inter-robot loop closures are merged into a global map. Hence, our framework also prevents the use of unreliable Byzantine robot's trajectories for map generation because our triangle validation mechanism ensures that Byzantine-generated loop closures remain invalidated, thus preventing their inclusion in the PGO and the global map.

Reputation Management Mechanism. Besides protecting the process from the injection of incorrect loop closures, our smart contract is also designed to identify and neutralise Byzantine robots. It does so through a reputation management mechanism based on crypto tokens which are scarce digital tokens stored in the blockchain. Each robot starts with a given amount of crypto tokens loaded in its blockchain wallet (accessed with standard public-key encryption). The blockchain stores the history of every token transaction between any wallet. As every robot maintains a synchronised copy of this ledger, the reputation (crypto tokens) of every robot is publicly known. In our implementation, we employ two types of crypto tokens: authorisation tokens and reputation tokens. The former are tokens that each robot needs to deposit to make a transaction through which a new loop closure is proposed (Fig. 2a-c). The deposited authorisation tokens are withheld by the smart contract and returned to the robot only once the loop closure is validated (Fig. 2d). This mechanism fixes the maximum number of unvalidated loop closures that a robot can have, preventing Byzantine robots from flooding the blockchain with incorrect loop closures as they will eventually deplete their authorisation tokens. In this way, robots submitting incorrect information are neutralised as they run out of authorisation tokens. Because only a set of designated wallets (one per robot) can receive authorisation tokens, the system is protected from Sybil attacks, in which a single robot could validate its own loop closures by using different identities.

The identification of Byzantine robots can be achieved through reputation tokens which are emitted by the smart contract every time a triangle identity validates three loop closures. Each of the three robots that submitted these loop closures receives one reputation token, which is publicly stored in the blockchain. Robots that do not increase their reputation for a long period may be identified as Byzantine.

Security Level Parameter. The security level indicates how many times an inter-robot loop closure has been validated by a different triplet of robots and hence, how secure it is. When a new loop closure is proposed, its security level is zero. When the loop closure becomes part of a validation triangle with other two loop closures proposed by two different robots, its security level is set to one. Each time the loop closure is included in other validation triangles involving different robot pairs, its security level increases by one. In our work, as soon as a loop closure reaches the security level one, it is included in the PGO and the smart contract returns the deposited authorisation tokens to the robot. This means that we secured the system from individual Byzantine robots, but not from colluding ones. However, the minimum security level can be increased to protect the system against potential collusion of Byzantine robots, albeit at the cost of a higher latency. The smart contract also gives a reputation token to the robots each time their loop closures lead to a security level increment. Therefore, robots that submit valid loop closures increase their reputation over time as more peers validate the loop closures, increasing their security level.

5 Results

We test our approach through a series of 40-minute-long simulation experiments. The simulations are run in the Gazebo simulator [24], with teams of 8 TurtleBots3-Waffle robots [1] collaboratively mapping an environment sized

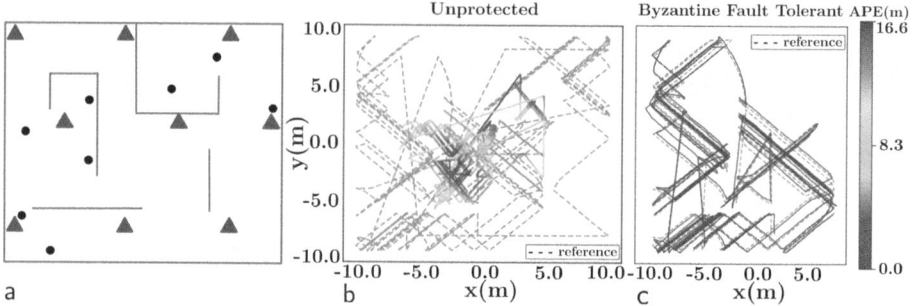

Fig. 3. (a) The tested environment with size $20\times22\,\text{m}^2$ and 9 scenes (purple triangles). The walls are depicted as black lines and the 8 TurtleBots3 robots as black dots. (b-c) Comparison between the ground truth (dashed grey lines) and the aggregated robot trajectories resulting from PGO of a representative run with 3 Byzantine robots that introduce loop closures with a constant error of 10 m. The trajectories are colour-coded (see right colour bar) indicating the error (APE) of each point. Swarm-SLAM without our blockchain-based protection layer produces trajectories with high error, whereas the error is close to zero when loop closures are validated by the smart contract. (Colour figure online)

$20\times22\,\text{m}^2$ with various walls separating the space and 9 scenes used for inter-robot loop closures (see Fig. 3a). Robots move through a random walk consisting of straight motion interrupted by random turns when an obstacle is detected at a distance smaller than 0.5 m; two robots can communicate within a maximum range of 5 m. Each robot acts as a blockchain node which maintains the blockchain and generates new blocks following the Proof-of-Authority consensus protocol. We employ the Toychain blockchain [37], a simple Python-based blockchain designed for scientific research. We integrated Swarm-SLAM using a smart contract that handles the security checks and crypto token distribution. The complete simulation software is open-source and available in two packages that are accessible from our project repository[1].

To measure the system's robustness, we compute two metrics using the *evo* software [16]: the Absolute Positional Error (APE) as the Euclidean distance between each point in the robot trajectories and the ground truth, and the Root Mean Square Error (RMSE) of these distances. We consider two types of Byzantine robots that differ in the error they apply to the loop closures. In both cases, Byzantine robots add a value—represented by a vector (ℓ_x, ℓ_y)—to the correct loop closure before broadcasting it to the other robots. In Fig. 4, we report the results for the case when Byzantine robots add a *constant* value $+10$ m to each vector component. In the supplementary material [32], we report the results for Byzantine robots that add a *random* value drawn from a uniform distribution $\mathcal{U}[-9, 9]$ m to each vector component (ℓ_x, ℓ_y). In both cases, the system without our protection layer suffers large errors as soon as one Byzantine robot is intro-

---

[1] https://github.com/clmoro/Blockchain-Based-BFT-Swarm-SLAM.git

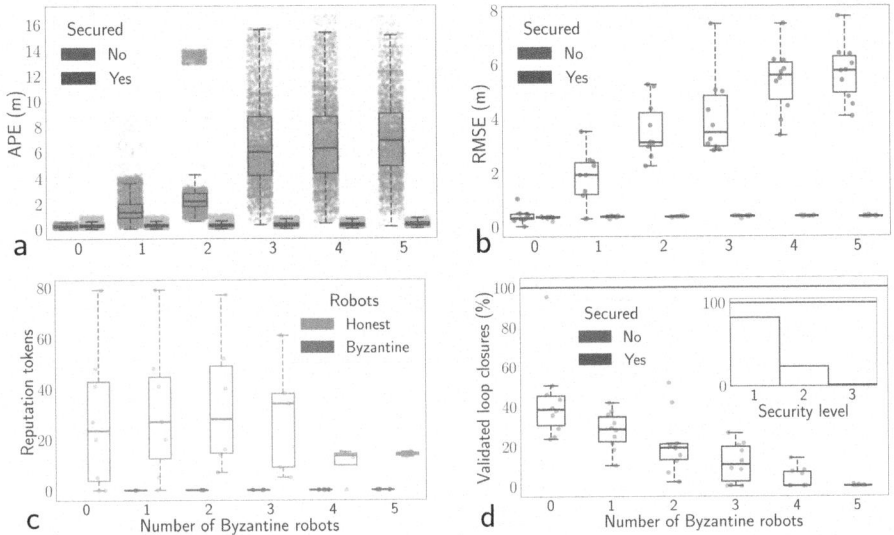

Fig. 4. (a-b) Comparison of the error in the aggregated robot trajectories after PGO for different numbers of Byzantine robots (x-axis) in a swarm of 8 robots using the original Swarm-SLAM (unsecured) or the Byzantine fault tolerant Swarm-SLAM (secured with smart contract). The boxplots in (a) show the APE for one representative simulation experiment while in (b) the RMSE is computed over 10 simulation runs (in the same environment but with different starting conditions). (c) Reputation tokens at the end (minute 40) of one representative experiment for Byzantines and non-Byzantine robots. The red boxes (on the left of each green box) are always flat at zero. (d) Proportion of validated loop closures that are used in the Swarm-SLAM's PGO (results for 10 simulation runs for each condition). The red line indicates that 100% of the proposed loop closures are used in PGO in the original Swarm-SLAM. In all panels, we show both the raw data (individual points) and the aggregated data distribution as boxplots (with interquartile range IQR box, median line, and whiskers to data within 1.5 IQR). The inset in d shows the proportion of loop closures reaching security level 1, 2, or 3 in the absence of Byzantine robots. The data for the inset are collected from one 45-min long run, independent from the main figure. (Colour figure online)

duced in the swarm, and the error increases with the number of Byzantine robots (Figs. 4a-b). Instead, when our blockchain-based smart contract secures the system, the error remains close to zero even when 5 of the 8 robots are Byzantine. The error is, however, never exactly zero because robots are subject to odometry noise. This odometry noise is smaller when there are more non-Byzantine robots (which compensate for each other's noise). Figures 3b-c show two examples of the accumulated APE on the aggregated robot trajectories with 3 Byzantine robots.

We also measure how the reputation tokens are distributed between Byzantine and non-Byzantine robots. Figure 4c shows that the number of reputation tokens at the end of the simulation allows distinguishing between the two types of robots as the Byzantine robots never receive any reputation token (red boxplot is flat at zero) because, in our experiments, they always submit incorrect loop closures that never get validated. Figure 4d shows that our security layer comes at the cost of reduced speed as the number of loop closures processed by the PGO is lower than half even in the absence of any Byzantine robots. We recall that the PGO only processes loop closures validated by the smart contract (y-axis of Fig. 4d). After 40 min, only a portion of loop closures are validated (and thus processed by the PGO), meaning that the other pending loop closures will be validated at a later time (thus adding some latency between the creation of an inter-robot loop closure and its use in the PGO). As the number of Byzantine robots increases, the number of validated loop closures decreases because fewer non-Byzantine robots contribute with correct data and forming loop-closure triangles becomes slower. Increasing the security level (i.e., number of validations before using the loop closure in the PGO) is another factor negatively impacting the system efficiency, yet protecting the system against Byzantine robot collusion. Figure 4d's inset shows that most loop closures achieve security level 1, however reaching levels 2 and 3 is less frequent in a swarm of 8 robots as each loop closure must be validated by more than half of the swarm.

6 Discussion and Conclusion

Swarm-SLAM [25] is a promising framework to enable robot swarms to perform decentralised collaborative mapping of unknown environments. However, our analysis shows that Swarm-SLAM is highly vulnerable to the presence of even a single Byzantine robot that shares incorrect loop closures or tampers with the pose graph optimisation (PGO) process. Through a blockchain-based smart contract that uses geometric constraints among loop closures to check their validity before using them to build the collective map (PGO step), we considerably improve Swarm-SLAM's security against Byzantine robots. However, this increased security comes at the cost of an increased latency between the computation of a loop closure and its use in the PGO, reducing the system's mapping speed. The proposed method allows both the identification and the neutralisation of Byzantine robots through the use of two types of crypto tokens that assign reputation (for identification) and rights to participate (for neutralisation).

Future research should investigate situations in which (i) Byzantine robots collude with each other to validate incorrect loop closures and (ii) Byzantine robots dynamically change their behaviour, proposing a mix of correct and incorrect loop closures. While we expect that dynamic Byzantines are harder to identify and neutralise, we expect that our solution will still prevent them from corrupting the map generation. Future research should also extend the smart contract to protect the system against PGO tampering through peer-reviewing of each other contributions, similar to our proposed loop closure peer validation.

Acknowledgements. We thank Miquel Kegeleirs, David Garzón Ramos, and Guillermo Legarda Herranz for the helpful discussions. V.S. and M.D. acknowledge support from the Belgian F.R.S.-FNRS. A.R. acknowledges support from DFG under Germany's Excellence Strategy - EXC 2117-422037984.

References

1. Amsters, R., Slaets, P.: Turtlebot 3 as a robotics education platform. In: Merdan, M., Lepuschitz, W., Koppensteiner, G., Balogh, R., Obdržálek, D. (eds.) RiE 2019. AISC, vol. 1023, pp. 170–181. Springer, Cham (2020). https://doi.org/10.1007/978-3-030-26945-6_16
2. Aswale, A., López, A., Ammartayakun, A., Pinciroli, C.: Hacking the colony: on the disruptive effect of misleading pheromone and how to defend against it. In: AAMAS 2022: Proceedings of the 21st International Conference on Autonomous Agents and Multiagent Systems, pp. 27–34. IFAAMAS, Richland, SC (2022)
3. Ayache, N., Faugeras, O.: Building, registrating and fusing noisy visual maps. Int. J. Robot. Res. **7**(6), 45–65 (1988). https://doi.org/10.1177/027836498800700605
4. Campos, M., Chanel, C., Chauffaut, C., Lacan, J.: Towards a blockchain-based multi-UAV surveillance system. Front. Robot. AI **8**, 557692 (2021). https://doi.org/10.3389/frobt.2021.557692
5. Chang, Y., et al.: LAMP 2.0: A robust multi-robot SLAM system for operation in challenging large-scale underground environments. IEEE Robot. Autom. Lett. 9175–9182 (2022). https://doi.org/10.1109/LRA.2022.3191204
6. Chen, W., et al.: Overview of multi-robot collaborative SLAM from the perspective of data fusion. Machines **11**(6), 653 (2023). https://doi.org/10.3390/machines11060653
7. Cieslewski, T., Choudhary, S., Scaramuzza, D.: Data-efficient decentralized visual SLAM. In: Proceedings of the 2018 IEEE International Conference on Robotics and Automation (ICRA), pp. 2466–2473. IEEE (2018).https://doi.org/10.1109/ICRA.2018.8461155
8. Cramariuc, A., et al.: maplab 2.0 – a modular and multi-modal mapping framework. IEEE Robot. Autom. Lett. **8**(2), 520–527 (2023). https://doi.org/10.1109/lra.2022.3227865
9. Crowley, J.L.: World modeling and position estimation for a mobile robot using ultrasonic ranging. In: Proceedings of the 1989 International Conference on Robotics and Automation (ICRA), vol. 2, pp. 674–680 (1989).https://doi.org/10.1109/ROBOT.1989.100062
10. Dorigo, M., Pacheco, A., Reina, A., Strobel, V.: Blockchain technology for mobile multi-robot systems. Nat. Rev. Electr. Eng. **1**(4), 264–274 (2024). https://doi.org/10.1038/s44287-024-00034-9
11. Dorigo, M., Theraulaz, G., Trianni, V.: Swarm robotics: past, present, and future [point of view]. Proc. IEEE **109**(7), 1152–1165 (2021). https://doi.org/10.1109/JPROC.2021.3072740
12. Durrant-Whyte, H., Bailey, T.: Simultaneous localization and mapping: part i. IEEE Robot. Autom. Mag. **13**(2), 99–110 (2006). https://doi.org/10.1109/MRA.2006.1638022
13. Dwork, C., Lynch, N., Stockmeyer, L.: Consensus in the presence of partial synchrony. J. ACM **35**(2), 288–323 (1988). https://doi.org/10.1145/42282.42283

14. Engel, J., Schöps, T., Cremers, D.: LSD-SLAM: large-scale direct monocular SLAM. In: Fleet, D., Pajdla, T., Schiele, B., Tuytelaars, T. (eds.) ECCV 2014. LNCS, vol. 8690, pp. 834–849. Springer, Cham (2014). https://doi.org/10.1007/978-3-319-10605-2_54
15. Fernandez-Cortizas, M., Bavle, H., Perez-Saura, D., Sanchez-Lopez, J.L., Campoy, P., Voos, H.: Multi S-graphs: an efficient distributed semantic-relational collaborative SLAM. IEEE Robot. Autom. Lett. **9**(6), 6004–6011 (2022). https://doi.org/10.1109/LRA.2024.3399997
16. Grupp, M.: EVO: python package for the evaluation of odometry and SLAM. https://github.com/MichaelGrupp/evo (2017)
17. Guerrero-Bonilla, L., Prorok, A., Kumar, V.: Formations for resilient robot teams. IEEE Robot. Autom. Lett. **2**, 841–848 (2017). https://doi.org/10.1109/LRA.2017.2654550
18. Hamann, H.: Swarm Robotics: A Formal Approach. Springer, Cham (2018). https://doi.org/10.1007/978-3-319-74528-2
19. Higgins, F., Tomlinson, A., Martin, K.M.: Survey on security challenges for swarm robotics. In: 2009 Fifth International Conference on Autonomic and Autonomous Systems, pp. 307–312. IEEE (2009). https://doi.org/10.1109/ICAS.2009.62
20. Huang, Y., Shan, T., Chen, F., Englot, B.: DiSCo-SLAM: distributed scan context-enabled multi-robot LiDAR SLAM with two-stage global-local graph optimization. IEEE Robot. Autom. Lett. **7**(2), 1150–1157 (2022). https://doi.org/10.1109/LRA.2021.3138156
21. Hunt, E., Hauert, S.: A checklist for safe robot swarms. Nat. Mach. Intell. **2**, 420–422 (2020). https://doi.org/10.1038/s42256-020-0213-2
22. Kegeleirs, M., Garzón Ramos, D., Birattari, M.: Random walk exploration for swarm mapping. In: Althoefer, K., Konstantinova, J., Zhang, K. (eds.) TAROS 2019. LNCS (LNAI), vol. 11650, pp. 211–222. Springer, Cham (2019). https://doi.org/10.1007/978-3-030-25332-5_19
23. Kegeleirs, M., Grisetti, G., Birattari, M.: Swarm SLAM: challenges and perspectives. Front. Robot. AI **8**, 618268 (2021). https://doi.org/10.3389/frobt.2021.618268
24. Koenig, N., Howard, A.: Design and use paradigms for Gazebo, an open-source multi-robot simulator. In: 2004 IEEE/RSJ International Conference on Intelligent Robots and Systems (IROS), vol. 3, pp. 2149–2154 (2004). https://doi.org/10.1109/IROS.2004.1389727
25. Lajoie, P.Y., Beltrame, G.: Swarm-slam: sparse decentralized collaborative simultaneous localization and mapping framework for multi-robot systems. IEEE Robot. Autom. Lett. **9**(1), 475–482 (2024). https://doi.org/10.1109/LRA.2023.3333742
26. Lajoie, P.Y., Hu, S., Beltrame, G., Carlone, L.: Modeling perceptual aliasing in SLAM via discrete–continuous graphical models. IEEE Robot. Autom. Lett. **4**(2), 1232–1239 (2019). https://doi.org/10.1109/lra.2019.2894852
27. Lajoie, P.Y., Ramtoula, B., Chang, Y., Carlone, L., Beltrame, G.: DOOR-SLAM: distributed, online, and outlier resilient SLAM for robotic teams. IEEE Robot. Autom. Lett. **5**(2), 1656–1663 (2020). https://doi.org/10.1109/lra.2020.2967681
28. Lajoie, P.Y., Ramtoula, B., Wu, F., Beltrame, G.: Towards collaborative simultaneous localization and mapping: a survey of the current research landscape. Field Robot. **2**(1), 971–1000 (2022). https://doi.org/10.55417/fr.2022032
29. Lamport, L., Shostak, R., Pease, M.: The Byzantine Generals Problem. ACM Trans. Program. Lang. Syst. **4**(3), 382–401 (1982). https://doi.org/10.1145/357172.357176

30. Lourakis, M., Argyros, A.: SBA: a software package for generic sparse bundle adjustment. ACM Trans. Math. Softw. **36**(1), 1–30 (2009). https://doi.org/10.1145/1486525.1486527
31. Macenski, S., Foote, T., Gerkey, B., Lalancette, C., Woodall, W.: Robot operating system 2: design, architecture, and uses in the wild. Sci. Robot. **7**(66), eabm6074 (2022). https://doi.org/10.1126/scirobotics.abm6074
32. Moroncelli, A., Pacheco, A., Strobel, V., Lajoie, P.Y., Dorigo, M., Reina, A.: Supplementary material for the paper: byzantine fault detection in swarm-SLAM using blockchain and geometric constraints (2024). https://sites.google.com/view/bft-swarm-slam
33. Mur-Artal, R., Montiel, J., Tardos, J.: ORB-SLAM: a versatile and accurate monocular SLAM system. IEEE Trans. Rob. **31**(5), 1147–1163 (2015). https://doi.org/10.1109/TRO.2015.2463671
34. Nakamoto, S.: Bitcoin: a peer-to-peer electronic cash system. electronic document. http://www.bitcoin.org (2008)
35. Nerurkar, E.D., Roumeliotis, S.I., Martinelli, A.: Distributed maximum a posteriori estimation for multi-robot cooperative localization. In: Proceedings of the 2009 IEEE International Conference on Robotics and Automation (ICRA), pp. 1402–1409. IEEE (2009). https://doi.org/10.1109/ROBOT.2009.5152398
36. Newcombe, R.A., Lovegrove, S.J., Davison, A.J.: DTAM: dense tracking and mapping in real-time. In: 2011 International Conference on Computer Vision, pp. 2320–2327 (2011). https://doi.org/10.1109/ICCV.2011.6126513
37. Pacheco, A., Denis, U., Zakir, R., Strobel, V., Reina, A., Dorigo, M.: Toychain: a simple blockchain for research in swarm robotics (2024). https://doi.org/10.48550/arXiv.2407.06630, arXiv preprint:2407.06630 [cs.RO]
38. Pacheco, A., Strobel, V., Dorigo, M.: A blockchain-controlled physical robot swarm communicating via an Ad-Hoc network. In: Dorigo, M., et al. (eds.) ANTS 2020. LNCS, vol. 12421, pp. 3–15. Springer, Cham (2020). https://doi.org/10.1007/978-3-030-60376-2_1
39. Pacheco, A., Strobel, V., Reina, A., Dorigo, M.: Real-time coordination of a foraging robot swarm using blockchain smart contracts. In: Dorigo, M., et al. Swarm Intelligence, ANTS 2022, LNCS, vol. 13491, pp. 196–208. Springer, Cham (2022). https://doi.org/10.1007/978-3-031-20176-9_16
40. Placed, J.A., et al.: A survey on active simultaneous localization and mapping: state of the art and new frontiers. IEEE Trans. Robot. **39**(3), 1686–1705 (2022). https://doi.org/10.1109/TRO.2023.3248510
41. Prorok, A., Malencia, M., Carlone, L., Sukhatme, G.S., Sadler, B.M., Kumar, V.: Beyond robustness: A taxonomy of approaches towards resilient multi-robot systems (2021). https://doi.org/10.48550/arXiv.2109.12343, arXiv preprint:2109.12343 [cs.RO]
42. Queralta Peña, J., Qingqing, L., Zou, Z., Westerlund, T.: Enhancing autonomy with blockchain and multi-access edge computing in distributed robotic systems. In: 2020 Fifth International Conference on Fog and Mobile Edge Computing (FMEC), pp. 180–187. IEEE (2020). https://doi.org/10.1109/FMEC49853.2020.9144809
43. Reina, A.: Robot teams stay safe with blockchains. Nat. Mach. Intell. **2**, 240–241 (2020). https://doi.org/10.1038/s42256-020-0178-1
44. Rodriguez-Losada, D., Matia, F., Jimenez, A.: Local maps fusion for real time multirobot indoor simultaneous localization and mapping. In: Proceedings of the 2024 IEEE International Conference on Robotics and Automation (ICRA), vol. 2, pp. 1308–1313. IEEE (2004). https://doi.org/10.1109/ROBOT.2004.1308005

45. Saeedi, S., Trentini, M., Seto, M., Li, H.: Multiple-robot simultaneous localization and mapping: a review. J. Field Robot. **33**(1), 3–46 (2016). https://doi.org/10.1002/rob.21620
46. Schmuck, P., Ziegler, T., Karrer, M., Perraudin, J., Chli, M.: COVINS: visual-inertial SLAM for centralized collaboration. In: 2021 IEEE International Symposium on Mixed and Augmented Reality Adjunct (ISMAR-Adjunct), pp. 171–176. IEEE Computer Society, Los Alamitos, CA, USA (2021). https://doi.org/10.1109/ISMAR-Adjunct54149.2021.00043
47. Smith, R.C., Cheeseman, P.: On the representation and estimation of spatial uncertainty. The Int. J. Robot. Res. **5**(4), 56–68 (1986). https://doi.org/10.1177/027836498600500404
48. Strobel, V., Castelló Ferrer, E., Dorigo, M.: Managing byzantine robots via blockchain technology in a swarm robotics collective decision making scenario. In: Proceedings of 17th International Conference on Autonomous Agents and MultiAgent Systems, AAMAS 2018, pp. 541–549. IFAAMAS, Richland, SC (2018)
49. Strobel, V., Castelló Ferrer, E., Dorigo, M.: Blockchain technology secures robot swarms: a comparison of consensus protocols and their resilience to byzantine robots. Front. Robot. AI **7**, 54 (2020). https://doi.org/10.3389/frobt.2020.00054
50. Strobel, V., Pacheco, A., Dorigo, M.: Robot swarms neutralize harmful Byzantine robots using a blockchain-based token economy. Sci. Robot. **8**(79), eabm4636 (2023). https://doi.org/10.1126/scirobotics.abm4636
51. Tian, Y., Chang, Y., Herrera Arias, F., Nieto-Granda, C., How, J.P., Carlone, L.: Kimera-multi: robust, distributed, dense metric-semantic SLAM for multi-robot systems. IEEE Trans. Rob. **38**(4), 2022–2038 (2022). https://doi.org/10.1109/TRO.2021.3137751
52. Van Calck, L., Pacheco, A., Strobel, V., Dorigo, M., Reina, A.: A blockchain-based information market to incentivise cooperation in swarms of self-interested robots. Sci. Rep. **13**, 20417 (2023). https://doi.org/10.1038/s41598-023-46238-1
53. Wood, G.: Ethereum: a secure decentralized generalised transaction ledger. Ethereum Found. **151**, 1–41 (2014). https://ethereum.github.io/yellowpaper/paper.pdf
54. Zhang, Y., Wu, Y., Tong, K., Chen, H., Yuan, Y.: Review of visual simultaneous localization and mapping based on deep learning. Remote Sens. **15**(11), 2740 (2023). https://doi.org/10.3390/rs15112740
55. Zhao, H., et al.: A generic framework for byzantine-tolerant consensus achievement in robot swarms. In: Proceedings of the 2023 IEEE/RSJ International Conference on Intelligent Robots and Systems (IROS), pp. 8839–8846. IEEE (2023). https://doi.org/10.1109/IROS55552.2023.10341423
56. Zhong, S., Qi, Y., Chen, Z., Wu, J., Chen, H., Liu, M.: DCL-SLAM: a distributed collaborative LiDAR SLAM framework for a robotic swarm. IEEE Sens. J. **24**(4), 4786–4797 (2024). https://doi.org/10.1109/JSEN.2023.3345541

Collective Bayesian Decision-Making in a Swarm of Miniaturized Robots for Surface Inspection

Thiemen Siemensma[1]($^{\boxtimes}$), Darren Chiu[2], Sneha Ramshanker[3], Radhika Nagpal[3], and Bahar Haghighat[1]

[1] University of Groningen, Groningen, The Netherlands
{t.j.j.siemensma,bahar.haghighat}@rug.nl
[2] University of Southern California, Los Angeles, CA, USA
chiudarr@usc.edu
[3] Princeton University, Princeton, NJ, USA
{sr6848,rn1627}@princeton.edu

Abstract. Robot swarms can effectively serve a variety of sensing and inspection applications. Certain inspection tasks require a binary classification decision. This work presents an experimental setup for a surface inspection task based on vibration sensing and studies a Bayesian two-outcome decision-making algorithm in a swarm of miniaturized wheeled robots. The robots are tasked with individually inspecting and collectively classifying a 1 m × 1 m tiled surface consisting of vibrating and non-vibrating tiles based on the majority type of tiles. The robots sense vibrations using onboard IMUs and perform collision avoidance using a set of IR sensors. We develop a simulation and optimization framework leveraging the Webots robotic simulator and a Particle Swarm Optimization (PSO) method. We consider two existing information sharing strategies and propose a new one that allows the swarm to rapidly reach accurate classification decisions. We first find optimal parameters that allow efficient sampling in simulation and then evaluate our proposed strategy against the two existing ones using 100 randomized simulation and 10 real experiments. We find that our proposed method compels the swarm to make decisions at an accelerated rate, with an improvement of up to 20.52% in mean decision time at only 0.78% loss in accuracy.

1 Introduction

Over the last few decades, automated inspection systems have increasingly become a valuable tool across various industries [7,8,24,25]. Studies have addressed applications in agricultural and hull inspection as well as infrastructure and wind-turbine maintenance [9,17,19,20]. Vibration analysis is a valuable tool in these inspection processes. Different types of vibration analysis are used to detect the condition of infrastructure through structural properties such as modal shapes and eigenfrequencies [2,12,21]. A class of inspection tasks involves making a binary decision about a spatially distributed feature of the inspected system. This type of decisions can be effectively addressed by a swarm of robots

[26,30]. When compared to a single entity, swarms improve decision time and accuracy by leveraging collective perception [29,31]. Moreover, swarms eliminate the problem of sensor-placement and can provide a high-resolution map of the environment [5,6]. Collective decision-making algorithms often draw inspiration from nature, such as groups of ants and bees [27]. A more mathematical approach is found in Bayesian algorithms. Applications of Bayesian algorithms have been studied in sensor networks [1,22] as well as robot swarms [13,14,16,29,31]. In the study outlined in [13], a collective of agents must determine whether the predominant color of a checkered surface pattern is black or white. The robots function as Bayesian modellers, exchanging information based on two information sharing strategies. The robots either (i) continuously broadcast their ongoing binary observations (*no feedback*) or (ii) continuously broadcast their irreversible decisions once reached (*positive feedback*), pushing the swarm to consensus. A common problem with collective perception methods is slow convergence, i.e. difficulty reaching high belief (probability) about the predominant color. Spatial correlation of observations cause the belief to fluctuate, resulting in long decision times and low decision accuracies, as shown by [3,4,14].

In this work, we build on top of the work in [13] in two ways. First, we propose a novel information sharing strategy, named *soft-feedback*. In this approach, binary information is shared between the robots based on their current belief, similar to [4,28], with the addition of sample variance and random sampling. The resulting strategy is shown to enhance convergence compared to no feedback and positive feedback, without compromising the accuracy of decisions. Second, we move away from the agent-based simulation setup of [13] and present a real experimental setup built around 3-cm-sized vibration sensing wheeled robots. We utilize vibration signals in the presence of measurement noise in place of simulated binary floor color observations. We develop a new sensor board to allow the robots to perform collision avoidance. A digital twin of our experimental setup is developed in Webots, derived from the work done in [10]. We use this model for calibration of our robots and optimization of the algorithm parameters in a Particle Swarm Optimization (PSO) loop.

2 Problem Definition

We task a swarm of N robots to individually inspect and collectively classify a 2D tiled surface section. The surface comprises two types of tiles, vibrating and non-vibrating tiles. The swarm must determine whether the tiled surface is majority vibrating or majority non-vibrating. We denote the fill-ratio f as the proportion of vibrating tiles. Thus, fill-ratios close to 0.5 represent a hard surface inspection problem, as the amount of vibrating and non-vibrating tiles is almost equal. The robots each individually inspect the surface, share their information with the rest of the swarm, and collectively classify the surface. The inspection task ends when every robot in the swarm reaches a final decision, determining if the fill-ratio is above or below 0.5. An underlying real-world scenario could involve inspecting a surface section and determining whether the surface is in a majority *healthy* or in a majority *unhealthy* state.

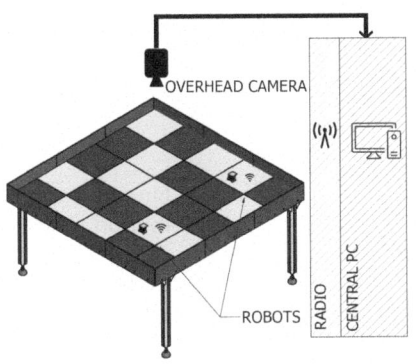
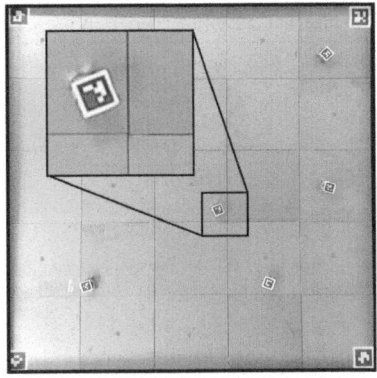

(a) Overall experimental setup.   (b) View from overhead camera.

Fig. 1. The experimental setup with a fill ratio of $f = \frac{12}{25} = 0.48$. (a) Schematic overview of the setup is shown. The central PC uses the radio and camera data for analysis. Vibration-motors are attached on the bottom side of white tiles. (b) A snapshot from the overhead camera with detailed view (black square). The red dot markings indicate vibrating tiles. Each robot carries a unique AruCo marker for tracking. AruCo markers in the corners of the environment mark the boundaries.

3 Inspection Algorithm

Algorithm 1 shows our collective Bayesian decision making algorithm. The robots individually estimate and classify the fill ratio f as above or below 0.5. Each robot acts as a Bayesian modeler integrating personal observations and information broadcast by other robots. We consider three information sharing strategies: the (i) no feedback (u^-) and (ii) positive feedback (u^+) strategies, previously studied in [13], and the (iii) soft feedback (u^s) strategy, which we propose in this work.

The robots make binary observations of the surface condition as black/white in simulation or vibrating/non-vibrating in the real setup as $O \in \{0,1\}$:

$$O \sim \text{Bernoulli}(f) \quad (1)$$

The fill ratio $f \in [0,1]$ is unknown and is modeled by a Beta-distribution:

$$f \sim \text{Beta}(\alpha, \beta) \quad (2)$$

The prior distribution of f is initialized as $\text{Beta}(\alpha_0 = 1, \beta_0 = 1)$. Upon sampling or receiving observations from other robots, the posterior of f is updated as:

$$f \mid O \sim \text{Beta}(\alpha + O, \beta + (1 - O)) \quad (3)$$

The robots perform a Levy-flight type random walk, moving forward for a time drawn from a Cauchy distribution with mean γ_0 and average absolute deviation γ followed by turning a uniform random angle $\phi \sim U(-\pi, \pi)$ in the direction of $\text{sign}(\phi)$ relative to the forward driving direction. The robots perform collision

Algorithm 1. Collective Bayesian Decision Making

Inputs: $u^-, u^+, u^s, T_{end}, \theta_o, \eta, \theta_c, \tau, p_c$
Initialize: $\alpha = 1, \beta = 1, d_f = -1, \text{robot } id, t_s = 0$
while $t < T_{end}$ **do**
  Perform random walk for τ time
  if $t - t_s > \tau$ **then**
    $O \leftarrow$ Observation   ▷ Get binary observation
    $t_s \leftarrow t$ ▷ Observation timestamp
    $\text{Beta}(\alpha, \beta) \leftarrow \text{Beta}(\alpha + O, \beta + (1 - O))$ ▷ Update modeling of f
    $p \leftarrow P(\text{Beta}(\alpha, \beta) < 0.5)$ ▷ Update belief on f
    $O_{count} \leftarrow O_{count} + 1$ ▷ Observation count
  end if
  if (u^s **Or** $d_f == -1$) **And** ($O_{count} > \theta_o$) **then**
    if $p > p_c$ **then**
      $d_f \leftarrow 0$
    else if $(1 - p) > p_c$ **then**
      $d_f \leftarrow 1$
    end if
  end if
  if u^s **then**
    $\Gamma \leftarrow \text{Var(Beta)}$
    $m \leftarrow \text{Bernoulli}\left((1-p)e^{-\eta\Gamma}(\frac{1}{2}-p)^2 + O(1 - e^{-\eta\Gamma}(\frac{1}{2}-p)^2)\right)$
    $\text{Broadcast}(m)$ ▷ Soft feedback
  else if (u^+ **And** $d_f \neq -1$) **then**
    $\text{Broadcast}(d_f)$ ▷ Positive feedback
  else
    $\text{Broadcast}(O)$ ▷ No feedback
  end if
  if Message in queue **then**
    $O \leftarrow$ Message   ▷ Receive message from swarm
    $\text{Beta}(\alpha, \beta) \leftarrow \text{Beta}(\alpha + O, \beta + (1 - O))$ ▷ Update modeling of f
  end if
end while

avoidance upon detecting an obstacle within a range of θ_c millimeters, by turning a random angle $\phi \sim U(-\pi, \pi)$ in the direction of $\text{sign}(\phi)$ relative to the forward driving direction.

Every τ milliseconds a robot samples a new observation O. In simulation, O is based on a binary floor color sampling. In experiments, O is calculated using a 500 millisecond vibration signal sample. The DC component of this sample is removed by employing a first-order high-pass filter with cutoff frequency $\omega_n = 40$ Hz. Given a sampling rate of 350 Hz, the filter parameters α_1, α_2, and α_3 are configured to values: 0.20, 0.60, and -0.60 respectively. We define the filtered signal at time step i as $\hat{a}_i$:

$$\hat{a}_i := \alpha_1 \hat{a}_{i-1} + \alpha_2 a_i + \alpha_3 a_{i-1} \tag{4}$$

where a_i is the magnitude of the IMU's raw acceleration data a_x, a_y, and a_z. The Root-Mean-Square (RMS) of $\hat{a}$ returns the energy of the signal as $\hat{E} = \sqrt{\frac{1}{n}\sum_{i=1}^{n} \hat{a}_i^2}$. Subsequently, the observation O is determined by comparing $\hat{E}$ with a threshold θ_E:

$$O = \begin{cases} 1 & \text{if } \hat{E} > \theta_E \\ 0 & \text{if } \hat{E} \leq \theta_E \end{cases} \quad (5)$$

We consider three information sharing strategies. (i) No feedback (u^-) considers sharing the latest observation O in any case. (ii) Positive feedback (u^+) is similar to (u^-) until reaching a final decision. From this point in time it broadcasts its irreversible final decision d_f. The intuition behind positive feedback is to push the swarm to consensus upon reaching a final decision by one robot. However, this is not very effective when no robot is able to reach a final decision. In this case, positive feedback is not different from no feedback. To resolve this, we propose soft-feedback (iii). Soft feedback (u^s) broadcasts a binary value sampled from a Bernoulli distribution. The corresponding probability is calculated through the soft feedback parameter $\eta \in \mathbb{R}^+$, the current observation $O \in \{0, 1\}$, and the current belief $p \in [0, 1]$ as below:

$$m \sim \text{Bernoulli}\left(\delta \cdot (1 - p) + (1 - \delta) \cdot O\right) \quad (6a)$$

$$\delta = e^{-\eta \Gamma} (\frac{1}{2} - p)^2 \quad (6b)$$

$$p = P(\text{Beta}(\alpha, \beta) < 0.5) \quad (6c)$$

where $m \in \{0, 1\}$ is the outgoing message, Γ is the variance of the Beta distribution and p is the robot's belief evaluated as the CDF of the Beta distribution at $f = 0.5$. The intuition behind soft-feedback is to broadcast information that is initially incorporating only the observation O, but gradually factors in more of the belief p. Namely, Eq. 6b depends on (i) a compelling component $e^{-\eta \Gamma} \in [0, 1]$ which increases the proportion of the current belief in messages as Γ decreases, and (ii) a stabilizing component $(\frac{1}{2} - p)^2 \in [0, 0.25]$ which is the squared distance of p from the indecisive state $p = 0.5$, increasing the proportion of robot's belief in the Bernoulli sampled message to enhance accuracy. This prevents the robot from prematurely making a decision with low confidence.

Upon reaching a minimum of θ_o number of observations, a robot considers making a final decision based on its belief p. If above the credibility threshold $p > p_c$, the robot's final decision d_f is set to 0. Conversely, if $(1 - p) > p_c$, d_f is set to 1. The inspection task ends when all robots have made a final decision.

4 Real Experimental Setup

Our experimental setup, shown in Fig. 1, is built around (i) a tiled surface section of size $1\,\text{m} \times 1\,\text{m}$, and (ii) a swarm of 3 cm-sized vibration-sensing wheeled robots that traverse and inspect the tiled surface section. The surface section consists

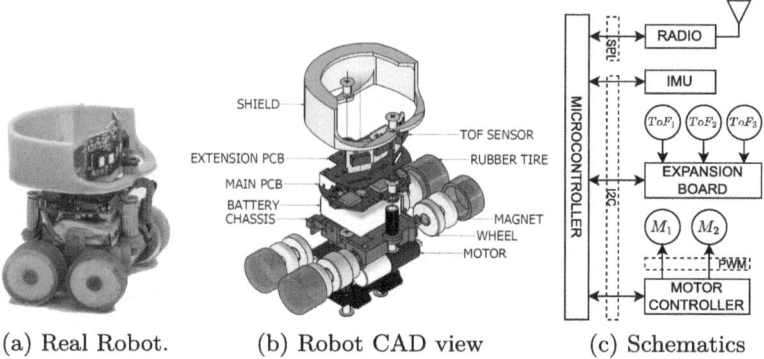

(a) Real Robot.   (b) Robot CAD view   (c) Schematics

Fig. 2. We use a revised and extended version of the original Rovable robot [9]. (a) The extended robot with IR sensor board. (b) Exploded 3D CAD view of the extended robot. (c) Electronic block diagram of the extended robot. The microcontroller (Atmel SAMD21G18) interfaces with the IMU (MPU6050), 2.4 GHz radio (nRF24L01+), motor-controllers (DRV8835), and ToF IR sensors (VL53L1X).

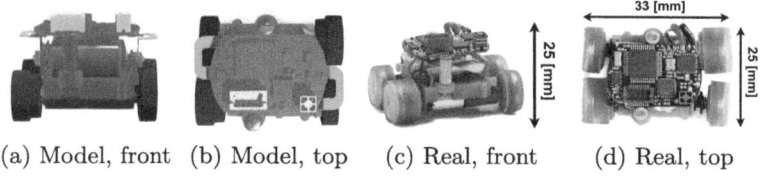

(a) Model, front   (b) Model, top   (c) Real, front   (d) Real, top

Fig. 3. We use a simpler robot model in simulation, with IR sensors directly simulated on the main PCB. Our simulated collision avoidance closely matches reality.

of 25 tiles, each of size 20 cm × 20 cm, that are laid out in a square grid with five tiles on each side. There are two types of tiles on the surface, vibrating and non-vibrating tiles. The vibrating tiles are excited using two miniature vibration motors mounted on top of one another underneath the tile at its center (ERM 3V Seeed Technology motors). All tiles are secured to an aluminium frame using 2 cm × 1 cm pieces of magnetic tape around the corners. The frame consists of four strut profiles in the middle and four others along the edges of the arena. We use an overhead camera (Logitech BRIO 4k) and ArUco markers for visual tracking of the robots.

We use a revised and extended version of the Rovable robot originally presented in [11]. As shown in Fig. 2, each robot measures 25mm × 33mm × 35mm and carries two customized Printed Circuit Boards (PCBs): (i) a main PCB hosting the micro-controller, an IMU, motor controllers, power circuitry and radio and (ii) an extension PCB hosting IR sensors for collision avoidance. The main PCB has essentially the same design as the one in [11], and was only revised and remade for updated components. The extension PCB is new and hosts three small Time-of-Flight (ToF) IR sensor boards, each facing a direction of 0°, 25°, and −25° relative to the forward driving direction of the robot. Each sensor has a field of view of 27° and a range of up to 1m. A 3D printed shield is mounted around the extension PCB to enhance the visibility of the robot when perceived by the IR sensors of other robots. The robot has four magnetic wheels. Only two wheels, one on the front and one on the back, are driven by PWM operated motors. At 100% PWM, the robot drives forward at around 5 cm per second.

5 Simulation and Optimization Framework

Our simulation framework provides a virtual environment where we can study the operation of our robot swarm. Within Webots, we set up two main components: (i) a realistic model of our robot, and (ii) a tiled surface that the robots inspect, with a black and white projected floor pattern. We use the black and white tiles in simulation as a proxy for vibrating and non-vibrating tiles in our real experimental setup. In simulation, we assume noise-free binary sampling of the surface and zero loss on inter-robot communication. Figure 3 shows the simulated and the real robots side by side. In simulation, the ToF sensor board is absent, but simulated IR sensors retain comparable range and positioning. We recreate mechanical differences that exist between real robots by adding randomized offsets to the simulated left and right motor speed commands:

$$M_l^s \leftarrow M_l^s r_v (1 - r_a) \tag{7a}$$

$$M_r^s \leftarrow M_r^s r_v (1 + r_a) \tag{7b}$$

where M_l^s and M_r^s are the left and right motor speeds, and r_v and r_a are drawn from empirically chosen uniform distributions $U \sim (0.95, 1.05)$ and $U \sim (-0.125, 0.125)$, respectively. We calibrate our simulation empirically considering three characteristic features: (i) sample distribution over the experimental setup, (ii) time between consecutive samples, and (iii) distance between consecutive samples. To obtain data for our calibrations, we run experiments using five robots for 3×20 min with algorithm parameters $[\gamma, \gamma_0, \tau, \theta_c] = [5000, 2000, 1500, 60]$ using the u^- information sharing strategy (Sect. 3). The Pearson correlation coefficients for the obtained data (partly shown in Fig. 4) corresponding to the three features mentioned above are calculated as 0.990, 0.984, and 0.735, respectively. These values confirm the similarity between simulated and real swarm behaviors, enabling optimizing real experiments using simulation.

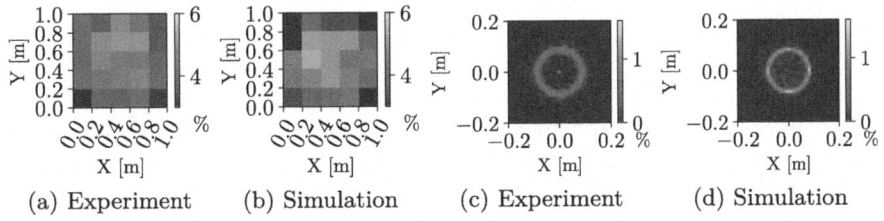

Fig. 4. Distribution of samples across tiles in real (a) and simulated (b) setups. The displacement between consecutive samples in real (c) and simulated (d) setups.

Our optimization framework involves two components: (i) our calibrated simulation and (ii) a noise-resistant PSO method. Throughout the PSO iterations, every particle is evaluated multiple times on randomized floor patterns with the same fill-ratio. The velocity and position of particle i are updated at iteration k as:

$$\mathbf{v}_i^{k+1} = \omega \cdot \mathbf{v}_i^k + \omega_p \cdot \mathbf{r_1}(\mathbf{p}_{b_i} - \mathbf{p}_i^k) + \omega_g \cdot \mathbf{r_2}(\mathbf{g}_b - \mathbf{p}_i^k) \tag{8a}$$

$$\mathbf{p}_i^{k+1} = \mathbf{p}_i^k + \mathbf{v}_i^{k+1} \tag{8b}$$

where $\mathbf{v}_i^k$ and $\mathbf{p}_i^k$ are the velocity and position vector of particle i at iteration k. $\mathbf{p}_{b_i}$ and $\mathbf{g}_b$ correspond to the position vector of the personal best and global best evaluations for particle i, respectively. We set the PSO weights for inertia, personal best, and global best as $[\omega \ \omega_p \ \omega_g] = [0.75 \ 1.5 \ 1.5]$, balancing local and global exploration [15,18,23]. The values $\mathbf{r_1}$ and $\mathbf{r_2}$ are drawn from a uniform distribution $U \sim (0,1)$ each iteration.

6 Experiments and Results

We use a swarm of five robots, all employing $p_c = 0.95$, and floor patterns with a fill ratio of $f = 0.48$ to conduct simulation and real experiments. We evaluate on decision time and accuracy for the strategies u^-, u^+, and u^s. We consider decision time as the time the last robot that makes a final decision. We average the beliefs (Eq. 6c) of the robots at this decision time to calculate a corresponding decision accuracy. We do not incorporate any base-line method, as this was already shown in [13].

Table 1. The PSO optimization parameters and bounds. P_0 is the empirical best guess particle. P^* is the resulting best particle with respect to our cost-function.

Parameter	γ_0[ms]	γ[ms]	τ[ms]	θ_c[mm]	θ_o
P_0	2000	5000	2000	60	50
$\min_i$	2000	0	1000	50	50
$\max_i$	15000	15000	3000	100	200
P^*	7565	15000	2025	50	85

6.1 Simulation Experiments

Five algorithmic parameters determine the sampling behavior of the swarm. These include mean (γ_0) and mean absolute deviation (γ) of the Cauchy distribution characterizing the robots' random walk, the sampling interval (τ), the collision avoidance threshold (θ_c), and the observations threshold (θ_o). Table 1 lists the boundaries of our optimization search space. The lower bounds for τ and θ_c are set to allow smooth pause-sample-move and collision avoidance maneuvers. The bounds on γ and γ_0 are set such that a robot is able to cross the arena in one random walk step. The bounds on the observation threshold θ_o are established empirically. The particles in the PSO swarm are initialized randomly within the bounded search space, with the exception of one particle P_0 set to an empirically chosen location. Each particle is evaluated multiple times to mitigate randomness. For a particle i, we define the performance cost $\mathcal{C}_i$ as:

$$\mathcal{C}_i = \mu\left(\begin{bmatrix} c_1 & c_2 & \ldots & c_{N_e} \end{bmatrix}^\top\right) + 1.1 \cdot \sigma\left(\begin{bmatrix} c_1 & c_2 & \ldots & c_{N_e} \end{bmatrix}^\top\right) \tag{9}$$

where N_e is the number of re-evaluations and c_j is the outcome of the evaluation j:

$$\epsilon_i(t, d_f) = \begin{cases} \epsilon_f \cdot t/T_{end} & d_f = d_f^* \\ \epsilon_f \cdot \epsilon_d & d_f \neq d_f^* \\ \epsilon_f & d_f = -1 \end{cases} \tag{10a}$$

$$c_j = \sum_{i=1}^{N_r} \epsilon_i \tag{10b}$$

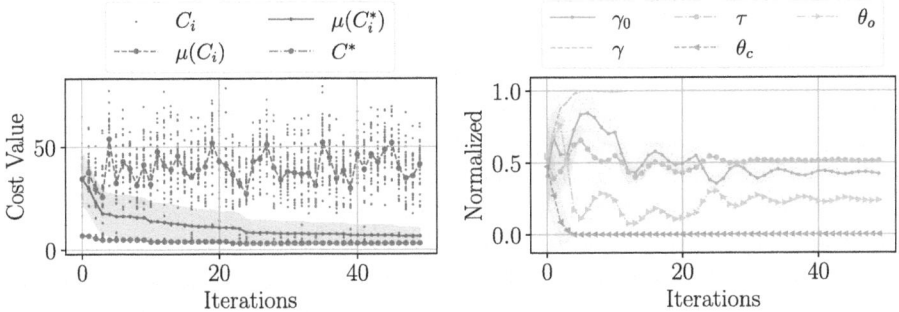

(a) Cost performance over iterations. (b) Average values of particle dimensions.

Fig. 5. We use 30 particles, each re-evaluated 16 times, over 50 iterations using Eq. 9. (a) Progression of cost performance of each particle (C_i), personal best cost performance of each particle (C_i^*), and the global best cost performance (C^*). (b) Progression of mean and standard deviation of parameters in Table 1.

where $N_r = 5$ is the number of robots, ϵ_i is the performance cost of robot i for which the robot's final decision d_f made at time t is compared with the correct decision d_f^*. A wrong decision is penalized by a factor of $\epsilon_d = 5$. The value $\epsilon_f = 1 + |\hat{f} - f^*|/\epsilon_t$ is calculated using the absolute difference between a robot's current estimate of the fill ratio $f = \alpha/(\alpha + \beta)$ and the correct fill ratio f^*, divided by a normalizing factor $\epsilon_t = 0.04$ that corresponds to the contribution of one tile in the overall 25-tile setup.

To find the optimal parameters for our inspection algorithm, we first consider running the algorithm with the u^- information sharing strategy through our optimization framework. Our intuition is that an optimal parameter set for

u^- should allow the swarm to obtain a well-representative sample of the environment in a time-efficient manner, thus, the same parameters should also perform optimally for u^+ and u^s. Using this parameter set, we then run a systematic search to find an optimal value for the soft feedback parameter η that characterizes the u^s information sharing strategy.

We consider the u^- strategy first. For the PSO optimizations, we use 30 particles, 50 iterations, and 16 re-evaluations. Each particle is evaluated for $T_{end} = 1200s$ or until all robots in the swarm have reached a decision. The optimization results are shown in Fig. 5. It can be seen that the average personal best performance of the particles converges to the performance of the global best particle P^*, which is listed in Table 1.

Using P^*, we then run a systematic search for the soft-feedback parameter η. We consider five candidate values for η based on prior empirical tests and run 100 randomized simulations to evaluate the performance of u^s against u^- and u^+. Figure 6 illustrates the results. We see that u^s consistently outperforms u^- and u^+ in decision time. Regarding accuracy, u^s closely approaches the performance of u^+ and u^- at $\eta = 1000$. Specifically, for $\eta = 1000$, the u^s achieves a 20.52% reduction in mean decision time at a 0.78% loss in accuracy, compared with u^+. When compared with u^-, u^s achieves a reduction of 22.10% in mean decision time at a 1.22% loss in accuracy.

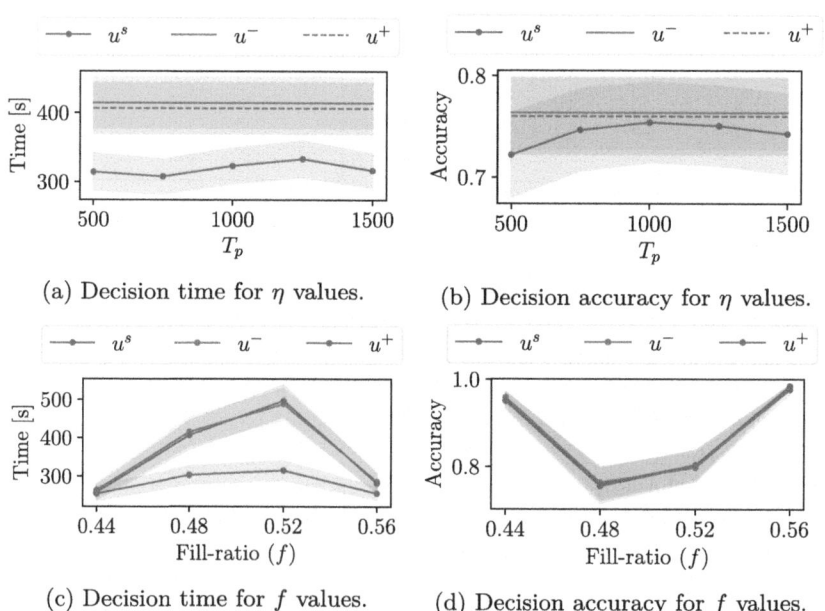

(a) Decision time for η values.

(b) Decision accuracy for η values.

(c) Decision time for f values.

(d) Decision accuracy for f values.

Fig. 6. Decision time (a) and decision accuracy values (b) of 100 randomized simulation experiments using systematic search for the soft-feedback parameter η of u^s, compared against u^- and u^+. Using $\eta = 1000$ we run 100 randomized simulations for different $f \in [0.44, 0.48, 0.52, 0.56]$ and compare with u^- and u^+. The resulting decision times and accuracies are shown in (c) and (d), respectively.

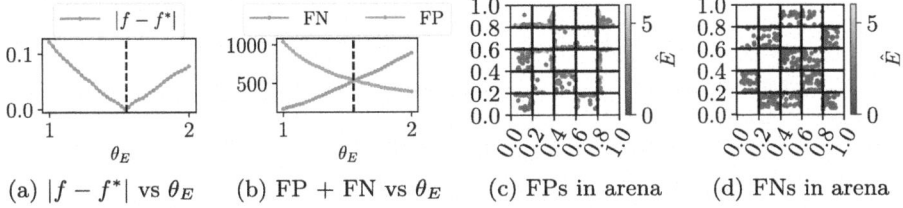

Fig. 7. The threshold parameter θ_E determines binary observations of vibration data in real experiments (see Eq. 5). (a) The fill ratio error $|f - f^*|$ for different values of θ_E (0.025 grid). (b) False observations for different values of θ_E (c) Spatial distribution of false positive and (d) false negative observations.

To assess the generalizability of our findings, we run 100 randomized simulations across fill-ratios of $f \in [0.44, 0.48, 0.52, 0.56]$ to compare u^-, u^+ and u^s (with $\eta = 1000$) based on decision time and accuracy. Each simulation ends upon reaching T_{end} or when all robots have reached a decision. For a fair comparison, we fix the random seeds used to generate floor patterns across the simulation instances. Figure 6c shows that u^s outperforms the other two strategies in decision time. Due to incorporating beliefs in messages, the swarm is compelled to make a decision rapidly, reducing mean and variation in decision times. This is particularly beneficial in harder environments where the fill ratio is close to $f = 0.5$, facilitating reaching the credibility threshold p_c.

6.2 Real Experiments

We validate our simulation results by real experiments in 10 trials for u^-, u^+, and u^s. We first tune the sample threshold θ_E using data from one hour of swarm operation employing u^- and algorithm parameters $[\gamma, \gamma_0, \tau, \theta_c] = [5000, 2000, 1000, 60]$, gathering a total of 6975 samples. Our evaluation criteria are the number of false observations and the fill-ratio error $|f - f^*|$. Figure 7 shows that we obtain $|f - f^*| \approx 0$ at $\theta_E = 1.55$. Employing $\theta_E = 1.55$ results in an equal amount of False Positives (FP) and False Negatives (FN), balancing the modeling error on the Beta distribution. Furthermore, we note that false observations appear mostly along edges of the tiles. This is expected as robots may sample close to the tile edges while they are in contact with two tiles.

We conduct 10 experimental trials on our experimental setup with $f^* = 0.48$ for assessing u^-, u^+ and u^s. The real experiments confirm our findings in simulation and reveal that the utilization of u^s notably decreases the swarm's decision time. Employing u^s compels the swarm to reach decisions at an accelerated rate compared to u^+ and u^-. The decision time and accuracy data from real experiments is shown in Fig. 8. We can see that u^s demonstrates inherently less variance in decision times. Moreover, we encounter fewer indecisive trial outcomes with u^s compared to u^+ and u^-.

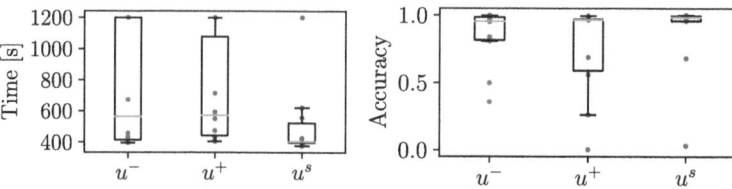

(a) Decision time for strategies. (b) Decision accuracy for strategies.

Fig. 8. Decision time and accuracy in real experiments with a fill-ratio of $f = 0.48$ for different information sharing strategies. If no decision is made within $t = T_{end}$, we refer to $T_{end} = 1200s$ for decision time and the corresponding belief at the end of the simulation experiment for accuracy. We see that u^s is faster and equally accurate as u^- and u^+ in real experiments, confirming our findings in simulation experiments.

7 Conclusion and Future Work

In this work, we presented an experimental setup for studying a surface inspection task using a swarm of vibration sensing robots and explored the application of a Bayesian decision-making algorithm. We developed a simulation framework leveraging the physics based Webots robotic simulator and a PSO method to optimize the parameters shaping the robots' sampling performance. The resulting optimal parameter values were assessed for three information sharing strategies in randomized simulations across different environments based on the swarm's decision time and accuracy. We observed that our proposed soft feedback strategy yields a significant decrease in decision time without a major compromise in decision accuracy, compared to two previously studied strategies. Furthermore, hardware experimental trials validated our simulation findings. In real experiments, no drop in the decision accuracy was observed, demonstrating the adaptability and robustness of the decision-making processes to noise. In our future work, we plan to increase the complexity of our experiments in several ways, considering (i) performing inspection of complex structures such as 3D surfaces or obstacle-dense environments, (ii) classifying time-varying fill-ratios on our experimental setup, and (iii) studying the effect of the swarm size on the inspection performance.

References

1. Alanyali, M., Venkatesh, S., Savas, O., Aeron, S.: Distributed bayesian hypothesis testing in sensor networks. In: Proceedings of the American Control Conference, vol. 6, pp. 5369–5374. Institute of Electrical and Electronics Engineers Inc. (2004). https://doi.org/10.23919/acc.2004.1384706
2. Deraemaeker, A., Worden, K.: New trends in vibration based structural health monitoring. Springer Vienna (2010)
3. Bartashevich, P., Mostaghim, S.: Benchmarking collective perception: new task difficulty metrics for collective decision-making. In: Moura Oliveira, P., Novais, P.,

Reis, L.P. (eds.) EPIA 2019. LNCS (LNAI), vol. 11804, pp. 699–711. Springer, Cham (2019). https://doi.org/10.1007/978-3-030-30241-2_58
4. Bartashevich, P., Mostaghim, S.: Multi-featured collective perception with evidence theory: tackling spatial correlations. Swarm Intell. **15**(1–2), 83–110 (2021). https://doi.org/10.1007/s11721-021-00192-8
5. Bayat, B., Crasta, N., Crespi, A., Pascoal, A.M., Ijspeert, A.: Environmental monitoring using autonomous vehicles: a survey of recent searching techniques (2017). https://doi.org/10.1016/j.copbio.2017.01.009
6. Bigoni, C., Zhang, Z., Hesthaven, J.S.: Systematic sensor placement for structural anomaly detection in the absence of damaged states. Comput. Methods Appl. Mech. Eng. **371** (2020). https://doi.org/10.1016/j.cma.2020.113315
7. Bousdekis, A., Apostolou, D., Mentzas, G.: Predictive maintenance in the 4th industrial revolution: benefits, business opportunities, and managerial implications. IEEE Eng. Manage. Rev. **48**(1), 57–62 (2020). https://doi.org/10.1109/EMR.2019.2958037
8. Brem, C., Siemens: Senseye Predictive Maintenance - Whitepaper True Cost Of Downtime 2022 (2023)
9. Carbone, C., Garibaldi, O., Kurt, Z.: Swarm robotics as a solution to crops inspection for precision agriculture. KnE Eng. **3**(1), 552 (2018). https://doi.org/10.18502/keg.v3i1.1459
10. Chiu, D., Nagpal, R., Haghighat, B.: Optimization and evaluation of multi robot surface inspection through particle swarm optimization. In: ICRA, pp. 8996–9002 (2024)
11. Dementyev, A., et al.: Rovables: miniature on-body robots as mobile wearables. In: UIST 2016 - Proceedings of the 29th Annual Symposium on User Interface Software and Technology, pp. 111–120. Association for Computing Machinery, Inc (2016). https://doi.org/10.1145/2984511.2984531
12. Doebling, S., Farrar, C., Prime, M., Shevitz, D.: Damage identification and health monitoring of structural and mechanical systems from changes in their vibration characteristics: a literature review. Technical Report (1996)
13. Ebert, J.T., Gauci, M., Mallmann-Trenn, F., Nagpal, R.: Bayes bots: collective bayesian decision-making in decentralized robot swarms. In: ICRA (2020). https://doi.org/10.1109/ICRA40945.2020.9196584
14. Ebert, J.T., Gauci, M., Nagpal, R.: Multi-feature col-lective decision making in robot swarms. In: Proceedings of the 17th International Conference on Autonomous Agents and Multiagent Systems (AAMAS), vol. 9 (2018)
15. Gad, A.G.: Particle swarm optimization algorithm and its applications: a systematic review. Arch. Comput. Methods Eng. **29**(5), 2531–2561 (2022). https://doi.org/10.1007/s11831-021-09694-4
16. Haghighat, B., Ebert, J., Boghaert, J., Ekblaw, A., Nagpal, R.: A swarm robotic approach to inspection of 2.5 d surfaces in orbit (2022)
17. Halder, S., Afsari, K.: Robots in inspection and monitoring of buildings and infrastructure: a systematic review (2023). https://doi.org/10.3390/app13042304
18. Innocente, M.S., Sienz, J.: Coefficients' settings in particle swarm optimization: insight and guidelines. Mecánica Comput. Comput. Intell. Tech. Optim. Data Model. **XXIX**, 9253–9269 (2010)
19. Lee, A.J., Song, W., Yu, B., Choi, D., Tirtawardhana, C., Myung, H.: Survey of robotics technologies for civil infrastructure inspection. J. Inf. Intell. Resilience **2**(1), 100018 (2023). https://doi.org/10.1016/j.iintel.2022.100018
20. Liu, Y., Hajj, M., Bao, Y.: Review of robot-based damage assessment for offshore wind turbines (2022). https://doi.org/10.1016/j.rser.2022.112187

21. Magalhães, F., Cunha, A., Caetano, E.: Vibration based structural health monitoring of an arch bridge: from automated OMA to damage detection. Mech. Syst. Signal Process. **28**, 212–228 (2012). https://doi.org/10.1016/j.ymssp.2011.06.011
22. Makarenko, A., Durrant-Whyte, H.: Decentralized bayesian algorithms for active sensor networks. Inf. Fusion **7**(4 SPEC. ISS.), 418–433 (2006). https://doi.org/10.1016/j.inffus.2005.09.010
23. Poli, R., Kennedy, J., Blackwell, T.: Particle swarm optimization. Swarm Intell. **1**(1), 33–57 (2007). https://doi.org/10.1007/s11721-007-0002-0
24. PwC: PdM 4.0. Technical Report (2017)
25. Roda, I., Macchi, M., Fumagalli, L.: The future of maintenance within industry 4.0: an empirical research in manufacturing. In: Moon, I., Lee, G.M., Park, J., Kiritsis, D., von Cieminski, G. (eds.) APMS 2018. IAICT, vol. 536, pp. 39–46. Springer, Cham (2018). https://doi.org/10.1007/978-3-319-99707-0_6
26. Schranz, M., Umlauft, M., Sende, M., Elmenreich, W.: Swarm robotic behaviors and current applications (2020). https://doi.org/10.3389/frobt.2020.00036
27. Seeley, T.D., Buhrman, S.C.: Group decision making in swarms of honey bees. Behav. Ecol. Sociobiol. **45**, 19–31 (1999)
28. Shan, Q., Mostaghim, S.: Discrete collective estimation in swarm robotics with distributed Bayesian belief sharing. Swarm Intell. **15**(4), 377–402 (2021). https://doi.org/10.1007/s11721-021-00201-w
29. Valentini, G., Brambilla, D., Hamann, H., Dorigo, M.: collective perception of environmental features in a robot swarm **9882** (2016). https://doi.org/10.1007/978-3-319-44427-7
30. Valentini, G., Ferrante, E., Dorigo, M.: The best-of-n problem in robot swarms: formalization, state of the art, and novel perspectives (2017). https://doi.org/10.3389/frobt.2017.00009
31. Valentini, G., Ferrante, E., Hamann, H., Dorigo, M.: Collective decision with 100 Kilobots: speed versus accuracy in binary discrimination problems collective decision with 100 kilo-bots: speed versus accuracy in binary discrimination problems. Auton. Agent. Multi-Agent Syst. **30**(3), 553–580 (2016). https://doi.org/10.1007/s10458-015-9323-3

Extinguishing Wildfires in Large Scale Scenarios Using Swarms of UAVs

Georgios Tzoumas(✉), Lucio Salina, Alex McConville, Tom Richardson, and Sabine Hauert

Department of Engineering Mathematics and Bristol Robotics Laboratory, University of Bristol, Bristol, UK
georgios.tzoumas@bristol.ac.uk

Abstract. The climate crisis induces the appearance of wildfires. Identifying and mitigating them at an early stage is crucial to control them successfully. To achieve this, we present algorithms to mitigate different types of wildfires using swarms of high-payload UAVs. In our experiments, a swarm of 30 UAVs monitors and suppresses wildfires in an area as large as California using a newly developed algorithm called Dynamic Space Partition for Firefighting (DSPF). We test the algorithm in two different environmental scenarios from low to high-difficulty fire conditions. We created DSPF with coordination (DSPFC) to enable multiple UAVs to engage larger fires. Using this algorithm, the aircraft that identified the firefront communicates with the two closest aircraft to self-organise and engage the wildfire. We developed a metric named 'fire mitigation effectiveness' (FME) to compare the different algorithms. Our results show that the DSPF was able to mitigate an average of 82% of the wildfires and achieve an FME of 61% at a low-difficulty scenario. When facing a high-difficulty scenario the DSPF strategy mitigated an average of 18% of fires, achieving an FME of 18%. The DSPFC achieved a better performance compared to the DSPF in the high-difficulty scenario mitigating an average of 73% of fires and achieving an FME of 50%.

1 Introduction

The effects of the climate crisis are seen more evidently, as temperatures and carbon emissions continue to rise [16,20]. Stopping destructive wildfires can reduce global emissions, as the Canadian wildfires of 2023 created the same amount of emissions as the yearly emissions of the global airline industry [31]. Early detection and mitigation of wildfires are key to controlling them more effectively. The sooner they are mitigated, the higher the chances of their successful extinguishment [9]. Therefore, fire services around the world require novel technologies to suppress wildfires at an early stage. UAVs are now frequently used to identify fires and to relay information to firefighters [1,2,8,14]. Previous research works, present systems to mitigate wildfires [15,24]. In those efforts, the use of smaller UAVs such as quadcopters has been seen but that type of aircraft usually do not have the payload capacity to mitigate wildfires. For this reason, high-payload

© The Author(s), under exclusive license to Springer Nature Switzerland AG 2024
H. Hamann et al. (Eds.): ANTS 2024, LNCS 14987, pp. 71–83, 2024.
https://doi.org/10.1007/978-3-031-70932-6_6

UAVs have been investigated as a potential tool to help in fire identification and mitigation [29]. In our previous work, the Dynamic Space Partition (DSP) algorithm was presented to organise a swarm of high payload UAVs to identify wildfires in large environments such as California [28]. In this work, we extend our DSP algorithm to improve the fire identification capabilities and to organise the swarm to mitigate identified wildfires. A new algorithm which adjusts the swarm autonomously in monitoring and firefighting states is presented named DSPF (Dynamic Space Partition for Firefighting). The algorithm is tested in different fire conditions and adjustments are made to allow the swarm to perform coordinated engagement on wildfires. We present the new algorithms, our experimental scenarios, and the effectiveness of the proposed strategies and we conclude with future work.

2 Related Work

The scalability and robustness of swarm mechanisms show potential in developing a tool for fire identification and mitigation [3,28]. Ausonio et al. have shown a conceptual swarm of UAVs that can be deployed to mitigate firefronts. They focus on small-scale scenarios with a firefront propagation model [5]. Innocente et al. presented quadcopters self-organising using the particle swarm optimisation (PSO) algorithm. A large focus is given to the wildfire expansion element and the effect that water payload drops have on the front [15]. Furthermore, the work of Seraj and Gombolay [24,25] has shown how swarms of UAVs can monitor propagating firefronts whilst taking into account the location of firefighters. Alsammak used PSO and Levi flights whilst considering the locations of fire spots and energy consumption [4].

For a swarm of UAVs to mitigate fires successfully it is necessary to identify fires as early as possible. This becomes a monitoring task as seen in many applications in robotics such as search and rescue and space exploration [19]. Sharma et al. show a nature-inspired algorithm using a clustering-based distribution factor (CBDF) for deterministic movement and nature-inspired algorithms like bacteria foraging optimisation (BFO), and bat algorithm (BA) for exploration [26]. Other researchers have used techniques, such as triangulation, to partition the area based on the number of available agents in the swarm [6]. Furthermore, Nair et al. focus on a combination of two Voronoi partition methodologies to ensure the continuity of the robots' trajectories [18].

Previously, we presented a system to monitor large areas for potential wildfires and created the dynamic space partition (DSP) algorithm [28]. Our algorithm is a bio-inspired mechanism that allows UAVs to re-organise dynamically and autonomously when the environment or the state of the UAVs changes inspired by Spears et al. [27,28]. Studies were completed to create dynamic fire models and to simulate the effect of the payload drop of the aircraft on the firefronts. The work of Hansen [10] has shown the required amount of water that is required to control wildfires in different conditions. Ausonio et al. have also used similar methodologies as Hansen to calculate how many UAVs would be needed to control a wildfire [5].

3 Methodology

3.1 Test Scenario and Aircraft

The scenario that we used to test our system is focused on the case of California which is highly affected by wildfires [30]. The simulation scenario requires the UAVs to explore a square area the size of California in 24 h of simulation, to identify and engage with 5 wildfires of different sizes. The swarm is pre-deployed to explore and monitor the required search area for fires to identify and mitigate them at an early stage. The fires are deployed 3.5 h after the simulation initialises. The fires start from the southerly side of the region and are spread based on a northerly wind. The aircraft have no prior knowledge of this scenario. Two airports are considered to be available for the aircraft to be deployed and to reload which are uniformly distributed within the search area. To develop and test our system, a custom-built software is used that is developed in Python 3.9. The developed tool, allows us to test algorithms in real-time using digital twin technologies of a swarm of fixed-winged aircraft. The system is described in our previous work [22]. We tested our systems 50 times per simulation scenario. The type of aircraft that was used in simulations is the ULTRA[1] platform which is produced by Windracers Ltd. The aircraft is a fixed-wing platform that can carry 100 kgs for 1000 kms. The minimum turn radius of the platform is 150 m and its cruise speed is 144 km/h. The kinematic model of the ith UAV is described using the following equations:

$$\begin{cases} \dot{x}_i = v \times \cos \theta_i \\ \dot{y}_i = v \times \sin \theta_i \\ \dot{\theta}_i = \omega_i \end{cases} \quad (1)$$

The agents are controlled by changing their heading. Aircraft are assumed to operate at the same altitude. The same model was used in our previous work and similar models can be seen in previous works [11,12,28].

3.2 Dynamic Space Partition for Firefighting (DSPF)

The DSP algorithm distributes UAVs over an area to identify wildfires. This is achieved with the creation of virtual waypoints named DSP points. Each UAV has its own DSP point. The UAVs communicate the location of their DSPs with each other. The DSP points operate by spreading over an area and maintaining a dispersion distance between each other. This is performed by generating attractive and repulsive forces amongst the points [28]. The UAVs are initially tasked to follow their DSP point. After they reach them, they perform Archimedean spirals as seen in the works of Saputra et al. and Hauert et al. [13,23]. The spiral trajectory is generated by taking into account the fire sensory range to avoid overlaps. After fires are identified, some adaptation needs to take place

---
[1] ULTRA (Unmanned Low-cost TRAnsport) platform developed by Windracers, https://windracers.org/.

to the algorithm to engage the identified firefronts. Thus, we created the DSP for firefighting (DSPF) algorithm. DSPF works in three stages: identification of wildfires; fire engagement; and return to base. The algorithm is shown in a flowchart in Fig. 1.

Identification of Wildfires. To identify wildfires, we use the assumption that the aircraft are equipped with a gimbaled multispectral camera and a visual camera. We assume that the aircraft can identify wildfires in a 6 km range. Fires are considered identified if the Euclidean distance between the aircraft and the location of the fire is smaller than the sensory range of the aircraft.

Fire Engagement. The agent that has identified the wildfire, changes its mode of operation from a 'DSP Search' mode to a 'Fire mitigation' mode. Thus, the search swarm has a reduced number of aircraft to identify wildfires. The aircraft that identified the fire, removes the DSP point from the rest of the group which affects the other points as they experience new forces. Since they do not have a force exerted from the removed DSP point, they will need to move to maintain their balance. The aircraft moves towards the location of the firefront and deposits its payload.

Return to Base. After a fire has been engaged, the aircraft identifies where the closest available base is and it moves towards that location. Once it reaches the base, it stops to refuel and then it resets its behaviour back to the DSP search algorithm. A turnaround time of 5 min is used as an approximation of the duration of the landing procedures and reloading of the aircraft. During their return to base, the aircraft is sensing if a fire is detected. If it has, it saves that information to a queue of fire locations and returns to attack the newly identified fires.

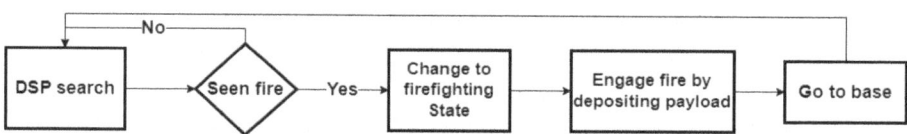

Fig. 1. DSPF Flowchart. An agent performs its DSP search and once it identifies a fire it changes its state to a firefighting state. The agent stops the DSP searching pattern and moves directly to the fire to engage with it performing an active firefighting drop. After the drop is performed, the aircraft goes back to the nearest base to reload and then resumes its DSP search.

3.3 DSP for Coordinated Fire Engagement (DSPFC)

In certain cases, larger fires can appear where one payload drop is not enough to mitigate them. To control them, coordinated strategies need to be implemented. Hence, the DSPF with coordination was created. With this algorithm, when an aircraft identifies a fire, it shares the location of the identified fire with the two closest aircraft in the swarm. It is assumed that each agent can communicate with other aircraft using a decentralised communications system in a specific communications range. Each aircraft identifies which aircraft are closest to its location and sends the location of the fire it identifies. The agents then change their state from a DSP searching agent to a firefighting agent, moving towards the shared location to deposit their payload. If they do not identify a fire location in that area or if the fire was already mitigated then they return to DSP search.

3.4 Fire Point Theory and Water to Control a Fire

Wildfire behaviours can be very complex and unpredictable as they are affected by various parameters like ground elevation and wind effects [17]. The work of Hansen et al. [10] has shown the water requirements to control a fire subject to different environmental conditions. Following their methodology, the fire point theory was used to calculate the critical flow rate of water that is needed to set a fire under control and the water that is needed to control them [7,21]. The maximum active area size is the area that the firefront covers. The larger the active fire area the larger the water requirement to control it. Please refer to the works of Hansen and Ausonio for more information on the water required to control a fire [5,10]. A summary of the parameters used in the low and high difficulty scenarios, can be seen in Table 1.

Table 1. Fire classification based on different environmental conditions and payload requirements to set them under control.

Fire classification	Low	High
Wind effect (m/s)	1	5
Maximum active area (m^2)	20	100
Water needed (L)	9–26	79–131
Payload drops to control fire	1	2
Time to reach maximum active area	30	50

Every fire entity in our simulation is modelled as a circle and consists of the following elements: Position; diameter; current area; maximum active area; water required to control the fire; and status: alive or extinguished. The wind is modelled as having a constant speed and direction and the fires move based on this effect. The kinematic model of the jth fire is described using the following equations:

$$\begin{cases} \dot{x}_j = WindSpeed \times \cos(WindDirection) \\ \dot{y}_j = WindSpeed \times \sin(WindDirection) \end{cases} \quad (2)$$

The active area firefront starts at zero when a fire entity is generated. The area increases until the fire reaches its maximum active area based on the fire classification that has been defined. On the low-difficulty and high-difficulty fires, we assume that 30 and 50 min respectively, are needed to reach the maximum active area. These variables are shown in Table 1. An example of the fire development is shown in Fig. 2.

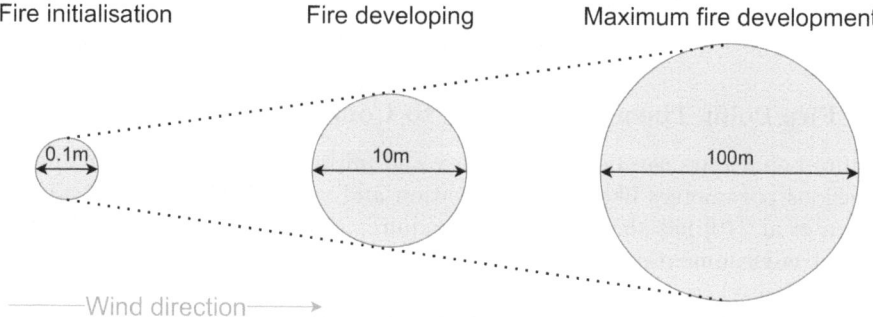

Fig. 2. The development of the fire is depicted in the figure showing its development in three stages assuming that there is a westerly wind. a) shows the initiation of the fire entity b) The centre of the fire moves towards the wind and its size increases from 0.1 m to 10 m after 10 min, c) The fire continues to move with the wind effect but stops increasing in size as it has reached its maximum active area size.

Performance Metrics. The performance metrics that are used to assess the effectiveness of the swarm are area coverage, fires mitigated and fire mitigation effectiveness (FME). The first one assesses how much of the area the aircraft have explored using the fire sensory range of the aircraft. This is assessed by summing the areas explored by all UAVs:

$$coverage = \frac{\bigcup_{n=1}^{N} AC}{Total\ world\ area} \quad (3)$$

where in Eq. 3 N is the number of aircraft in the swarm, AC *(Area Covered)* is the total area that was explored by a single aircraft.

The mitigation is assessed on how many of the existing fires were mitigated by the swarm:

$$mitigation = \frac{\sum_{n=1}^{N} FM}{Total\ fires\ in\ the\ world} \quad (4)$$

In Eq. 4 FM *(Fires Mitigated)* is the total number of fires that the aircraft have mitigated.

The FME takes into account not only how many fires have been mitigated but also how fast this operation was performed. To calculate the FME we calculate the area of the time series mitigation of the test runs. An example of that graph can be seen in Fig. 3d.

$$FME = \int_{t_1}^{t_2} mitigation \times dt \qquad (5)$$

In Eq. 5, *mitigation* refers to the percentage of mitigated fires. Time steps t_1 and t_2 refer to the change in time when a new fire is mitigated. For example, when the first fire is mitigated, t_1 is equal to zero referring to the start of the simulation and t_2 is the time step that the fire mitigation was achieved.

The FME of each run was then compared to the best operating system which can achieve 100% of fire mitigation right at the initiation of the fires. This is called the potential FME (PFME) which can be calculated as:

$$Potential\ FME = (t_{total\ simulation} - t_{initation\ of\ fires}) \times Best\ mitigation \qquad (6)$$

where, $t_{total simulation}$ is 86400 referring to the duration of the simulation in seconds, $t_{initation of fires}$ is 12600 and *Best mitigation* is 100%.

4 Results

4.1 Low-Difficulty Fires

The low-difficulty fires can be mitigated if one payload drop is made from an aircraft. The total average of the mitigated fires for this system is 82% as can be seen in Fig. 4a. In every test run, a time series graph is produced allowing us to calculate the FME of each run. The FME for the mitigation of low-difficulty fires reached 61% as it is shown in Fig. 4b.

4.2 High-Difficulty Fires

The high-difficulty fires require at least 2 drops to be performed within a time frame of half an hour to be considered mitigated. Initially, the UAVs were controlled with the DSPF algorithm, meaning that they were not communicating to other aircraft the location of each fire. As it is seen in Fig. 4a the system was able to mitigate an average of 18% of fires. The FME reached 18% which is a significant drop from the previous scenario. As this is a more difficult scenario, DSPFC was used to allow coordination between aircraft. With this algorithm, the agent that identified the wildfire sends a message to the two closest agents to stop their DSP search behaviour and go to that location to mitigate the identified wildfire. The system was able to mitigate an average of 73% of the fires and reached an FME of 50%.

4.3 Swarm Behaviour in Fire Mitigation

The operation of the system detecting and mitigating 5 high-difficulty fires using DSPFC is shown in Fig. 3. Each agent's trajectory is plotted in Fig. 3a. The exploration strategy shows agents performing trajectories that fit Archimedean spirals. Boundary restrictions limit some agents from performing full spirals which makes them explore the edges of the world. Certain agents perform two different spirals, which take place when agents reset their DSP search behaviour. This occurs because they are tasked to stop the exploration and go back to their DSP point to monitor the already explored area, or because they have deposited a payload on a wildfire and then they go back to the DSP search. This is seen more clearly at the bottom right of Fig. 3a.

In Fig. 3b we see the trajectories of each DSP point of the agents. The points are spread out to expand on the required search space. Some points expand outwards without returning to any base indicating the aircraft has not switched from 'DSP search' to 'Fire mitigation'. This is seen in the top left corner of Fig. 3b with the dark and light green trajectories. On the bottom left side of Fig. 3b, we see the purple trajectory of a DSP point which moves initially towards the boundaries of the search area and then returns to the base at the centre. This means that the corresponding agent has performed a payload drop and then it is tasked to reset the searching behaviour at the base after it refuelled. That can also be seen at the bottom right part of Fig. 3b.

Figure 3c shows the fire movement, with five fires generated in the southern region affected by a northerly wind. The grey and green fires are extinguished quickly, while rest develop further. The agent's state alteration at the bottom left corresponds to the mitigation of the grey fire in that region, as seen in Fig. 3c.

The first fire detection and the first mitigation took place after 3 and 6 min respectively, demonstrating rapid engagement and mitigation of the firefront. This is illustrated in Fig. 3d showing the coverage and mitigation percentage of the test runs. The swarm mitigated all fires after 11 h of simulation.

5 Discussion

The swarm system was used to address two environmental scenarios, demonstrating the aircraft's ability to identify and mitigate wildfires. The agents' capacity to autonomously switch between searching and mitigating behaviours without compromising area coverage or effectiveness highlights the resilience and adaptability of the swarm system. The results showed the swarm's potential to mitigate fires early, with the first firefront mitigation occurring just 6 min after the fire started. This underscores the system's potential if aircraft are pre-deployed. While the DSPFC algorithm proved more effective in high-difficulty scenarios, environmental factors must be considered to ensure simulation accuracy. These scenarios underscore the importance of early firefront mitigation.

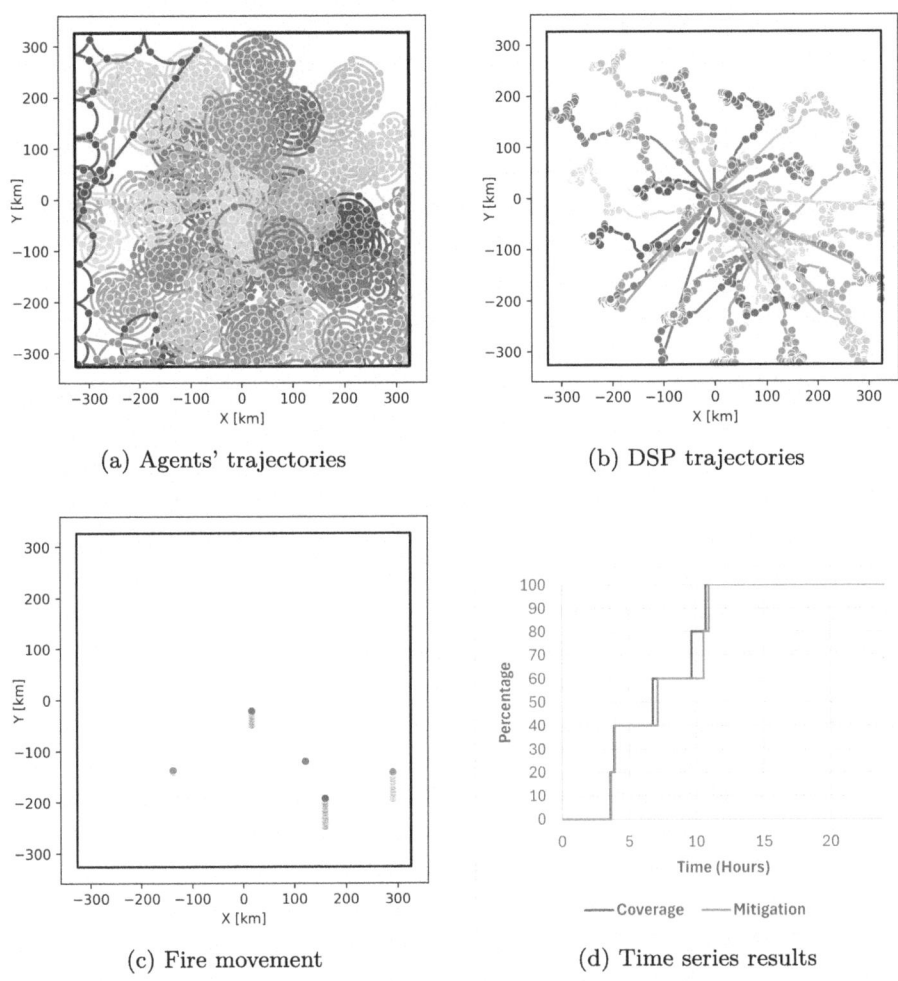

Fig. 3. This figure shows the data output of each test run. In this case a high-difficulty fire mitigation scenario with the use of DSPFC. In Fig. 3a the trajectories of the agents are shown plotted over the duration of the simulation. In the figure, the Archimedean spiral exploration can be seen. In Fig. 3b we can see the trajectories of the DSP points. The DSP point trajectories can show if an aircraft performed a payload drop on a fire. On the right-hand side, the pink trajectory shows that the DSP starts its position from the starting point indicating a reset in the DSP search behaviour of the aircraft. In Fig. 3c the movement of fires can be seen. The green and grey fires are extinguished at an early stage, showing the system's potential to detect and engage wildfires before they become uncontrollable. The other fires are also mitigated but at a later stage of the operation. Figure 3d shows the performance of the system. It is shown that the system completed the first fire mitigation after 3 h and 36 min of simulation took place.

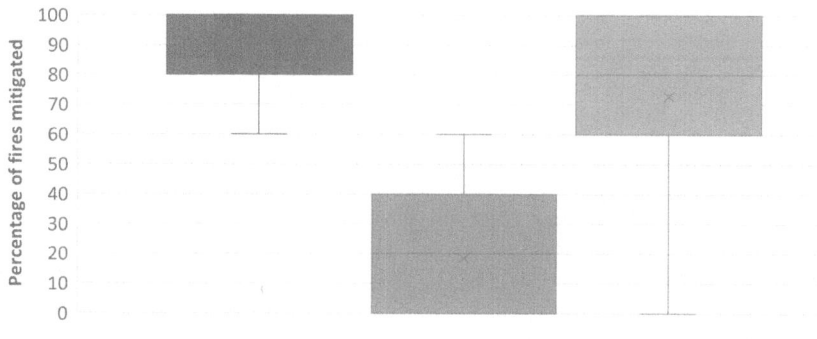

■ DSPF - Low difficulty fires   ■ DSPF - High Difficulty fires   ■ DSPFC - High Difficulty fires

(a) This figure shows the wildfires that were mitigated by the swarm system in three different scenarios. The system consisted of 30 aircraft and in all cases, 5 fires were deployed and needed to be mitigated by the system. In the low-difficulty scenario using the DSPF, an average of 82% fires were mitigated. In the high-difficulty case with the DSPF, the system was able to mitigate an average of 18% fires which is a significant drop due to the higher difficulty of fire mitigation. In the high-difficulty case using DSPFC an increase in performance is seen with an average of 73% fires were mitigated.

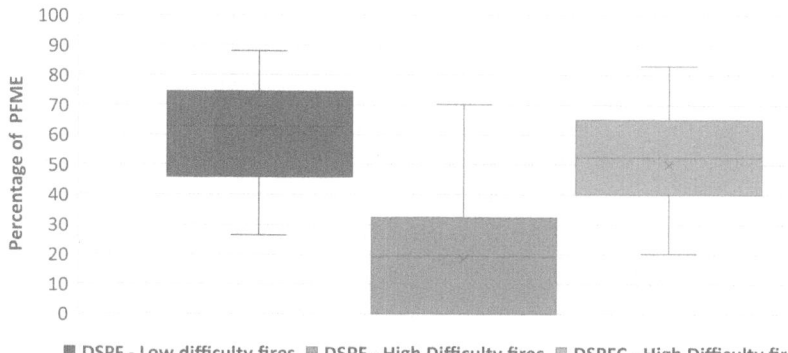

■ DSPF - Low difficulty fires   ■ DSPF - High Difficulty fires   ■ DSPFC - High Difficulty fires

(b) This figure shows the fire mitigation effectiveness (FME) achieved by the system in three different scenarios. The system consisted of 30 aircraft and in all cases, 5 fires were deployed and needed to be mitigated by the system. The FME takes into account how many fires have been mitigated and how fast this was performed. The potential FME (PFME) corresponds to the ideal performing system where the system could mitigate instantly all fires when they appear. We use this to compare the results of the FME of each run. In the blue bar plot, low-difficulty fires were mitigated using the DSPF. The system achieved an average FME of 61%. The orange bar plot corresponds to the mitigation of high-difficulty fires with the DSPF algorithm. The system achieved an FME of 18%. The grey bar plot corresponds to the mitigation of high-difficulty fires using the DSPFC algorithm. Here, the system managed to achieve an FME of 50%.

Fig. 4. Bar charts displaying the performance of the systems.

6 Conclusions

When wildfires are identified, it is crucial for their successful suppression, to engage with them as soon as possible. High-payload UAVs provide the opportunity to identify fires and mitigate them directly after recognition. To achieve this, different types of algorithms need to be developed. The DSP for firefighting (DSPF) allows the swarm to explore an area and when a fire is identified to move towards the fire location and to deposit a fire extinguishing payload without disrupting the operations of the rest of the swarm. We tested our algorithms in an effort to identify and mitigate 5 fires in an area as large as California. To assess how effective our controllers are, we developed the fire mitigation effectiveness (FME) metric where not only the number of mitigated fires is taken into account but also how fast this was achieved. Using this strategy the system was able to mitigate on average 82% of fires. The system achieved an FME of 61% when compared to an ideal system. When facing more challenging scenarios, where larger fires can appear, the system has shown some difficulties in mitigating them. In this case, an average of 18% of fires were mitigated and an FME of 18% was achieved. The algorithm was then adjusted to allow coordinated mitigation of firefronts, creating DSPFC. This algorithm has shown improved performance, mitigating in the high-difficulty scenario an average of 73% of fires and achieving an FME of 50%. The swarm was also able to mitigate a wildfire 16 min after its initiation showing the capability of the swarm for early detection and mitigation. These are promising results showing how swarms of UAVs can perform identification and mitigation of firefronts.

7 Future Work

Although the paradigm of a swarm of 30 aircraft can be a realistic set-up to monitor large areas the size of California, it is possible that more aircraft can improve the performance of the system. As more agents can be used to monitor the area more frequently, they will be able to deliver extinguishing material more rapidly thus engaging fires faster and controlling them more easily. Other elements can be explored further, such as the effect of non-perfect fire detection. In the field, there is a high chance that detection systems such as optical cameras or infrared sensors might not detect the fire perfectly, thus this will need to be incorporated in these studies as well. Additionally, the payload drop effect on the fire needs to be researched further since a perfect payload drop will be impossible to achieve in the field. Wind perturbations and localisation of the fire fronts are quite dynamic thus the effect of the extinguishing payload on the fire will need to be investigated further.

Acknowledgements. This work was supported by Windracers ltd. and by Innovate UK grant 10023377.

References

1. Akhloufi, M.A., Castro, N.A., Couturier, A.: UAVs for wildland fires. In: Autonomous Systems: Sensors, Vehicles, Security, and the Internet of Everything, vol. 10643, pp. 134–147. SPIE (2018)
2. Alon, O., Rabinovich, S., Fyodorov, C., Cauchard, J.R.: Drones in firefighting: a user-centered design perspective. In: Proceedings of the 23rd International Conference on Mobile Human-Computer Interaction, pp. 1–11 (2021)
3. Alsammak, I.L.H., Mahmoud, M.A., Aris, H., AlKilabi, M., Mahdi, M.N.: The use of swarms of unmanned aerial vehicles in mitigating area coverage challenges of forest-fire-extinguishing activities: a systematic literature review. Forests **13**(5), 811 (2022)
4. Alsammak, I.L.H., Mahmoud, M.A., Gunasekaran, S.S., Ahmed, A.N., AlKilabi, M.: Nature-inspired drone swarming for wildfires suppression considering distributed fire spots and energy consumption. IEEE Access **11**, 50962–50983 (2023). https://doi.org/10.1109/ACCESS.2023.3279416
5. Ausonio, E., Bagnerini, P., Ghio, M.: Drone swarms in fire suppression activities: a conceptual framework. Drones **5**(1), 17 (2021)
6. Baranzadeh, A.: Decentralized autonomous navigation strategies for multi-robot search and rescue. arXiv preprint arXiv:1605.04368 (2016)
7. Beyler, C.: A unified model of fire suppression by. J. Fire Prot. Eng. **4**(1), 5–16 (1992). https://doi.org/10.1177/104239159200400102
8. Bouguettaya, A., Zarzour, H., Taberkit, A.M., Kechida, A.: A review on early wildfire detection from unmanned aerial vehicles using deep learning-based computer vision algorithms. Sig. Process. **190**, 108309 (2022)
9. Carta, F., Zidda, C., Putzu, M., Loru, D., Anedda, M., Giusto, D.: Advancements in forest fire prevention: a comprehensive survey. Sensors **23**(14), 6635 (2023)
10. Hansen, R.: Corrigendum to: estimating the amount of water required to extinguish wildfires under different conditions and in various fuel types. Int. J. Wildland Fire **21**(6), 778 (2012)
11. Hao, C., Xiangke, W., Lincheng, S., Yirui, C.: Formation flight of fixed-wing UAV swarms: a group-based hierarchical approach. Chin. J. Aeronaut. **34**(2), 504–515 (2021). https://doi.org/10.1016/j.cja.2020.03.006
12. Hauert, S., et al.: Reynolds flocking in reality with fixed-wing robots: communication range vs. maximum turning rate. In: 2011 IEEE/RSJ International Conference on Intelligent Robots and Systems, pp. 5015–5020. IEEE (2011). https://doi.org/10.1109/IROS.2011.6095129
13. Hauert, S., Leven, S., Zufferey, J.C., Floreano, D.: Communication-based leashing of real flying robots. In: 2010 IEEE International Conference on Robotics and Automation, pp. 15–20. IEEE (2010)
14. Hossain, F.A., Zhang, Y.M., Tonima, M.A.: Forest fire flame and smoke detection from UAV-captured images using fire-specific color features and multi-color space local binary pattern. J. Unmanned Veh. Syst. **8**(4), 285–309 (2020). https://doi.org/10.1139/juvs-2020-0009
15. Innocente, M.S., Grasso, P.: Self-organising swarms of firefighting drones: Harnessing the power of collective intelligence in decentralised multi-robot systems. J. Computat. Sci. **34**, 80–101 (2019)
16. Jones, M.W., Smith, A., Betts, R., Canadell, J.G., Prentice, I.C., Le Quéré, C.: Climate change increases the risk of wildfires. ScienceBrief Rev. **116**, 117 (2020)

17. Marques, S., et al.: Characterization of wildfires in Portugal. Eur. J. Forest Res. **130**, 775–784 (2011)
18. Nair, V.G., Guruprasad, K.: GM-VPC: an algorithm for multi-robot coverage of known spaces using generalized Voronoi partition. Robotica **38**(5), 845–860 (2020)
19. Pang, B., Song, Y., Zhang, C., Yang, R.: Effect of random walk methods on searching efficiency in swarm robots for area exploration. Appl. Intell. **51**, 5189–5199 (2021)
20. Pausas, J.G., Keeley, J.E.: Wildfires and global change. Front. Ecol. Environ. **19**(7), 387–395 (2021). https://doi.org/10.1002/fee.2359
21. Rasbash, D., Drysdale, D., Deepak, D.: Critical heat and mass transfer at pilot ignition and extinction of a material. Fire Saf. J. **10**(1), 1–10 (1986). https://doi.org/10.1016/0379-7112(86)90026-3
22. Salinas, L.R., Tzoumas, G., Pitonakova, L., Hauert, S.: Digital twin technology for wildfire monitoring using UAV swarms. In: 2023 International Conference on Unmanned Aircraft Systems (ICUAS), pp. 586–593. IEEE (2023)
23. Saputra, O.D., Irfan, M., Putri, N.N., Shin, S.Y.: UAV-based localization for distributed tactical wireless networks using archimedean spiral. In: 2015 International Symposium on Intelligent Signal Processing and Communication Systems (ISPACS), pp. 392–396 (2015). https://doi.org/10.1109/ISPACS.2015.7432802
24. Seraj, E., Gombolay, M.: Coordinated control of UAVs for human-centered active sensing of wildfires. In: 2020 American Control Conference (ACC), pp. 1845–1852. IEEE (2020)
25. Seraj, E., Silva, A., Gombolay, M.: Safe coordination of human-robot firefighting teams. arXiv preprint arXiv:1903.06847 (2019)
26. Sharma, S., Shukla, A., Tiwari, R.: Multi robot area exploration using nature inspired algorithm. Biol. Inspired Cogn. Archit. **18**, 80–94 (2016)
27. Spears, W.M., Spears, D.F.: Physicomimetics: Physics-Based Swarm Intelligence. Springer, Cham (2012)
28. Tzoumas, G., Pitonakova, L., Salinas, L., Scales, C., Richardson, T., Hauert, S.: Wildfire detection in large-scale environments using force-based control for swarms of UAVs. Swarm Intell. **17**(1–2), 89–115 (2023)
29. Tzoumas, G., Tom, R., Hauert, S.: Aged care with socially assistive robotics under advance care planning. In: 2024 IEEE International Conference on Advanced Robotics and its Social Impacts (ARSO). IEEE (2024)
30. Wang, D., et al.: Economic footprint of California wildfires in 2018. Nat. Sustain. **4**(3), 252–260 (2021)
31. Wang, Z., et al.: Severe global environmental issues caused by Canada's record-breaking wildfires in 2023 (2023)

Grasshopper Optimization Algorithm (GOA): A Novel Algorithm or A Variant of PSO?

Negin Harandi[1,2(✉)], Arnout Van Messem[3], Wesley De Neve[1,4], and Joris Vankerschaver[1,2]

[1] Center for Biosystems and Biotech Data Science, Department of Environmental Technology, Food Technology and Molecular Biotechnology, Ghent University Global Campus, Yeonsu-gu, Incheon, South Korea
{negin.harandi,wesley.deneve,joris.vankerschaver}@ghent.ac.kr
[2] Department of Applied Mathematics, Computer Science and Statistics, Ghent University, Ghent, Belgium
[3] Department of Mathematics, Université de Liège, Liège, Belgium
Arnout.VanMessem@uliege.be
[4] Department of Electronics and Information Systems, Ghent University, Ghent, Belgium

Abstract. In the world of new optimization methods, there is a concern that various methods, despite having different names, are quite similar. This raises a crucial question: Does the introduction of a new source of inspiration justify assigning a new name to an optimization algorithm, especially when its functionality closely mirrors or simplifies an existing, well-known method? This paper takes a close look at the Grasshopper Optimization Algorithm (GOA), investigating its concepts and comparing them to different versions of Particle Swarm Optimization (PSO). Our findings lead to a noteworthy conclusion: GOA, despite its branding as a novel algorithm, is not a new algorithm, but can be viewed as a derivative of PSO.

1 Introduction

In the domain of computational optimization, the search for novel algorithms has led to the development of various methods inspired by the collective behavior of natural or artificial agents. Among these, the Grasshopper Optimization Algorithm (GOA) [28] is widely used and takes its inspiration from the swarming behavior of grasshoppers.

GOA claims to offer a novel approach to solving complex optimization problems by mimicking the social and movement dynamics of grasshoppers in nature. However, this study aims to critically analyze GOA, arguing that despite its apparent innovation, this algorithm is not a fundamentally novel algorithm, but rather a variant of the well-known Particle Swarm Optimization (PSO) algorithm. This perspective draws on the broader discourse of algorithmic novelty in computational research, as highlighted by Weyland [37,38], Camacho et al. [5–7], and Piotrowski et al. [25].

By focusing on GOA, this paper contributes to this ongoing debate by examining the repackaging of existing algorithms under the guise of novelty and its impact on the field of computational optimization. There are three main aspects to evaluate the similarity between optimization algorithms: structural, empirical, and behavioral similarity [2,14,15,32]. This paper focuses on the first two aspects. To achieve this objective, we start by reviewing PSO and its various modifications in Sect. 2, and GOA in Sect. 3. Section 4 discusses how GOA compares to PSO. Subsequently, we present findings indicating that there is no substantial difference between GOA and PSO variant in terms of effectiveness on the benchmark functions in Sect. 5. Section 6 outlines various directions for future research.

2 Particle Swarm Optimization

Particle swarm optimization (PSO) is a stochastic population-based optimization approach that was first introduced by Kennedy and Eberhart [17,19]. This algorithm is one of the most popular swarm-based optimization algorithms and is designed to find optimal regions in complex search spaces using the interaction of individuals in a population of particles, based on the metaphor of social interaction.

In the original PSO algorithm [19], at every iteration t, each particle $i \in \{1,\ldots,N\}$, with N the total number of particles in the swarm, is aware of its current position x_i^t, its velocity v_i^t, its personal best position p_i^t, and the position g^t of the overall best particle. At the next iteration $t+1$, each particle evaluates its new position x_i^{t+1}, calculated based on its previous velocity v_i^t and position x_i^t, using the objective function $F(x)$. For a minimization problem, if the new position x_i^{t+1} of particle i yields a lower value of the objective function than that of its personal best position p_i^t, i.e., if $F(x_i^{t+1}) < F(p_i^t)$, then the personal best is updated to this new position: $p_i^{t+1} = x_i^{t+1}$.

Simultaneously, all particles communicate their personal best positions to the swarm. If there exists a particle with lower fitness than the current global best, then the global best is updated to be the particle with the lowest fitness. This global best position represents the collective knowledge of the swarm and guides the search direction of all particles.

The rules for updating the velocity and position of a particle in the original PSO are:

$$v_i^{t+1} = v_i^t + \phi_1 \odot (p_i^t - x_i^t) + \phi_2 \odot (g^t - x_i^t) \tag{1}$$
$$x_i^{t+1} = x_i^t + v_i^{t+1}.$$

Here, the notation ⊙ refers to the Hadamard product, or component-wise multiplication of two vectors. Moreover, $\boldsymbol{\phi}_k$, $k = 1, 2$, represents a vector of random numbers which, at each iteration and for each particle, has been drawn from a uniform distribution $\mathcal{U}(0, c_k)$ to provide diversity to the particle's movement. The coefficients c_1 and c_2 are referred to as the cognitive and social coefficients, respectively, and modulate (through $\boldsymbol{\phi}_1$ and $\boldsymbol{\phi}_2$) the contribution of the personal best and the global best to velocity update equation. Collectively, c_1 and c_2 are also referred to as the acceleration coefficients.

The determination of these coefficients is a matter of balancing the algorithm's exploration (global search) and exploitation (local search) capabilities. While starting with the values such as $c_1 = c_2 = 2$ is a recommended value [29], fine-tuning these parameters through empirical experimentation or adaptive mechanisms can lead to improved performance in specific optimization tasks [1,8,13,26,33]. Last, after each velocity update, the components of $\boldsymbol{v}_i^{t+1}$ are clamped between a minimum velocity $\boldsymbol{v}_{\min}$ and a maximum velocity $\boldsymbol{v}_{\max}$, to ensure the velocity does not grow without bound.

2.1 Standard PSO

In a subsequent paper, Eberhart and Kennedy [11] introduced a new version of PSO in which each particle has information about its personal best position and the local best particle in a certain neighborhood around the particle. The personal best position $\boldsymbol{p}_i^t$ is the best position encountered by particle i up to iteration t, and the local best position $\boldsymbol{l}_i^t$ for the particle i is the position of the particle in its neighborhood that has the best fitness. In other words, particles move toward the points defined by their personal best and local best in their neighborhood, instead of moving toward the stochastic average of personal best and global best in the population.

In the original PSO, the velocity term is memoryless. Taking into account that the recommended value for c_1 and c_2 is the same and equal to 2, $\boldsymbol{\phi}_1$ and $\boldsymbol{\phi}_2$ values for both parts of the equation will have components that are one on average. Hence, the particles move towards the global best position until another particle achieves better fitness and takes over. Therefore, all the particles tend to come closer to the new global best. Note that although PSO aims to find the global optimum, there is no guarantee that this will happen (indeed, PSO may not even be a local optimizer [3]). In this regard, Shi and Eberhart [29,30] proposed a new parameter, called the inertia weight ω, to control the velocity in each iteration and to achieve a balance between exploitation and exploration. While a large inertia weight facilitates a global search, a small inertia weight facilitates a local search. This version is considered the standard PSO (SPSO) algorithm. The rules to update a particle's velocity and position in SPSO are given as:

$$\begin{aligned}\boldsymbol{v}_i^{t+1} &= \omega \boldsymbol{v}_i^t + \boldsymbol{\phi}_1 \odot (\boldsymbol{p}_i^t - \boldsymbol{x}_i^t) + \boldsymbol{\phi}_2 \odot (\boldsymbol{l}_i^t - \boldsymbol{x}_i^t) \\ \boldsymbol{x}_i^{t+1} &= \boldsymbol{x}_i^t + \boldsymbol{v}_i^{t+1}.\end{aligned} \quad (2)$$

The inertia weight can be static, remaining constant throughout the search process, or dynamic, where it changes according to the progress of the search. A common strategy for dynamic adjustment involves (linearly) decreasing ω gradually as the search progresses, with the aim of shifting the balance from exploration to exploitation.

2.2 Bare-Bones PSO

Over the years, many variants of PSO have been introduced to improve the algorithm. One of those variants relevant to our analysis is the 'Bare-bones' PSO (BBPSO) proposed by Kennedy [18] as a model of PSO dynamics [35], which belongs to a class of PSO algorithms known as velocity-free variants. The particle update rule in the velocity-free version is:

$$x_i^{t+1} = \frac{1}{2}(g^t + p_i^t) + |g^t - p_i^t|\gamma, \qquad (3)$$

where γ is a random number drawn from the standard normal distribution $\mathcal{N}(0,1)$. The notation $|g^t - p_i^t|$ refers to the component-wise absolute value of the vector $g^t - p_i^t$.

One BBPSO variant of note uses all the information in the swarm instead of just the global best g^t, and centers the particles around the average of the best positions [4]:

$$x_i^{t+1} = \frac{1}{N}\sum_{i=1}^{N} p_i^t + |g^t - p_i^t|\gamma. \qquad (4)$$

This is an example of a fully-informed PSO (FiPSO) algorithm (see Sect. 2.3) and will be referred to as Bare-bones fully-informed PSO (BBFiPSO).

Blackwell [4] generalized Bare-bones PSO so that each particle has a set B_i^t of neighboring particles, and where each particle is centered around the best position, g_i^t, in that neighborhood (instead of the global best). This idea is expressed by:

$$g_i^t = \mathrm{argmin}_{(j \in B_i^t)} F(p_j^t) \qquad (5)$$
$$x_i^{t+1} = g_i^t + \alpha m_i^t \gamma.$$

Here, α is a positive constant that affects the convergence rate. Note that for $\alpha = 1$, we obtain Kennedy's BBPSO. F is the objective function of a minimization problem. In this paper, we consider the global neighborhood and fully-connected swarm in which all particles are included and are connected. Therefore, m_i^t can be written as:

$$m_i^t = |g_i^t - p_i^t|. \qquad (6)$$

2.3 Fully Informed PSO

In 2002, Clerc and Kennedy [9] and Van Den Bergh [34] carried out the first formal stability analysis for PSO by simplifying the standard stochastic PSO to a deterministic dynamical system by treating the random coefficient vectors as constants as follows:

$$v_i^{t+1} = v_i^t + \phi(m_i^t - x_i^t) \tag{7}$$
$$m_i^t = \frac{\phi_1 p_i^t + \phi_2 g_i^t}{\phi_1 + \phi_2},$$

where $\phi = \phi_1 + \phi_2$. Furthermore, they introduced a multiplicative factor χ, referred to as the constriction coefficient, in the velocity update equation, which then becomes

$$v_i^{t+1} = \chi \left(v_i^t + \phi(m_i^t - x_i^t) \right). \tag{8}$$

At each iteration, the constriction coefficient reduces the magnitude of the velocity v_i^{t+1} by a constant amount, thus making the system take smaller steps in the position space. Clerc and Kennedy also derived certain relations between χ, ϕ_1, and ϕ_2 that are necessary for the system to be stable; parameter values that are often used in this version of the algorithm are $\chi = 0.7298$ and $\phi_1 = \phi_2 = 2.05$.

In 2004, Mendes et al. [23] generalized the definition for m_i^t in Eq. (7) to include the position of all particles and not just the best one, making the particles "fully informed":

$$m_i = \frac{\sum_{k \in B} W_{ik} \phi_k \odot p_k}{\sum_{k \in B} W_{ik} \phi_k}. \tag{9}$$

Here, p_k is the best position found by particle k in its neighborhood, and ϕ_k is a vector of uniformly randomly distributed numbers in the interval $[0, \phi_{\max}/|B|]$, where $|B|$ denotes the number of particles in the neighborhood B. The coefficients W_{ik} are weights determining the contribution of particle k to the velocity update of particle i. These weights can include any aspect of the particle that is deemed relevant, and Mendes et al. [23] experimented with the fitness of the neighboring particle, the distance between particles i and k, and the case of constant weights. We will revisit the choice of weights in Sect. 4, when we demonstrate the equivalence between GOA and a particular version of the fully informed PSO. There, we will choose the weight W_{ik} to be a function of the distance between particles i and k.

Later in 2006, Kennedy and Mendes [20] modified the canonical model, given in Eq. (2), and presented a generalized version (referred to as the FIPS model) so that the acceleration coefficient ϕ is distributed throughout the neighborhood rather than dividing it between two terms p_i^t and l_i^t.

$$v_i^{t+1} = \chi \left(v_i^t + \sum_{n=1}^{N} \frac{\phi(p_n^t - x_i^t)}{N} \right) \quad (10)$$

$$x_i^{t+1} = x_i^t + v_i^t.$$

This equation is one of the main equations that we will return to show the equivalency between GOA and one of the PSO variants.

3 Grasshopper Optimization Algorithm

The Grasshopper Optimization Algorithm (GOA) is a nature-inspired optimization algorithm introduced by Saremi et al. in 2017 [28]. Grasshoppers are considered pests due to their damaging effects on crop production and agriculture, and exhibit a unique behavior known as swarming. Despite being typically observed individually in nature, grasshoppers can form one of the largest swarms among all creatures. This swarm phenomenon is present in both nymphs and adult grasshoppers. In large numbers, nymph grasshoppers move synchronously resembling rolling cylinders while consuming vegetation along their path. As they mature into adults, they gather together in the air to form a migrating swarm that covers vast distances. The algorithm operates with a population of grasshoppers, each representing a potential solution within the search space, and models the movement and interaction of grasshoppers in a swarm to optimize solutions to various problems.

In the GOA algorithm, the positions of the particles in the swarm are updated as follows:

$$x_i^{t+1} = c_t^2 \frac{u-l}{2} \odot \left(\sum_{\substack{j=1 \\ j \neq i}}^{N} S\left(\|x_j^t - x_i^t\| \right) \frac{x_j^t - x_i^t}{d_{ij}} \right) + g^t. \quad (11)$$

Here, g^t is the best solution found by the grasshoppers at iteration t, and plays a similar role to the global best in PSO. Components of the position vector are limited to the range provided by the vectors u and l (upper bound and lower bound, respectively), and $d_{ij} = \|x_j^t - x_i^t\|$ is the Euclidean distance between the grasshoppers i and j. The social interaction function S is defined as:

$$S(d) = f \exp(-d/a) - \exp(-d), \quad (12)$$

and is applied component-wise to the vector $|x_j^t - x_i^t|$. The social interaction function modulates the attractive behaviour of the swarm (when $S(d) > 0$) versus the repelling behaviour (when $S(d) < 0$). In [28], the constants f and a are chosen to be $f = 0.5$ and $a = 1.5$ to balance the effects of attraction and repulsion. The multiplicative factor c_t depends on the iteration t and decreases linearly over T timesteps from a value $c_{\max}$ to $c_{\min}$:

$$c_t = c_{\max} - t \frac{c_{\max} - c_{\min}}{T}. \quad (13)$$

4 How Does GOA Compare to PSO?

Given that GOA is an algorithm without velocity term and with a global neighborhood, it bears a strong resemblance to BBPSO with global neighborhood. Recall from Sect. 2.2 that Blackwell [4] generalized Bare-bones PSO in such a way that the swarm centers around the best particle g_i in their neighborhood, which can be defined based on the problem, and the particle update equation can be written as follows:

$$x_i^{t+1} = g^t + \alpha m_i^t \gamma. \tag{14}$$

Here, the choice of m_i^t reflects the contribution of the other particles to the next position x_i^{t+1}. The FiPSO class of algorithms (see Sect. 2.3) uses information from all neighbors but with different weights. One variant of FiPSO in [23], which is useful to our study, is the wdFIPS algorithm. While the generalized version of FiPSO (Eq. (10)) considers the average distance of the particles' personal best to the target particle, the wdFIPS algorithm in particular weighs the contribution of each neighboring particle by its distance to the target particle, which is similar to the GOA approach. Therefore, the personal best in Eq. (10) is replaced with the position of the particles. As a result, m_i^t can be written as

$$m_i^t = \sum_{k=1}^{N} W_{ki}^t \frac{\phi(x_k^t - x_i^t)}{N}, \tag{15}$$

Substituting this expression into the particle update Eq. (14) gives

$$x_i^{t+1} = g^t + \alpha\gamma \sum_{k=1}^{N} W_{ki}^t \frac{\phi(x_k^t - x_i^t)}{N}. \tag{16}$$

To simplify the problem, the stochastic term can be replaced with a coefficient β, making the particle update equation a deterministic one:

$$x_i^{t+1} = g^t + \frac{\alpha\beta}{N} \sum_{k=1}^{N} W_{ki}^t (x_k^t - x_i^t). \tag{17}$$

Given that the weight W_{ki}^t typically represents the distance from the target particle i to the neighbor k, we propose using the social interaction function S, given by Eq. (12), for it:

$$W_{ki}^t = \frac{S\left(\|x_k^t - x_i^t\|\right)}{d_{ik}}, \tag{18}$$

The resulting wdFIPS update equation is then given by

$$x_i^{t+1} = g^t + \frac{\alpha\beta}{N} \sum_{k=1}^{N} S\left(\|x_k^t - x_i^t\|\right) \frac{(x_k^t - x_i^t)}{d_{ik}}. \tag{19}$$

Comparing this equation with the GOA equation (Eq. (11)) shows that GOA can be considered as a bare-bones fully informed PSO (BBFiPSO) with a global neighborhood.

5 Experimental Results

To further assess the equivalence between GOA and the different PSO algorithms discussed previously, we performed a computational analysis across a variety of unimodal and multimodal objective functions [10, 21, 24, 31, 39, 40]. The mathematical definitions of these benchmark functions are presented as a table available at the following GitHub repository: ANTS2024-Code. In addition to GOA, we also benchmarked: SPSO (Eq. (2)), BBPSO (Eq. (3)), FiPSO (Eq. (10)), BBFiPSO (Eq. (4)). Note that some optimization algorithms have a tendency to converge towards the origin, making their performance on benchmark functions that have their optimum at the origin appear more effective, a problem known as center bias [21]. To assess an optimization algorithm in an unbiased way, it is therefore common to evaluate it on objective functions that have been rotated, shifted, and scaled so that the optimum value is at a location other than the origin (see e.g. the objective functions in CEC'05, BBOB, etc.). However, as GOA and PSO have been shown (empirically) not to suffer from center bias (see [22]), we have chosen not to pursue this.

5.1 Experimental Setup

Each algorithm was subjected to an evaluation process involving 200 independent runs to ensure the reliability and stability of the results obtained. The experiments were conducted in a 30-dimensional search space, using 100 individuals (particles for PSO and grasshoppers for GOA) and 150 iterations per run (refer to Table 1 for more details on the parameter choices). The values of the parameters follow their definition in the papers that they were drawn from.

The complete source code for this study, including all Python scripts and related resources, is available for access and download at the following GitHub repository: https://github.com/negin17h/ANTS2024-Code.

The computational experiments were performed on a MacBook Pro with a 2.4 GHz 8-Core Intel Core i9 processor, 32 GB DDR4 memory with 2667 MHz frequency, Intel UHD Graphics 630 1536 MB, and Mac OS Sonoma 14.5, using Python 3.8 and several libraries, including NumPy 1.24.2 [12], Pandas 1.5.3 [27], Matplotlib 3.7.5 [16], and Seaborn 0.13.2 [36].

Table 1. Parameter Values of the Algorithms

Algorithm	Parameter	Value
	Topology	Fully Connected
	Neighborhood	Global
	Independent Runs	200
	Dimension	30
	Number of Individuals	100
	Number of Iterations	150
SPSO	Inertia Weight (ω)	Linearly decrease from 0.9 to 0.2
	Cognitive and Social Components (c_1, c_2)	2, 2
	Velocity Control	0.2
FiPSO	Constriction Coefficient (χ)	0.7298
	Cognitive and Social Components (c_1, c_2)	2.05, 2.05
	Velocity Control	0.2
BBPSO	–	–
BBFiPSO	Alpha (α)	1
GOA	Intencity of attraction (f)	0.5
	Attractive length scale (a)	1.5
	Decreasing coefficient (c)	Decrease from 1 to 0.00004

5.2 Quantitative Results and Discussion

Figures 1 and 2 show the comparison of the different algorithms for each benchmark function in the unimodal and multimodal test sets, respectively. Furthermore, to investigate whether GOA is significantly better than other algorithms, a one-sided Wilcoxon test has been performed at 5% significance level.

In most cases (15 out of 18 benchmarks), there is either no significant improvement in terms of effectiveness between GOA and BBFiPSO ($p = 1$), or BBFiPSO outperforms GOA. Taking into account that the interquartile range (IQR) for BBFiPSO is smaller than that of GOA in most cases, BBFiPSO shows better capabilities to explore the search space and converge towards optimal or near-optimal solutions, regardless of the nature of the benchmark function (unimodal or multimodal). This similarity in performance can be explained by the fact that both GOA and BBFiPSO are velocity-free and share all the information among grasshoppers or particles in the swarm. Note that for most of these functions, there are no substantial differences in terms of effectiveness between GOA and BBPSO, or between GOA and FiPSO, or even PSO variants outperform GOA.

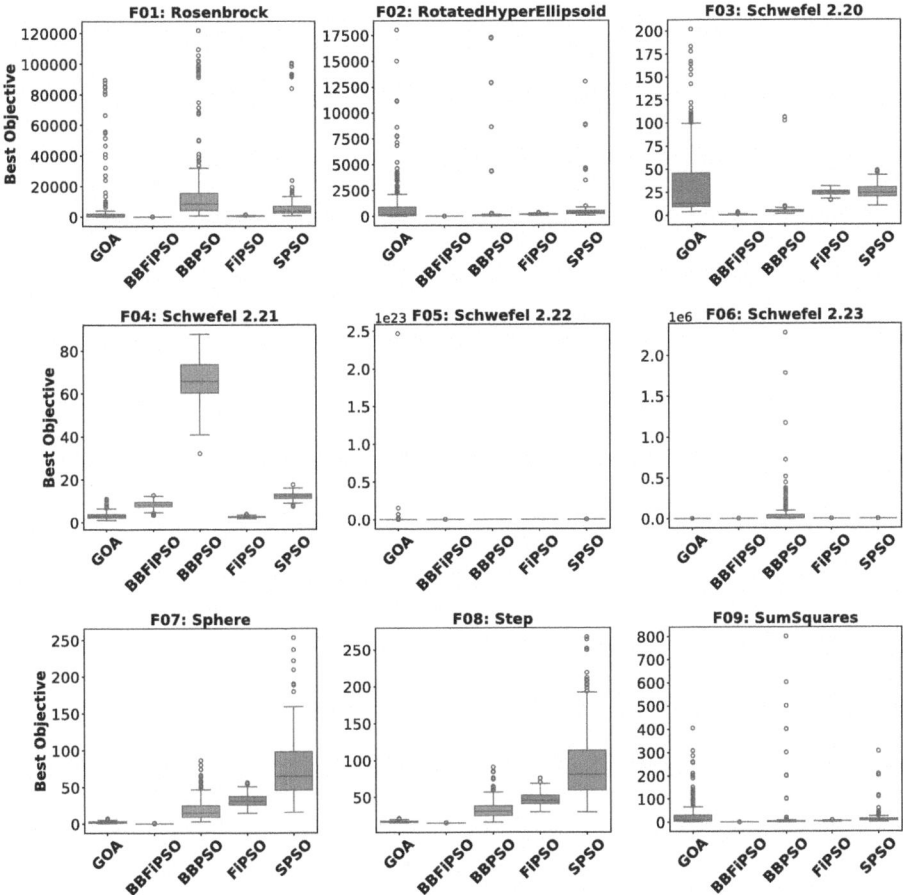

Fig. 1. Comparison between GOA and PSO variants on different unimodal benchmark functions based on 200 runs for each algorithm using 100 particles in a 30 dimensional space and 150 iterations.

Furthermore, for the remaining three benchmark functions in this list ($F04$, $F17$, $F18$), there is a significant improvement between GOA and BBFiPSO ($p < 2e-16$). However, no significant improvements in terms of effectiveness have been found between GOA and FiPSO for $F04$ ($p = 0.998$), and between GOA and BBPSO for functions $F17$ and $F18$ ($p = 1$), which is expected, given that the former is a particular example of a BBPSO-like or FiPSO-like algorithm, as discussed in Sect. 2.2.

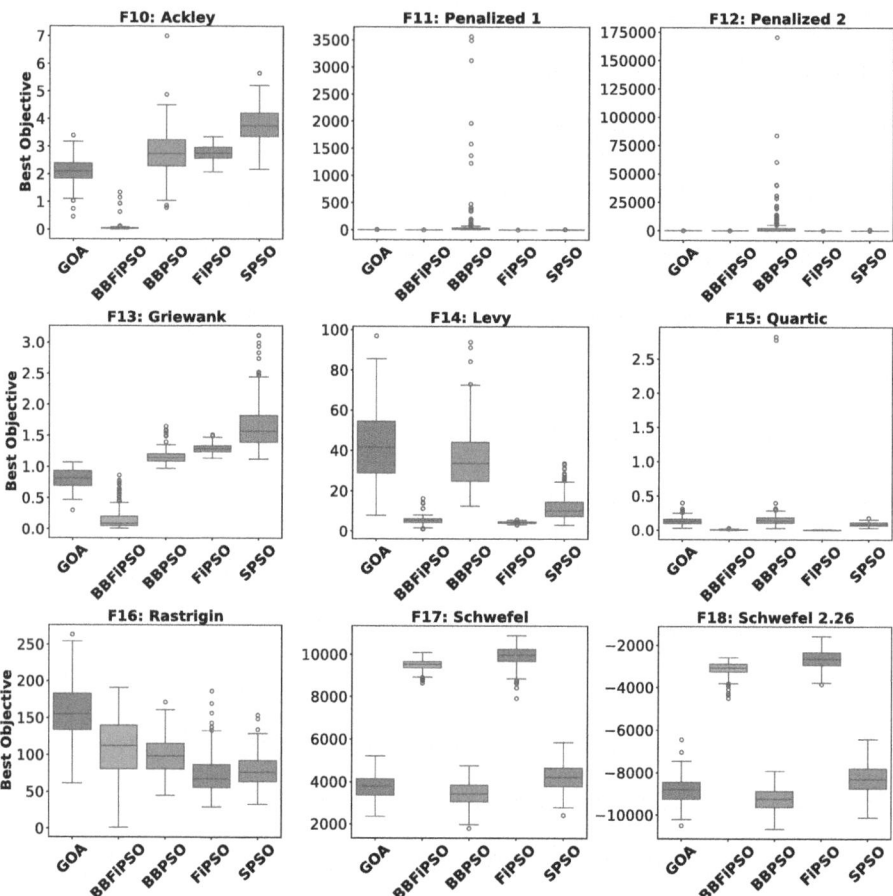

Fig. 2. Comparison between GOA and PSO variants on different `multimodal` benchmark functions based on 200 runs for each algorithm using 100 particles in a 30 dimensional space and 150 iterations.

6 Conclusions

In this paper, we have presented one of the highly cited swarm intelligence algorithms in terms of PSO. We performed a comparative analysis between GOA and PSO and its variants, which showed that the GOA and BBFiPSO share a similar approach. However, given that BBFiPSO has a more condensed IQR, it has greater capabilities to converge to an optimal or near-optimal solution. In terms of efficiency, considering that GOA calculates the distance among all grasshoppers to the target grasshopper, the computational cost is much higher compared to the PSO family.

Future research may investigate the conditions and parameters that influence the performance of these algorithms, potentially leading to the development of more refined and problem-specific optimization strategies.

Acknowledgements. The work of J.V. was supported by a grant from the Special Research Fund (BOF) of Ghent University (BOF/STA/202109/039).

References

1. Ardizzon, G., Cavazzini, G., Pavesi, G.: Adaptive acceleration coefficients for a new search diversification strategy in particle swarm optimization algorithms. Inf. Sci. **299**, 337–378 (2015). https://doi.org/10.1016/j.ins.2014.12.024
2. de Armas, J., Lalla-Ruiz, E., Tilahun, S.L., Voß, S.: Similarity in metaheuristics: a gentle step towards a comparison methodology. Nat. Comput. **21**(2), 265–287 (2022). https://doi.org/10.1007/s11047-020-09837-9
3. Van den Bergh, F., Engelbrecht, A.P.: A convergence proof for the particle swarm optimiser. Fund. Inform. **105**(4), 341–374 (2010). https://doi.org/10.3233/FI-2010-370
4. Blackwell, T.: A study of collapse in bare bones particle swarm optimization. IEEE Trans. Evol. Comput. **16**(3), 354–372 (2011). https://doi.org/10.1109/TEVC.2011.2136347
5. Camacho-Villalón, C.L., Dorigo, M., Stützle, T.: Why the intelligent water drops cannot be considered as a novel algorithm. In: Dorigo, M., Birattari, M., Blum, C., Christensen, A.L., Reina, A., Trianni, V. (eds.) ANTS 2018. LNCS, vol. 11172, pp. 302–314. Springer, Cham (2018). https://doi.org/10.1007/978-3-030-00533-7_24
6. Camacho-Villalón, C.L., Dorigo, M., Stützle, T.: The intelligent water drops algorithm: why it cannot be considered a novel algorithm: a brief discussion on the use of metaphors in optimization. Swarm Intell. **13**, 173–192 (2019). https://doi.org/10.1007/s11721-019-00165-y
7. Camacho Villalón, C.L., Stützle, T., Dorigo, M.: Grey wolf, firefly and bat algorithms: three widespread algorithms that do not contain any novelty. In: Dorigo, M., et al. (eds.) ANTS 2020. LNCS, vol. 12421, pp. 121–133. Springer, Cham (2020). https://doi.org/10.1007/978-3-030-60376-2_10
8. Chen, K., Zhou, F., Yin, L., Wang, S., Wang, Y., Wan, F.: A hybrid particle swarm optimizer with sine cosine acceleration coefficients. Inf. Sci. **422**, 218–241 (2018). https://doi.org/10.1016/j.ins.2017.09.015
9. Clerc, M., Kennedy, J.: The particle swarm - explosion, stability, and convergence in a multidimensional complex space. IEEE Trans. Evol. Comput. **6**(1), 58–73 (2002). https://doi.org/10.1109/4235.985692
10. Digalakis, J.G., Margaritis, K.G.: On benchmarking functions for genetic algorithms. Int. J. Comput. Math. **77**(4), 481–506 (2001). https://doi.org/10.1080/00207160108805080
11. Eberhart, R., Kennedy, J.: A new optimizer using particle swarm theory. In: Proceedings of the Sixth International Symposium on Micro Machine and Human Science, MHS 1995, pp. 39–43. IEEE (1995). https://doi.org/10.1109/MHS.1995.494215
12. Harris, C.R., et al.: Array programming with NumPy. Nature **585**(7825), 357–362 (2020). https://doi.org/10.1038/s41586-020-2649-2

13. Harrison, K.R., Engelbrecht, A.P., Ombuki-Berman, B.M.: Self-adaptive particle swarm optimization: a review and analysis of convergence. Swarm Intell. **12**, 187–226 (2018). https://doi.org/10.1007/s11721-017-0150-9
14. Hayward, L., Engelbrecht, A.: Determining metaheuristic similarity using behavioural analysis. IEEE Trans. Evol. Comput. (2023). https://doi.org/10.1109/TEVC.2023.3346672
15. Hayward, L., Engelbrecht, A.: How to tell a fish from a bee: constructing metaheuristic search behaviour characteristics. In: Proceedings of the Companion Conference on Genetic and Evolutionary Computation, pp. 1562–1569 (2023). https://doi.org/10.1145/3583133.3596338
16. Hunter, J.D.: Matplotlib: a 2D graphics environment. Comput. Sci. Eng. **9**(3), 90–95 (2007). https://doi.org/10.1109/MCSE.2007.55
17. Kennedy, J.: The particle swarm: social adaptation of knowledge. In: Proceedings of 1997 IEEE International Conference on Evolutionary Computation (ICEC 1997), pp. 303–308. IEEE (1997). https://doi.org/10.1109/ICEC.1997.592326
18. Kennedy, J.: Bare bones particle swarms. In: Proceedings of the 2003 IEEE Swarm Intelligence Symposium, SIS 2003 (Cat. No. 03EX706), pp. 80–87. IEEE (2003). https://doi.org/10.1109/SIS.2003.1202251
19. Kennedy, J., Eberhart, R.: Particle swarm optimization. In: Proceedings of ICNN 1995-International Conference on Neural Networks, vol. 4, pp. 1942–1948. IEEE (1995). https://doi.org/10.1109/ICNN.1995.488968
20. Kennedy, J., Mendes, R.: Neighborhood topologies in fully informed and best-of-neighborhood particle swarms. IEEE Trans. Syst. Man Cybern. Part C (Appl. Rev.) **36**(4), 515–519 (2006). https://doi.org/10.1109/TSMCC.2006.875410
21. Kudela, J.: A critical problem in benchmarking and analysis of evolutionary computation methods. Nat. Mach. Intell. **4**(12), 1238–1245 (2022). https://doi.org/10.1038/s42256-022-00579-0
22. Kudela, J.: The evolutionary computation methods no one should use. arXiv preprint arXiv:2301.01984 (2023)
23. Mendes, R., Kennedy, J., Neves, J.: The fully informed particle swarm: simpler, maybe better. IEEE Trans. Evol. Comput. **8**(3), 204–210 (2004). https://doi.org/10.1109/TEVC.2004.826074
24. Molga, M., Smutnicki, C.: Test functions for optimization needs. Test Functions for Optimization Needs **101**, 48 (2005)
25. Piotrowski, A.P., Napiorkowski, J.J., Rowinski, P.M.: How novel is the "novel" black hole optimization approach? Inf. Sci. **267**, 191–200 (2014). https://doi.org/10.1016/j.ins.2014.01.026
26. Ratnaweera, A., Halgamuge, S.K., Watson, H.C.: Self-organizing hierarchical particle swarm optimizer with time-varying acceleration coefficients. IEEE Trans. Evol. Comput. **8**(3), 240–255 (2004). https://doi.org/10.1109/TEVC.2004.826071
27. Reback, J., McKinney, W., et al.: Pandas: powerful Python data analysis toolkit. pandas.pydata.org (2020). https://pandas.pydata.org/
28. Saremi, S., Mirjalili, S., Lewis, A.: Grasshopper optimisation algorithm: theory and application. Adv. Eng. Softw. **105**, 30–47 (2017). https://doi.org/10.1016/j.advengsoft.2017.01.004
29. Shi, Y., Eberhart, R.: A modified particle swarm optimizer. In: 1998 IEEE International Conference on Evolutionary Computation Proceedings. IEEE World Congress on Computational Intelligence (Cat. No. 98TH8360), pp. 69–73. IEEE (1998). https://doi.org/10.1109/ICEC.1998.699146

30. Shi, Y., Eberhart, R.C.: Empirical study of particle swarm optimization. In: Proceedings of the 1999 Congress on Evolutionary Computation-CEC99 (Cat. No. 99TH8406), vol. 3, pp. 1945–1950. IEEE (1999). https://doi.org/10.1109/CEC.1999.785511
31. Surjanovic, S., Bingham, D.: Virtual library of simulation experiments: test functions and datasets. http://www.sfu.ca/~ssurjano. Accessed 13 Mar 2024
32. Swan, J., et al.: A research agenda for metaheuristic standardization. In: Proceedings of the XI Metaheuristics International Conference, pp. 1–3. Citeseer (2015)
33. Tripathi, P.K., Bandyopadhyay, S., Pal, S.K.: Multi-objective particle swarm optimization with time variant inertia and acceleration coefficients. Inf. Sci. **177**(22), 5033–5049 (2007). https://doi.org/10.1016/j.ins.2007.06.018
34. Van Den Bergh, F.: An Analysis of Particle Swarm Optimizers. University of Pretoria (South Africa) (2001)
35. Wang, D., Tan, D., Liu, L.: Particle swarm optimization algorithm: an overview. Soft. Comput. **22**, 387–408 (2018). https://doi.org/10.1007/s00500-016-2474-6
36. Waskom, M.L.: Seaborn: statistical data visualization. J. Open Source Softw. **6**(60), 3021 (2021). https://doi.org/10.21105/joss.03021
37. Weyland, D.: A rigorous analysis of the harmony search algorithm: how the research community can be misled by a "novel" methodology. Int. J. Appl. Metaheuristic Comput. (IJAMC) **1**(2), 50–60 (2010). https://doi.org/10.4018/jamc.2010040104
38. Weyland, D.: A critical analysis of the harmony search algorithm—how not to solve sudoku. Oper. Res. Perspect. **2**, 97–105 (2015). https://doi.org/10.1016/j.orp.2015.04.001
39. Yang, X.S.: Test problems in optimization. arXiv preprint arXiv:1008.0549 (2010)
40. Yao, X., Liu, Y., Lin, G.: Evolutionary programming made faster. IEEE Trans. Evol. Comput. **3**(2), 82–102 (1999). https://doi.org/10.1109/4235.771163

Group-Level Behavioral Switch in a Robot Swarm Using Blockchain

Himank Gupta[1,2](✉), Volker Strobel[1], Alexandre Pacheco[1], Eliseo Ferrante[3], Enrico Natalizio[2,4], and Marco Dorigo[1]

[1] IRIDIA, Université Libre de Bruxelles, Brussels, Belgium
{himank.gupta,volker.strobel,alexandre.pacheco}@ulb.be, mdorigo@ulb.ac.be
[2] Technology Innovation Institute, Abu Dhabi, UAE
enrico.natalizio@tii.ae
[3] Vrije Universiteit Amsterdam, Amsterdam, The Netherlands
e.ferrante@vu.nl
[4] CNRS, LORIA, Université de Lorraine, Villers-lés-Nancy, France

Abstract. In this paper, we introduce the concept of group-level behavioral switch (GLBS) in a robot swarm. We consider two distinct types of GLBS that differ in whether or not the individual robots in the group need to switch their behavior at the same time: the Synchronous GLBS (S-GLBS) and the Asynchronous GLBS (A-GLBS). To implement these GLBSs, we propose a blockchain-based solution built on the Ethereum platform. We then study its performance in terms of required time and success rate in a series of simulation experiments.

1 Introduction

Swarm robotics is the discipline that studies how a large number of robots with relatively limited capabilities plan and coordinate so as to complete complex tasks that are difficult or impossible for individual robots [4,7,9,10]. Swarm robotics is considered as one of the most promising research directions in robotics [37] and it is envisioned that in the future robot swarms will be used in a wide range of possible applications, including environmental monitoring, search and rescue, precision farming, surveillance missions, and extinguishing fires.

Studies in the field have mainly focused on how to implement individual collective behaviors in robot swarms such as aggregation [1], pattern formation [2], self-assembly [14,15,21,27], collective exploration [11,20,39], coordinated motion [13,32], and collective transport [16]. However, for complex real-world applications it is also necessary for the robot swarm to be able to transition from one of the aforementioned behaviors to the next. For instance, during a search and rescue mission, a robot swarm is required to transition between two distinct collective behaviors: first it performs collective exploration to locate casualties and search for survivors, and then it transitions to coordinated transport to bring survivors to a safe place. In this paper, we study how to implement these types of transitions, that we call "group-level behavioral switches", or GLBSs for short, using blockchain technology. Most of the literature in swarm robotics, with the notable exceptions of [6,25,28], has studied collective behaviors in isolation and a versatile way to switch from one collective behavior to another is missing. In GLBS a certain number, or percentage, of the robots in the swarm have to switch their behavior. Note that, if all the robots in the swarm need to perform the switch then we talk of a swarm-level behavioral switch, or SLBS for short.

A straightforward approach to GLBS would be to develop a centralized controller that monitors the whole swarm and instructs a group of robots to switch behavior when certain conditions are met. However, a centralized controller is a single point of failure for the swarm and is not scalable to larger swarm sizes.

Another way to implement a GLBS involves using a distributed mechanism, such as the one presented in [25] that utilizes a hash-table for disseminating information within the swarm. However, this solution has two drawbacks. First, every individual robot in the swarm can input information into the hash-table at any time, and propagate it to other robots. This can potentially cause a high number of conflicts in information, especially for larger swarm sizes. Second, in their original formulation, there is no explicit mechanism to handle the presence of malicious information, whereas blockchain-based approaches have already been shown to be effective against this issue [31].

In this paper, our objective is to examine the use of blockchain technology to achieve GLBSs in robot swarms. In fact, recent advances in swarm robotics have demonstrated that blockchain technology can be leveraged to achieve fast and conflict-free consensus, even in the presence of Byzantine (faulty or non-cooperating) robots [29,31], and to give instructions to individual robots in the swarm [23]. However, it was not yet studied how to instruct a group of robots to change its behavior in a coordinated way.

The main contributions of this paper are the following:

1. We introduce the group-level behavioral switch (GLBS) notion and classify it into two categories, i.e., asynchronous behavioral switch (A-GLBS) and synchronous behavioral switch (S-GLBS), which differ in whether or not the individual robots need to switch their behavior at the same time.
2. We develop a blockchain-based solution, based on the Ethereum platform, that showcases how the blockchain mitigates the issues of centralized and distributed approaches in the decision-making process for GLBS.
3. We evaluate our blockchain-based solution by conducting experiments on an example scenario chosen so that it includes both A-GLBS and S-GLBS: a simulated fire extinguishing application.[1]

The subsequent sections of the paper are structured in the following manner: Sect. 2 provides a brief review of relevant literature related to decision-making in swarm robotics and the utilization of blockchain technology. Section 3 provides a concise introduction to blockchain technology, smart contracts, and consensus algorithms. In Sect. 4, we present the experimental scenario and the experiment timeline, and provide some details on the experimental setup. Section 5 defines the performance metrics that we use to evaluate the effectiveness of our solution

---

[1] Note that this is mainly a representative scenario that well illustrates situations in which GLBSs might be needed.

and describes and discusses the results obtained. Finally, conclusions and future works are highlighted in Sect. 6.

2 Related Work

The work presented in this paper is closely related to research on collective decision-making and on swarm-level finite state machines. In fact, the GLBS requires a group of robots in the swarm to make a collective decision to transition to a new state (e.g. of a finite state machine), in response to some event.

Collective decision-making, which is the first step in implementing a GLBS, has been thoroughly studied in the classic, non-blockchain-based robot swarm literature [3,12,33–35]. A limitation of these works when used as the first step for GLBS is that, after convergence, the robots in the swarm do not have explicit knowledge about the swarm status, which is fundamental for implementing a GLBS. Recent advances in the use of blockchain technology to reach consensus in a robot swarm for collective decision-making [8,17,23,31] provide a possible solution to this problem. Indeed, the blockchain, which is maintained in a distributed way by the robots in the swarm, provides a distributed database that allows each of the robots in the swarm to know about the convergence status of any other robot in the swarm. This knowledge can be used to implement our GLBS. However, GLBS was not explicitly studied in that work.

Our implementation is built on the concept of a swarm-level or group-level finite state machine in a robot swarm. We could find only two studies focusing on this concept. The first one introduced the Buzz programming language [25,28], which uses the concept of virtual stigmergy to facilitate coordination within the swarm. This concept is exploited as an approach for executing transitions between different behaviors at the swarm level using a finite state machine (FSM). However, as explained in the Introduction, this approach does not handle well conflicting information nor does it account for the presence of Byzantine robots. The second one proposes a distributed controller also based on a FSM for the allocation of sub-swarms to predefined tasks through a bidding process [6]. However, the study assumes that the swarm is always connected, which makes it unsuitable when this cannot be guaranteed because of low swarm density.

3 Blockchain Fundamentals

A blockchain is a decentralized and distributed ledger that securely records transactions in a transparent and tamper-resistant manner. The primary objective of the original blockchain introduced by Nakamoto [19] was to establish a digital currency system that enables peer-to-peer transactions (i.e., transfers of crypto assets) without the need for external authorities. A blockchain is composed of a sequence of blocks, each of which is a data structure consisting of at least two elements: i) the hash value of the previous block, used to establish the connection and order of the blocks within the blockchain, and ii) a list of transactions.

When a node wants to add a transaction to the blockchain, it sends it to a pool of unconfirmed transactions. To generate a new block, nodes gather a subset of the transactions in the pool, verify each transaction's validity, and bundle the valid transactions in a new candidate block. A consensus algorithm is responsible for deciding if and in which order the new candidate blocks should be added to the blockchain. (Note that the consensus algorithm is always the cause of a delay between the moment a node sends a transaction and the moment in which this transaction is incorporated in the blockchain and disseminated in the network.) Recent research [22,31,38] has demonstrated the feasibility of executing blockchain technology in (real) robot swarms using a lightweight consensus algorithm called proof-of-authority (PoA[2]) [26], which is therefore the consensus algorithm that we use in this work.

In a PoA-based blockchain, the first block, also known as the genesis block, incorporates parameters of the consensus protocol, such as the block interval period, the initial crypto-asset allocation per node, and a list of the permitted sealers (i.e., nodes that are allowed to generate new blocks). With PoA, before being able to propose a new block, sealers need to wait for a certain amount of time (specified by the block interval period) from the time the last block was added to the blockchain. In addition, in a blockchain consisting of N sealers, after a sealer has proposed a new block, it needs to wait for at least $(N/2 + 1)$ blocks before being able to propose another block. Therefore, at any time, there are no more than $(N - (N/2 + 1))$ nodes permitted to propose a new block. Forks may occur in the blockchain when more nodes propose a block at the same time. When this happens, the GHOST protocol [36] is used to resolve the forks.

Alongside the utilization of the PoA algorithm, we also make use of smart contracts for the decision making in GLBS. A smart contract is a software program stored in the blockchain and executed by all the nodes in the blockchain network [5]. This mechanism enables decentralized networks to agree on the code, input, and output of software programs. In the context of swarm robotics, smart contracts can act as distributed controllers that allow robots to execute actions based on a consensus in the blockchain network.

4 Methods

4.1 The Experimental Scenario

To evaluate the performance of our blockchain-based approach to the GLBS, we consider a "fire extinguishing" scenario which is representative of any scenarios that require a GLBS (Fig. 1). In this scenario, the robots in the swarm perform a sequence of three behaviors: (i) `detect fires`, (ii) `navigate toward fires`, and (iii) `extinguish fires`. During the first behavior, i.e., fires detection, a robot swarm patrols the environment in search for fires. Upon detection of a fire, in the second behavior, a subset of the robot swarm navigates towards it. Once this group of robots reaches the vicinity of the fire, the robots in the group

---

[2] https://github.com/ethereum/EIPs/issues/225.

Fig. 1. Our simulation scenario implemented using the ARGoS simulator [24].

switch to the third behavior and start extinguishing the fire. The third behavior in particular has a strong *group-level requirement*, meaning that fire can only be extinguished through the collective action of a group of robots.

In order to complete the sequence of behaviors, the group of robots has to perform two GLBSs: first an asynchronous GLBS (A-GLBS) and then a synchronous GLBS (S-GLBS). The A-GLBS is the switch from detect fire to navigate toward fire, which starts after the fire has been detected. The S-GLBS is the switch from navigate toward fire to extinguish fire, which is activated when the required number of robots have reached the fire location.

In the A-GLBS (i.e., from detect fire to navigate toward fire), the involved robots are not required to switch their behavior simultaneously, as they do not need to coordinate their movement toward the fire. In the S-GLBS (i.e., from navigate toward fire to extinguish fire), a minimum number of robots need to start shooting water simultaneously for their fire-extinguishing action to be effective. This number, that we call R_{est}, is a parameter of the problem.

In this paper, we consider a specific instance of the generic problem described above; in particular, we assume that: (i) only one fire is present in the experimental arena, (ii) during the fire detection behavior the robots perform a random walk, (iii) the robots share a common frame of reference and a shared clock, and (iv) the minimum cardinality of the set R_{est} of robots necessary to extinguish the fire is $|R_{est}| = 4$. In a more general scenario one could relax some of these assumptions, for example the number of fires and the number of robots necessary to extinguish them could be variable. However, in this paper, our main goal is to show the suitability of our blockchain-based decision making for GLBS: we leave therefore the analysis of more general scenarios for future research.

Even though the swarm could, at least in principle, extinguish the fire using just $|R_{est}|$ robots, a larger swarm has a number of benefits: it reduces the time required to explore the environment to find the fire location, and allows to increase the probability that enough robots are available to execute the GLBSs. In practice, we set the swarm size to $N > |R_{est}|$, and we require that the number of robots involved in each GLBSs is greater than $|R_{est}|$. In particular:

- Let R_{sync} be the set of robots involved in the S-GLBS. We require at least $|R_{sync}| = |R_{est}| + \Delta R_{sync}$ robots, where $|R_{est}|$ is the number of robots necessary for extinguishing the fire and ΔR_{sync} is the number of additional robots necessary to increase the probability of a successful S-GLBS execution.

– Let R_{async} be the set of robots involved in the A-GLBS. We require at least $|R_{async}| = |R_{sync}| + \Delta R_{async}$ robots, where $|R_{sync}|$ is the number of robots necessary for implementing the S-GLBS and ΔR_{async} is the number of additional robots necessary to increase the probability of a successful execution of the A-GLBS.

Therefore, in our solution:

- N robots perform the `detect fire` behavior,
- $|R_{async}|$ robots are involved in the A-GLBS and in the `navigate toward fire` behavior, and
- $|R_{sync}|$ robots are involved in the S-GLBS and $|R_{est}|$ of these robots are active during the `extinguish fire` behavior,

where $|R_{est}| < |R_{sync}| < |R_{async}| < N$.

4.2 The Experiment Timeline

The experiment timeline (see Fig. 2) consists of a sequence of three swarm behaviors: (i) `detect fire`, (ii) `navigate toward fire`, and (iii) `extinguish fire`, as described in the following.

Detect Fire Behavior. The experiment starts at time t_0 when the swarm of N robots begins exploring the arena through a random walk, with the goal of locating the region affected by the fire (`detect fire` behavior in Fig. 2). At time

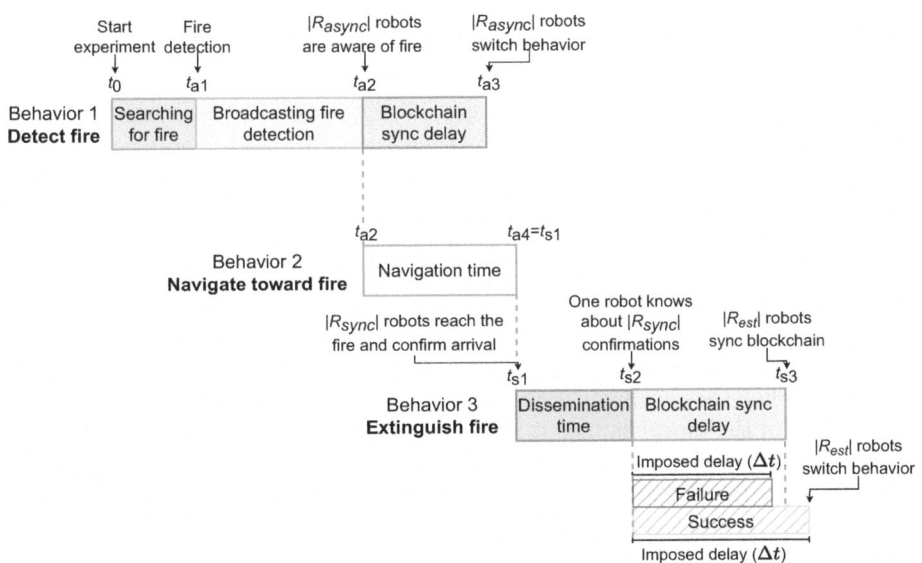

Fig. 2. Timeline of execution of the three behaviors used by the robot swarm in the fire extinguishing scenario—see Sect. 4.2 for more details.

t_{a1} (subscript a stands for asynchronous), for the first time one of the robots identifies the presence of a fire and sends a transaction—that is, the transaction is sent to a pool of unconfirmed transactions, as described in Sect. 3—containing information about the presence of a fire and its location (this transaction will be later added to the blockchain). The remaining robots become aware of the fire by synchronizing with the blockchain through peers when they are in communication proximity (i.e., they are at a distance of 40 cm) during random walk. When a robot receives information regarding fire detection, it confirms reception by sending a transaction, which activates the A-GLBS. The robots that are already aware of the fire continue the random walk to favor information dissemination until the smart contract has received $|R_{async}|$ confirmations (at t_{a2}). At this point, the robots in R_{async} get informed (by querying the smart contract) that it is time to switch their behavior from `fire detection` to `navigate toward fire`. The robots in R_{async} that first receive the blockchain version containing all the $|R_{async}|$ confirmations switch their behavior instantly at t_{a2}. The remaining robots in R_{async} switch their behavior later, as soon as they synchronize with the blockchain of the robots that have already switched their behavior. At time t_{a3} all the robots in R_{async} have switched their behavior and the A-GLBS are t_{a3}, therefore completed. This synchronization delay (from t_{a2} to t_{a3}) is due to the blockchain synchronization delay inherent to blockchain technology.

Navigate Toward Fire Behavior. During the `navigate toward fire` behavior, the robots in R_{async} move toward the area affected by the fire. The transition from `detect fire` to `navigate toward fire` happens gradually: the first robot (r_x) switches at t_{a2} and the others up to time t_{a3}. Navigation time is the time interval between t_{a2} and t_{a4}, i.e., the time elapsed from the moment the first robot $r_x \in R_{async}$ switches its behavior until $|R_{sync}|$ robots reach the fire location.

Extinguish Fire Behavior. At time $t_{s1} = t_{a4}$ (subscript s stands for synchronous), at least $|R_{sync}|$ robots have reached the fire and they have all sent their transactions confirming their arrival. Between t_{s1} and t_{s2}, called dissemination time, the blocks containing these transactions are generated and disseminated by the robots until, at time t_{s2}, at least one robot r_y has received the blockchain version containing all the $|R_{sync}|$ confirmations. Therefore, at t_{s2}, one of the robots would in principle be ready to start extinguishing the fire because it knows that the required number of robots are positioned around the fire. However, as we want the robots to start extinguishing the fire at the same time, it is necessary that they wait until at least $|R_{est}|$ of the robots in R_{sync} have the same version of the blockchain as r_y. This happens at time t_{s3}. Unfortunately, as the robots that synchronize their blockchain with the blockchain of robot r_y do not know the value of t_{s3}, they cannot decide how long to wait before starting to extinguish the fire. Our solution to achieve synchronized fire extinguishing is to impose a delay with respect to t_{s2} for switching to the extinguish fire behavior that needs to be large enough so that the S-GLBS is activated at or after t_{s3}. When selecting the value Δt to give to this imposed delay, one should consider that there is a trade-off between the success rate of the S-GLBS and its duration:

Table 1. Variables and parameters used in experiments

Variables	Value
Arena size (m^2)	7, 14, 28
Density (robots/m^2)	1, 2, 3, 4, 5, 6
Δt (s)	2.0, 2.1, 2.2, ..., 29.9, 30.0
Parameters	**Value**
R_{est}	4
ΔR_{sync}	2
ΔR_{async}	1
Fire patch diameter (cm)	60
Communication range (cm)	40
Block interval period (s)	2

higher values of Δt will increase both the S-GLBS duration and its probability of success, whereas lower values will make the S-GLBS faster but at the cost of a lower probability of success.

4.3 Experimental Setup

All experiments are conducted using the ARGoS [24] robot simulator, in which we have designed our fire extinguishing scenario (discussed in Sec. 4.1). As a robot model, we use the e-puck robot plugin, available in ARGoS. The e-puck robot [18] is equipped with a total of eight proximity sensors, used to detect and avoid obstacles as well as other robots. Additionally, we use the three ground sensors of the robot to simulate fire detection (black ground color indicates the presence of fire in Fig. 1). To estimate the presence of neighboring robots, the e-puck robot uses a range-and-bearing sensor. Finally, the robot is equipped with a pair of wheels, enabling locomotion and environmental exploration.

Every single robot is programmed to function as a node in the Ethereum blockchain by using the ARGoS-blockchain interface [30,31]. Robots communicate using range-and-bearing sensors only when they are in line-of-sight and the spatial distance between them does not exceed 40 cm. In each experiment, the number of robots is given by the product of the arena size times the swarm density (see Table 1). This yields experiments with up to 168 robots. A total of 30 iterations of each experiment are performed for every combination of density, imposed delay and arena size, as indicated in Table 1. The remaining experimental parameters are provided in Table 1.

5 Results

In this section, we analyze the performance of the two GLBS mechanisms considered in this paper, A-GLBS and S-GLBS, in terms of time needed to perform the switch and of success rate. The switch time is defined as follows:

- A-GLBS time is the time interval between t_{a1} (at least one robot detects the fire) and t_{a3} (the $|R_{async}|$ robots have switched their behavior); and
- S-GLBS time is the sum of the dissemination time $t_{s1} \to t_{s2}$ and the imposed delay Δt (see Fig. 2). For an experiment to be successful, S-GLBS time must be larger or equal to $t_{s1} \to t_{s3}$ (i.e., the blockchain synchronization delay must be smaller than the imposed delay).

We discuss the performance of A-GLBS in terms of A-GLBS time in Sect. 5.1, where we specifically study the effect of changing the swarm density and the arena size. In Sect. 5.2, we discuss the performance of S-GLBS in terms of time and success rate of the experiment. As explained in Sect. 4.2, the S-GLBS performance is expected to strongly depend on the imposed delay (Δt). For this reason, the imposed delay will be the main factor we consider while discussing these results, along with the density and the arena size.

5.1 Asynchronous Behavioral Switch (A-GLBS)

A-GLBS time is shown in Fig. 3 for the different arena sizes considered and for different swarm densities. The A-GLBS requires the dissemination of the information about fire detection and reaching consensus on this information on the blockchain. We expect that an increase of swarm density causes a decrease of A-GLBS time because the robots are likely to have a higher connectivity and can therefore disseminate information faster. We can observe that this expectation is met in all the different arena sizes considered (see Fig. 3). In addition, we observe that for all arena sizes variability tends to decrease for increasing densities. This is because when the density increases the probability of having communication delays caused by disconnected robots decreases and therefore the time required to complete the A-GLBS is less variable.

5.2 Synchronous Behavioral Switch (S-GLBS)

To analyze the performance of the S-GLBS, first we note that we consider the S-GLBS successful only if the blockchain synchronization (which completes at t_{s3},

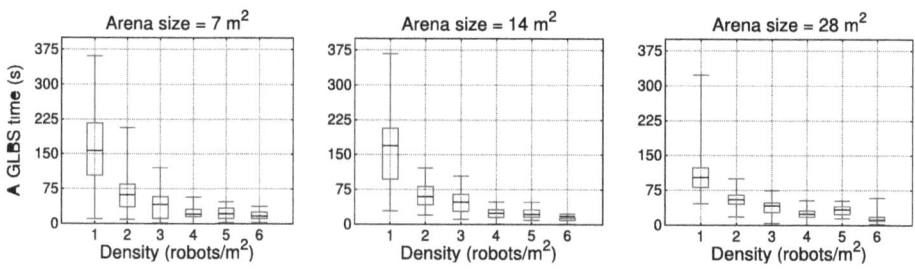

Fig. 3. Asynchronous GLBS: A-GLBS time ($t_{a1} \to t_{a3}$) as a function of the swarm density. Experiments are repeated 30 times. In box-plots, whiskers indicate the minimum and maximum values and horizontal black line indicates the median.

see Fig. 2) occurs before the end of the imposed delay Δt. In Fig. 4, we analyze the success rate in terms of percentage of failed experiments (top row) and S-GLBS time (middle and bottom row) as a function of Δt for varying swarm densities and arena sizes. When reporting the value of S-GLBS time corresponding to each imposed delay, we only considered the successful runs.

Results presented in Fig. 4 (first row) show that when the imposed delay increases the number of failed experiments decreases. However, this comes at the cost of an increased S-GLBS time (last two rows).

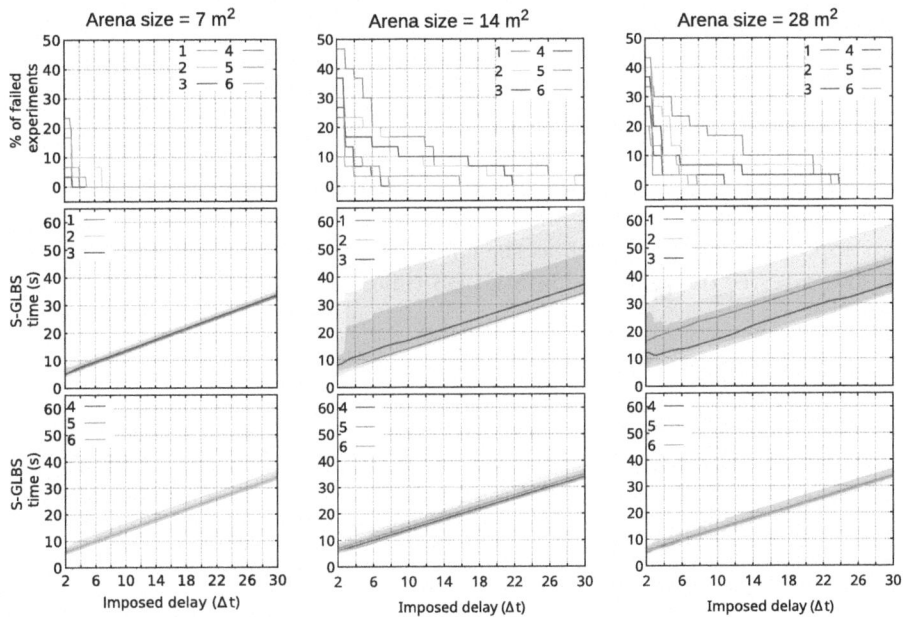

Fig. 4. Synchronous GLBS: Percentage of failed experiments (top row) and S-GLBS time (middle and bottom rows) as a function of imposed delay Δt. Experiments are repeated 30 times; S-GLBS time is computed only for successful experiments. The top row plots the percentage of failed experiments for six different swarm densities, the middle and bottom rows plot S-GLBS time for densities (robots/m^2) 1, 2, 3 and densities 4, 5, 6, respectively. In the middle and bottom rows, the lines represent the median and the shaded areas depict the inter-quartile range. Results show that in the small arena or when the density is ≥ 4 robots/m^2 there is no significant difference in S-GLBS time.

We also observe significant variations in the S-GLBS time for lower densities (≤ 3 robots/m^2) in medium (14 m^2) and large (28 m^2) arenas as depicted in the middle row of Fig. 4. On the other hand, in the small arena (7 m^2) the degree of variation is lower. This can be explained as follows. The time required for the S-GLBS is influenced by two key factors: the block interval period and the block propagation time (because transactions that contain the confirmations about the robots' arrivals at the fire need to be included in blocks and propagated among the robots). Because the number $|R_{sync}|$ of robots required for the S-GLBS does not change across arena sizes, it represents a larger proportion of the swarm in the small arena than in the larger ones. Therefore, it is highly likely that, in the small arena, either one of the robots $\in R_{sync}$ or one of the robots within the communication range (note that the communication range is constant across arena sizes but leads to a higher connectivity in the small arena) is capable of acting as a sealer for the PoA consensus protocol (see Sect. 3). This leads to a block production time with low variability. Hence, the variability of S-GLBS time is lower for the smaller arena. In contrast, when considering medium and large arenas, the number $|R_{sync}|$ of robots no longer constitutes a relatively significant proportion of the robot swarm. As a result, it is possible that there are longer delays before the blockchain transactions created by the robots $\in R_{sync}$ are incorporated into a block.

Conversely, in case of higher densities (≥ 4 robots/m^2) (see bottom row of Fig. 4), enhanced connectivity among robots in the swarm facilitates the finding of an appropriate sealer as well as the efficient propagation of blocks. This leads to a lower degree of variability in S-GLBS time.

6 Conclusions and Future Work

This paper has presented the notion of group-level behavioral switch (GLBS) and its categorization into two distinct types: the asynchronous GLBS and the synchronous GLBS. In order to let a robot swarm reach consensus for a GLBS, a solution has been proposed based on blockchain technology, developed on the Ethereum platform, and evaluated using a fire extinguishing scenario as testbed. The results demonstrate that the time required for the A-GLBS is influenced by the density of the swarm: the higher the density, the lower the A-GLBS time. We also assessed the time required for the S-GLBS and its success rate as a function of the imposed delay Δt. The findings indicate that there exists a trade-off between the S-GLBS success rate and its time for completion. Our proposed solution is versatile, as it allows a user to choose the imposed delay depending on specific requirements of the given scenario. It should however be noted that the choice of the imposed delay in principle depends also on the underlying communication infrastructure which influences the speed with which the robots in the swarm synchronize with the most recent blockchain. In future work, we will investigate the exact form of this dependency as well as other approaches for synchronizing the behavioral switch without imposing the delay externally. Furthermore, in the presented research, we assumed that all robots in the swarm use a shared clock in order to achieve synchronicity in the case of S-GLBS. However, in our future work, we want to accomplish synchronization without relying on a common clock. We also intend to study the effects that the presence of Byzantine robots have on decision-making for GLBSs.

Acknowledgements. V. Strobel and M. Dorigo acknowledge support from the Belgian F.R.S.-FNRS, of which they are a Postdoctoral Researcher and a Research Director respectively.

References

1. Bahçeci, E., Şahin, E.: Evolving aggregation behaviors for swarm robotic systems: a systematic case study. In: Proceedings IEEE Swarm Intelligence Symposium, pp. 333–340 (2005)
2. Bahçeci, E., Soysal, O., Şahin, E.: A review: pattern formation and adaptation in multi-robot systems. Technical report. CMU-RI-TR-03-43, Carnegie Mellon University, Pittsburgh, PA (2003)
3. Bartashevich, P., Mostaghim, S.: Multi-featured collective perception with evidence theory: tackling spatial correlations. Swarm Intell. **15**, 1–28 (2021). https://doi.org/10.1007/s11721-021-00192-8
4. Brambilla, M., Ferrante, E., Birattari, M., Dorigo, M.: Swarm robotics: a review from the swarm engineering perspective. Swarm Intell. **7**(1), 1–41 (2013). https://doi.org/10.1007/s11721-012-0075-2
5. Buterin, V.: A next-generation smart contract and decentralized application platform. Ethereum project white paper. Technical report, Ethereum Foundation (2014). https://ethereum.org/en/whitepaper/. Accessed 02 July 2024
6. Chen, J., Sun, R., Kress-Gazit, H.: Distributed control of robotic swarms from reactive high-level specifications. In: 2021 IEEE 17th International Conference on Automation Science and Engineering (CASE), pp. 1247–1254 (2021). https://doi.org/10.1109/CASE49439.2021.9551578
7. Dorigo, M., Birattari, M., Brambilla, M.: Swarm robotics. Scholarpedia **9**(1), 1463 (2014). https://doi.org/10.4249/scholarpedia.1463
8. Dorigo, M., Pacheco, A., Reina, A., Strobel, V.: Blockchain technology for mobile multi-robot systems. Nat. Rev. Electr. Eng. **1**(4), 264–274 (2024). https://doi.org/10.1038/s44287-024-00034-9
9. Dorigo, M., Théraulaz, G., Trianni, V.: Reflections on the future of swarm robotics. Sci. Robot. **5**(49) (2020). https://doi.org/10.1126/scirobotics.abe4385
10. Dorigo, M., Théraulaz, G., Trianni, V.: Swarm robotics: past, present and future. Proc. IEEE **109**(7), 1152–1165 (2021). https://doi.org/10.1109/JPROC.2021.3072740
11. Ducatelle, F., Di Caro, G., Pinciroli, C., Gambardella, L.M.: Self-organized cooperation between robotic swarms. Swarm Intell. **5**, 73–96 (2011). https://doi.org/10.1007/s11721-011-0053-0
12. Ebert, J.T., Gauci, M., Mallmann-Trenn, F., Nagpal, R.: Bayes bots: collective Bayesian decision-making in decentralized robot swarms. In: 2020 IEEE International Conference on Robotics and Automation (ICRA), pp. 7186–7192 (2020). https://doi.org/10.1109/ICRA40945.2020.9196584
13. Ferrante, E., Turgut, A.E., Mathews, N., Birattari, M., Dorigo, M.: Flocking in stationary and non-stationary environments: a novel communication strategy for heading alignment. In: Schaefer, R., Cotta, C., Kołodziej, J., Rudolph, G. (eds.) PPSN 2010. LNCS, vol. 6239, pp. 331–340. Springer, Heidelberg (2010). https://doi.org/10.1007/978-3-642-15871-1_34
14. Groß, R., Bonani, M., Mondada, F., Dorigo, M.: Autonomous self-assembly in swarm-bots. IEEE Trans. Rob. **22**(6), 1115–1130 (2006). https://doi.org/10.1109/TRO.2006.882919

15. Groß, R., Dorigo, M.: Self-assembly at the macroscopic scale. Proc. IEEE **96**(9), 1490–1508 (2008)
16. Groß, R., Dorigo, M.: Towards group transport by swarms of robots. Int. J. Bio-Inspired Comput. **1**(1–2), 1–13 (2009). https://doi.org/10.1504/IJBIC.2009.022770
17. Keramat, F., Peña Queralta, J., Westerlund, T.: Partition-tolerant and Byzantine-tolerant decision making for distributed robotic systems with IOTA and ROS2. IEEE Internet Things J. **10**(14), 12985–12998 (2023). https://doi.org/10.1109/JIOT.2023.3257984
18. Mondada, F., et al.: The e-puck, a robot designed for education in engineering. In: Gonçalves, P.J.S., et al. (eds.) Proceedings of the 9th Conference on Autonomous Robot Systems and Competitions, pp. 59–65. IPCB: Instituto Politècnico de Castelo Branco, Portugal (2009)
19. Nakamoto, S.: Bitcoin: a peer-to-peer electronic cash system (2008)
20. Nouyan, S., Campo, A., Dorigo, M.: Path formation in a robot swarm: self-organized strategies to find your way home. Swarm Intell. **2**(1), 1–23 (2008). https://doi.org/10.1007/s11721-007-0009-6
21. O'Grady, R., Groß, R., Christensen, A.L., Dorigo, M.: Self-assembly strategies in a group of autonomous mobile robots. Auton. Robot. **28**(4), 439–455 (2010). https://doi.org/10.1007/s10514-010-9177-0
22. Pacheco, A., Strobel, V., Dorigo, M.: A blockchain-controlled physical robot swarm communicating via an ad-hoc network. In: Dorigo, M., et al. (eds.) ANTS 2020. LNCS, vol. 12421, pp. 3–15. Springer, Cham (2020). https://doi.org/10.1007/978-3-030-60376-2_1
23. Pacheco, A., Strobel, V., Reina, A., Dorigo, M.: Real-time coordination of a foraging robot swarm using blockchain smart contracts. In: Dorigo, M., et al. (eds.) ANTS 2022. LNCS, vol. 13491, pp. 196–208. Springer, Cham (2022). https://doi.org/10.1007/978-3-031-20176-9_16
24. Pinciroli, C., et al.: ARGoS: a modular, parallel, multi-engine simulator for multi-robot systems. Swarm Intell. **6**(4), 271–295 (2012). https://doi.org/10.1007/s11721-012-0072-5
25. Pinciroli, C., Beltrame, G.: Swarm-oriented programming of distributed robot networks. Computer **49**, 32–41 (2016). https://doi.org/10.1109/MC.2016.376
26. Polge, J., Robert, J., Le Traon, Y.: Permissioned blockchain frameworks in the industry: a comparison. ICT Express **7**(2), 229–233 (2021). https://doi.org/10.1016/j.icte.2020.09.002
27. Rubenstein, M., Cornejo, A., Nagpal, R.: Programmable self-assembly in a thousand-robot swarm. Science **345**(6198), 795–799 (2014). https://doi.org/10.1126/science.1254295
28. St-Onge, D., Varadharajan, V.S., Švogor, I., Beltrame, G.: From design to deployment: decentralized coordination of heterogeneous robotic teams. Front. Robot. AI **7** (2020). https://doi.org/10.3389/frobt.2020.00051
29. Strobel, V., Castelló Ferrer, E., Dorigo, M.: Managing Byzantine robots via blockchain technology in a swarm robotics collective decision making scenario. In: Dastani, M., Sukthankar, G., André, E., Koenig, S. (eds.) Proceedings of the 17th International Conference on Autonomous Agents and Multiagent Systems, pp. 541–549. AAMAS 2018. International Foundation for Autonomous Agents and Multiagent Systems, Richland, SC (2018)
30. Strobel, V., Castelló Ferrer, E., Dorigo, M.: Blockchain technology secures robot swarms: a comparison of consensus protocols and their resilience to Byzantine robots. Front. Robot. AI **7**(54) (2020). https://doi.org/10.3389/frobt.2020.00054

31. Strobel, V., Pacheco, A., Dorigo, M.: Robot swarms neutralize harmful Byzantine robots using a blockchain-based token economy. Sci. Robot. **8**(79), eabm4636 (2023). https://doi.org/10.1126/scirobotics.abm4636
32. Trianni, V., Nolfi, S., Dorigo, M.: Hole avoidance: experiments in coordinated motion on rough terrain. In: Groen, F., Amato, N., Bonarini, A., Yoshida, E., Kröse, B. (eds.) Intelligent Autonomous Systems 8 - IAS 8, pp. 29–36. IOS Press, Amsterdam (2004)
33. Valentini, G., Ferrante, E., Dorigo, M.: The best-of-n problem in robot swarms: formalization, state of the art, and novel perspectives. Front. Robot. AI **4**(9) (2017). https://doi.org/10.3389/frobt.2017.00009
34. Valentini, G., Ferrante, E., Hamann, H., Dorigo, M.: Collective decision with 100 kilobots: speed versus accuracy in binary discrimination problems. Auton. Agent. Multi-Agent Syst. **30**(3), 553–580 (2016). https://doi.org/10.1007/s10458-015-9323-3
35. Valentini, G., Hamann, H., Dorigo, M.: Self-organized collective decision making: the weighted voter model. In: Proceedings of 13th International Conference on Autonomous Agents and Multiagent Systems (AAMAS 2014), pp. 45–52. International Foundation for Autonomous Agents and Multiagent Systems, Richland, SC (2014)
36. Wood, G.: Ethereum: a secure decentralised generalised transaction ledger- EIP-150 revision. Technical report, Ethereum Foundation (2017)
37. Yang, G.Z., et al.: The grand challenges of science robotics. Sci. Robot. **3**(14) (2018). https://doi.org/10.1126/scirobotics.aar7650
38. Zhao, H., et al.: A generic framework for Byzantine-tolerant consensus achievement in robot swarms. In: Proceedings of the 2023 IEEE/RSJ International Conference on Intelligent Robots and Systems (IROS 2023), pp. 8839–8846. IEEE (2023). https://doi.org/10.1109/IROS55552.2023.10341423
39. Zlot, R., Stentz, A., Dias, M.B., Thayer, S.: Multi-robot exploration controlled by a market economy. In: Proceedings of the 2002 IEEE International Conference on Robotics and Automation, pp. 3016–3023 (2002). https://doi.org/10.1109/ROBOT.2002.1013690

Heterogeneity Can Enhance the Adaptivity of Robot Swarms to Dynamic Environments

Raina Zakir[1](✉) [iD], Mohammad Salahshour[2] [iD], Marco Dorigo[1] [iD], and Andreagiovanni Reina[2,3] [iD]

[1] IRIDIA, Université Libre de Bruxelles, Brussels, Belgium
`raina.zakir@ulb.be, mdorigo@ulb.ac.be`
[2] Department of Collective Behaviour, Max Planck Institute of Animal Behavior, Konstanz, Germany
`{msalahshour,areina}@ab.mpg.de`
[3] CASCB, Universität Konstanz, Konstanz, Germany

Abstract. We study how robot swarms can collectively adapt to dynamic environments by changing what they collectively select as the best among a set of n possible options. While the robots rely on local communication with one another, follow simple rules, and make estimates of the option's qualities subject to measurement errors, the swarm as a whole can infer the change in the environment to make accurate collective decisions. Most studies focusing on dynamic environments have achieved adaptive behaviour by including random noise or threshold-based approaches to continuously explore alternatives and prevent opinion stagnation once consensus is achieved. In this study, we investigate whether or not swarms of robots with heterogeneous behaviours can be more adaptive than homogeneous swarms. We consider two behaviours from the literature which robots use to update their opinions: the majority rule, where robots gather information from all neighbours, and the voter rule, where robots use information from a single neighbour. In static environments, swarms of majority-rule robots, by using a larger amount of social information, typically make quicker decisions than swarms of voter-rule robots. However, our multiagent and robot simulations show that including voter-rule robots within a swarm of majority-rule robots can increase the group's responsiveness to environmental changes. This result shows the potential benefits of mixing simpler and relatively more advanced robots in the same swarm.

1 Introduction

Collective decision-making is an essential capability to enable autonomy in swarm robotics. Swarm robotics is inspired by the self-organising behaviours observed in biological systems, particularly evident in eusocial insects [13]. Examples of collective decision-making are honeybees' choice of a site for their nest among various alternatives [46], or ants' selection of the shortest path from their nest to a food source [17]. These examples found in nature and characterised by the absence of a global coordinator serve as the foundation for developing collective decision-making algorithms in swarm robotics [39], especially in swarms of minimalistic robots that execute simple behaviours to reach collective decisions in diverse scenarios, such as the selection of an aggregation site [8,48], the coordination of motion in a common direction [29], or the identification of the most abundant environmental feature [6,58]. The use of minimalistic robots, constrained by limited memory, computational power, and communication capabilities, is often mandated by the application scenario. For instance, size is crucial for nano-robots navigating blood vessels, whereas budget constraints may lead to the use of inexpensive, disposable robots in hazardous environments [19], where the risk of robot loss is high. As a result of their limitations, minimalist robots often present a high level of sensory noise, which leads to uncertainty in environmental and social estimations, and ultimately poses a significant challenge to the process of making collective decisions.

An important category of collective decision-making is the "best-of-n problem" where a group needs to choose the best option among n alternatives that can differ in quality [55]. Given the noisy assessment of the quality of the options in minimalistic systems, undertaking best-of-n decisions can be challenging for a robot swarm. This challenge is further amplified in dynamic environments, where conditions change over time [34].

Making collective decisions in the best-of-n problem while requiring minimalism relies on voting algorithms governed by simple rules, typically studied in opinion dynamics [5], where each opinion shared by the robots is treated as a vote. Among the computationally simplest algorithms is the voter model (or voter rule) [7,18], where robots consider only one vote (opinion) from a randomly chosen neighbour. This model has been expanded into the weighted voter model in [53], wherein robots express their votes for a duration proportional to the quality of the communicated option, often resulting in the selection of the best option due to modulation of positive feedback. Another well-known approach to collective decision-making in robot swarms is the use of the local majority rule [11,16,23,28], where each robot chooses the option with the most votes from its neighbours. Compared to the voter rule, the local majority rule has higher computational costs (as robots need to process and aggregate multiple votes instead of randomly selecting one), allowing the robots to pool more accurately neighbour's opinions and, in turn, enabling quicker collective decisions [40,54]. The concept of majority rule, where options favoured by over 50% of neighbouring agents are selected, can be extended by employing various sub- and super-majority quorums [26], such as the k-unanimity rule [44] or the q-voter rule [27]. Besides selecting social information using methods like voter or majority rules, a robot must integrate new social information with its personal opinions. This integration can be done using the direct-switch rule to overwrite the opinion with new social information [53] or by temporarily dropping any personal opinion before adopting a new one, as in the cross-inhibition rule [37].

Besides a few exceptions, e.g., [31,34,49], most research efforts for solving the best-f-n problem in robot swarms have primarily examined scenarios in static environments, where environmental conditions and option qualities remain constant over time. Even more rare are studies on collective decision-making in environments with $n > 2$, i.e., more than two options [3,10,14,25,51]. In real-life applications (e.g. drug-delivering nanobots in human bodies or detection or clearing of chemical spills), robot swarms may need to operate in dynamic environments with several alternatives that change quality over time. Hence, an important aspect to consider in designing robot swarms is their adaptivity, i.e., the swarm's capability to reconsider its opinion in response to environmental changes, such as when the quality of an option diminishes, or when a new, more desirable option appears [22]. Survival in an uncertain environment, even for biological systems, requires the ability to infer and respond to changing environmental conditions [9,12,43]. For adaptivity to occur in robot swarms, it is imperative that the swarm does not become 'locked-in' on an outdated opinion. One way to enable the system to keep reconsidering decisions is through the introduction of noise, which can be either random noise [51], or noise introduced by periodically resetting the robots' personal opinions, in this way allowing the integration of new environmental evidence [31,49]. In general, relying too much on social information can prevent the system from adapting to changes [34,35,42], as indicated by previous work that highlighted the benefits of lower robot connectivity to keep social exchanges low and enable group adaptivity [2,51]. Other work avoided opinion stagnation by including in the swarm a group of robots that never change opinion (called stubborn or zealot robots) [34,35]. These robots occasionally influence the other robots to reconsider opinions that are not shared by the majority; although this solution enables adaptivity, it requires the designer to know in advance all the available options so that stubborn robots can be allocated accordingly.

The majority of studies investigating collective decision-making have considered behavioural homogeneity, where all the robots in the swarm follow the same rules for selecting and updating opinions, i.e., every robot runs the same algorithm. While behavioural heterogeneity can enhance the functionality of both robotic and natural swarms [24,30,57], models of swarms comprising robots with different behaviours are more difficult to analyse as even minor behavioural differences can trigger significant, often unpredictable changes in the collective dynamics. For instance, even unintentional differences, such as differences in actuation errors, can yield qualitatively different collective responses [36]. Although the literature on this topic remains relatively sparse, prior research has highlighted the huge potential of heterogeneity, showing performances surpassing those of homogeneous swarms [1,20,21].

In this paper, we exploit behavioural heterogeneity to improve the adaptivity of the robot swarm in dynamic environments. We focus on a best-of-5 problem in a collective perception scenario used in several collective decision-making studies [3,14,47,56,58]. In this scenario, the environment floor is covered by coloured tiles (see Fig. 1), where each colour represents an environmental attribute. The environment is dynamic due to the periodic change of colour of the tiles. The goal of the robot swarm is to collectively reach a consensus on the predominant floor colour (i.e., the most frequent tile colour) and to adapt the consensus when the environment changes. Because the robots can only make local noisy estimates of the environmental state, they rely on information exchanges between nearby robots to make accurate decisions. The robot algorithms are based on an extension of the agent-based model proposed in [42]. In Sect. 2, we present our agent-based model and in Sect. 3 we describe how our model can be relevant for and implemented in a swarm robotics application. In Sect. 4, through abstract multiagent simulations and realistic physics-based robot simulations, we show that heterogeneous swarms that comprise robots that use the majority rule and robots that use the voter rule can improve the speed response to adapt collectively to environmental changes, surpassing, under some conditions, the performance of a homogeneous swarm. Section 5 concludes the paper by indicating how the results of this study could be extended to characterise when either behavioural homogeneity or behavioural heterogeneity enables a better collective adaptivity to environmental changes.

2 The Model

Our model extends the agent-based model proposed in [42]. In our model, a population of individuals live in an uncertain time-varying environment which can be in one of n possible states. Each individual can access the n representations, corresponding to the environmental states, and can communicate these using n different types of messages to form an opinion about the state of the environment. In a simulation run, the environmental state changes every t time steps, transitioning into one of the other $n-1$ states chosen uniformly at random. In between each environmental change, each agent undergoes an update τ times. Individuals can make noisy observations of the environmental state, which constitutes their *personal information*. We model the observation noise through a single parameter, the error probability $\eta \in [0, 1]$. When an individual makes an observation, with probability $1 - \eta$ the observation is correct corresponding to the true environmental state and with probability η, the observation is incorrect corresponding to one of the wrong states (chosen uniformly at random). In addition to personal information, individuals can also access their *social information* which corresponds to the opinions of their neighbours. Two individuals are neighbours when they are directly connected in the communication network. Individuals combine social and personal information by weighting their personal information by a factor ω and using a simple decision-making rule.

We consider two decision-making rules: the majority rule and the voter rule. With the majority rule, the individual counts how many neighbours have their opinion in favour of each of the n environmental states. The individual adds the value ω to the count for the option corresponding to its personal information. In other words, the count for option i is $M_i + \omega M \delta_i$, where M is the number of neighbours, M_i is the number of neighbours with opinion i, and δ_i is one when

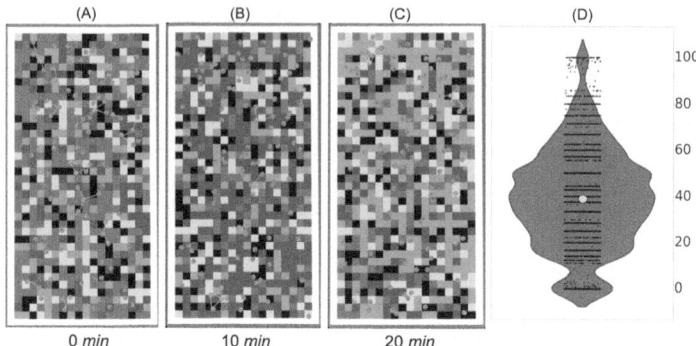

Fig. 1. (A–C) Snapshots of robotics simulations with 100 Kilobots in the tested collective perception scenario. The floor (i.e., the Kilogrid table) is composed of coloured tiles. The robots are tasked with selecting the predominant colour among five alternatives. (A) At the beginning of the experiments, the floor has 40% of red tiles and the remaining 60% are evenly split among blue, green, yellow, and black tiles. (B–C) The environment has changed to a majority for blue and green tiles, respectively. (D) A strip plot overlaid on a violin plot for 5,500 individual environmental observations when the true value is 40% (yellow dot). The simulated robots observe the true environmental state (i.e., the majority colour) on average 71% of the times, thus noise $\eta = 0.29$. (Colour figure online)

the individual's personal information is equal to i and is zero otherwise. Finally, the individual adopts the opinion with the highest count. If more than one state has the highest count, the individual adopts one of them chosen uniformly at random. With the voter rule, the individual chooses the state to adopt as its opinion with a probability proportional to the counts. This rule is equivalent to adopting the personal information with probability $\hat{\omega} = \omega/(1+\omega)$, and adopting the opinion of one randomly-chosen neighbour with probability $1 - \hat{\omega}$.

We form heterogeneous swarms comprising individuals employing one of the two decision-making rules, either voter or majority, where a fraction $k \in [0, 1]$ of the agents use the voter rule and the rest (i.e., fraction $1 - k$) use the majority rule. Thus, for values of $k = 0$ and $k = 1$, the swarm is homogeneous.

We consider asynchronous opinion updates, i.e., at each time step only one individual makes an environmental observation and updates its opinion following either of the two rules. Once it updates its opinion, the individual communicates it to all its neighbours. We consider two types of communication networks: structured and random. In a structured communication network, individuals live on a first-nearest-neighbour square lattice with periodic boundaries and von Neumann connectivity. In a random communication network, each individual is connected to four randomly chosen individuals to whom it transmits its signals. The random network is also dynamic as the neighbours of each individual are drawn randomly each t time steps.

3 Swarm Robotics Simulations

While abstract multiagent simulations offer quick computations of the system dynamics, they may not fully encompass the complexities and real-world constraints of robotic systems. Therefore, to provide a more comprehensive verification of transferability, we test the collective behaviour through physics-based simulations of a swarm of $N = 100$ autonomous robots.

Collective Perception Scenario. We test the collective decision-making algorithms in a collective perception scenario where the robots' objective is to collectively infer the environmental state, which comprises randomly distributed floor tiles of five different colours: red, blue, green, yellow, and black, see Fig. 1. This corresponds to an instance of the best-of-5 problem, where the correct environmental state corresponds to the colour that occurs most frequently in the floor tiles. In line with the decision-making literature [15,38,45,50], when scaling the problem to decisions between more than two options, there is one best option (i.e., the predominant colour representing the correct environmental state) and $n-1$, in our case four, remaining options with the same lower quality (i.e., four colours appear in the floor tiles in equal minority proportion).

The environment changes every t time steps, meaning that the tiles' colours change and, after every change, one of the colours that was in minority becomes the predominant one. In the robot simulations, t corresponds to 10 min. We run the experiments that last $3t = 30$ min, therefore we can test the ability of the robots to adapt to three environmental changes. At the start of each simulation, all robots are initialised at random locations and committed to the blue option while the environment is initialised in a red state (i.e., the red tiles are the most numerous). In this way, we simulate that at the beginning of the simulation, an environmental change has happened. Every 10 min, the environment changes, at minute 10, the blue colour becomes predominant and at minute 20, the green colour becomes predominant, as shown in Fig. 1.

Simulation Setup. In this analysis, we use Kilobots [41]—small-sized, minimalistic, and cost-effective robots that can broadcast infrared (IR) messages with 9-byte payload in a range of 10 cm, move at 1 cm/s, and have a control loop of approximately 33 ms. Because the Kilobots have limited sensing capabilities, we run our experiments putting the robots in a virtual environment, the Kilogrid [52], that allows them to make environmental readings otherwise impossible. The Kilogrid is an electronic table measuring $1 \times 2 \, \text{m}^2$, consisting of 800 cells that can display any RGB colour and interact with the Kilobots via IR messages. All cells, except white border cells (Fig. 1), continuously transmit IR messages with their ID and colour. We simulate the robot swarm using ARGoS [33], a state-of-the-art swarm robotics simulator, which has dedicated plugins to simulate accurately both the Kilobots [32] and the Kilogrid [2].

Robot Behaviour. Kilobots perform a random walk (alternating 10 s straight motion with 5 s rotation) to explore the environment and interact with other robots. Since Kilobots lack proximity sensors, Kilogrid cells transmit a binary 'wall flag' (high or low) to indicate proximity to a wall. Border cells and their adjacent non-white cells send a high wall flag, while other internal cells send a low wall flag. Upon receiving a high wall flag, a robot executes a simple obstacle avoidance routine to avoid collisions with the wall.

The robots repeatedly observe the environment, with each observation lasting about 800 robot control loops (rcl). Thus, a robot makes roughly $\tau = 22$ observations between environmental changes (every 10 min). During an observation, the robot counts the tiles (Kilogrid cells) of each colour it encounters. The robot can only read the colour of the Kilogrid cell it is on. Through a random walk, during each observation, the robot visits on average 8 tiles. After 800 rcl, the robot sets as its environmental observation the most observed colour (the tile counter is reset every observation cycle of 800 rcl). Because of the limited number of cells visited by the robot, the environmental observation is subject to noise η, i.e., the proportions of tile counts are often different than their true proportions. The value of η varies depending on the true proportions. We test two different proportions (η values) in the robot experiments. When the predominant colour appears in 40% of the tiles and each of the other four colours in 15% of them (as shown in Fig. 1A–C), the average observation error is $\eta = 0.29$ (Fig. 1D). When the predominant colour appears in 30% of the tiles and each of the other four colours in 17.5% of them, the average observation error is $\eta = 0.47$.

Throughout the experiment, the robots broadcast a message expressing their opinion (i.e., the environmental state they believe to be true) every 2 s on average. They also display their opinion to a human observer by lighting their LEDs in the same colour. The majority-rule robots process all received messages, grouping them by colour in a set r, whereas the voter-rule robots only store the colour indicated in the last message they receive, overwriting the content at each new message. After each environmental observation (every 800 rcl), the robot uses either the majority or voter rule for decision-making and updates its opinion. In the majority rule process, the robot counts messages received within r as m, calculates $\omega \times m$, and adds it to the count of the most observed tile colour in r. It then applies the majority rule on r to update its opinion. In the voter rule, the robot uses personal observation with probability $\hat{\omega}$ to switch to the most observed tile colour, or with probability $1 - \hat{\omega}$ uses social information to update its opinion. This process repeats throughout the experiment.

4 Results

We measure the system performance Λ as the proportion of robots that infer the correct environmental state on average throughout the experiment. To do so, we compute this proportion for each time step and divide it by the experiment length. We report results for different levels of personal observation noise η and different personal information weights ω (which is used in the decision rule).

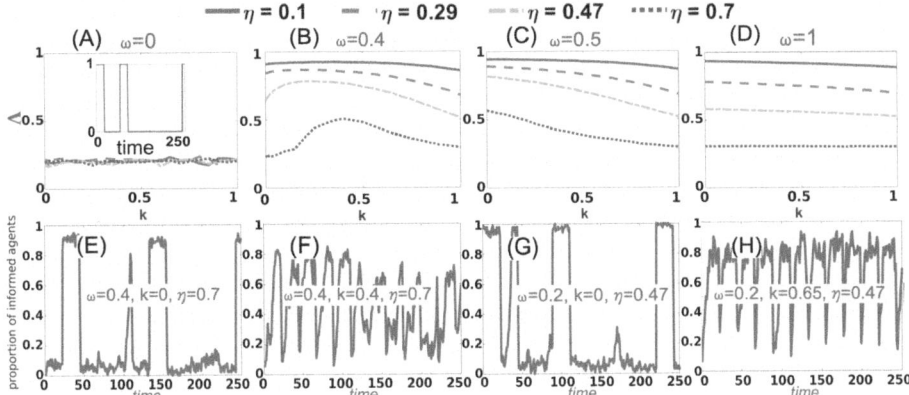

Fig. 2. Multiagent simulation results for swarms of 100 agents interacting on random time-varying communication network. (A–D) Time-averaged fraction of agents inferring the correct environmental state (Λ on the y-axis) for different swarm compositions (k on the x-axis), for various observation noise values $\eta \in \{0.1, 0.29, 0.47, 0.7\}$ and personal information weights $\omega \in \{0, 0.4, 0.5, 1\}$ (average of 11,000 time steps, with environmental changes every $\tau = 22$ time steps). The inset in panel A shows a snippet of the simulation with time on the x-axis and the proportion of correct agents on the y-axis. (E–H) The temporal evolution of some selected cases for a short segment of 250 time steps for $\omega = 0.4$ and $\eta = 0.7$ (in E, F), and $\omega = 0.2$ and $\eta = 0.47$ (in G, H).

4.1 Multiagent Simulation Results

The results of stochastic multiagent simulations with 100 agents are presented in Figs. 2 and 3. We ran one long simulation with 500 environmental switches (thus a total of $500t = 11,000$ time steps) per configuration. Since the simulation process is ergodic, the results are akin to running multiple simulations.

Figures 2A–D shows results for four values of $\omega \in \{0, 0.4, 0.5, 1\}$ and four personal observation noise $\eta \in \{0.1, 0.29, 0.47, 0.7\}$. Relying prevalently on social or personal information ($\omega = 0$ and $\omega = 1$, respectively) leads in most cases to poor results. For instance, when $\omega = 0$ in Fig. 2A, the agents only use social information, making the population blind to environmental changes and stuck in an immutable consensus for one option (also represented by the inset). Instead, for intermediate values of ω, the system is able to combine social and personal information achieving higher performance Λ. On the horizontal axis of Fig. 2A–D, we vary the swarm composition k indicating the proportion of voter-rule agents (where the other agents, proportion $1 - k$, use the majority rule). While Fig. 2C shows that a homogeneous swarm of majority-rule agents ($k = 0$) has the best performance Λ, Fig. 2B also shows that there are conditions when heterogeneous swarms ($0 < k < 1$) are superior to homogeneous ones. In particular, when observation noise is high and the personal information is weighted less than the social information ($\omega < 0.5$), combining the two types of agents can be beneficial.

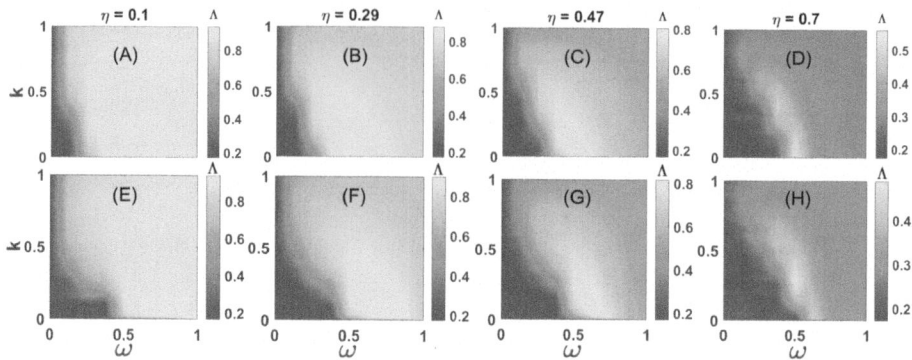

Fig. 3. Results from multiagent simulations with 100 agents on two network types: top row (A–D) shows random time-varying communication, bottom row (E–H) shows structured communication (lattice with von Neuman neighbourhood). The colour map indicates performance Λ, the average fraction of agents inferring the correct environmental state over time. The colour bar scale in each panel is different to better visualise the region of maximal performance in each condition. (Colour figure online)

The temporal evolution of the fraction of informed individuals shown in Figs. 2E–H helps to understand the dynamics of the system in different configurations. With majority-rule agents only (homogeneous swarm, $k = 0$), the swarm is rarely able to adapt to changes (see Figs. 2E and 2G), behaving in a way comparable to an observation-blind swarm (shown in the inset of Fig. 2A). In these figures, there are few and high peaks. This means that with $k = 0$ and low $\omega < 0.5$, the agents prevalently rely on social information, obtaining in this way very high levels of group agreement (high peaks), however at the expense of poor adaptability to changes (few peaks). These few peaks mostly happen when, by chance, the environment changes to the state that corresponds to the opinion in which the swarm is locked in. This situation of opinion stagnation can be improved by including in the swarm a proportion of voter-rule agents, as shown in Figs. 2F and 2H for $k = 0.4$ and $k = 0.65$. Here, the system reliably and rapidly adapts to change (periodic peaks at every environmental change), however at the cost of a smaller agreement (lower peaks).

The colour maps of Fig. 3 show a more complete exploration of the parameter space. The two rows show results for the two considered types of networks. The top row shows results for agents interacting on a random time-varying communication network, instead the bottom row shows results for swarms of agents interacting on a static structured communication network. For both types of networks, the trend is similar, however when agents communicate on a structured network, having a heterogeneous swarm can perform better than a homogeneous swarm in a larger range of parameters.

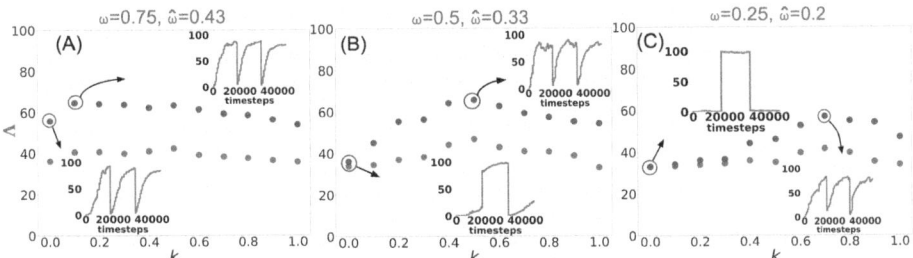

Fig. 4. Swarm robotics simulation results for swarms of 100 Kilobots. Time-averaged fraction of robots inferring the correct environmental state (Λ on the y-axis) for different swarm compositions (k on the x-axis), for various observation noise values (red points for $\eta = 0.475$ and blue points for $\eta = 0.29$) and personal information weights ω. The average is computed on 30 simulation runs per configuration of 30-min experiments with environmental changes every 10 min. Each panel's insets show the temporal evolution of the homogeneous swarm (with majority-rule robots only) and the heterogeneous swarm (with the best performance Λ) for $\eta = 0.29$.

4.2 Swarm Robotics Simulation Results

The swarm robotics results highlight the benefits of having a heterogeneous swarm as combining the two types of robots (majority-rule and voter-rule robots) leads to the best performance in a larger range of parameters. Figure 4 shows the results for two observation noise levels ($\eta = 0.475$ as red points and $\eta = 0.29$ as blue points) and three ω values (different panels). In line with the multiagent predictions of Sect. 4.1, the robot simulation results show that the benefits of having a heterogeneous swarm are larger when ω is low and the estimation noise is high. The insets in Fig. 4 show the temporal dynamics of some representative cases, confirming that swarms only composed of majority-rule robots ($k = 0$) are unable to rapidly adapt to changes (for high ω, adaptation is slow and for low ω, there is no sign of adaptation). As discussed in the previous section, for $k = 0$ and low ω, the occasional peaks in the temporal evolution are due to the environment transitioning to the state that matches the opinion in which the robots have been initialised. More precisely, this match happens after the first environment change after 10 min of simulation, when the predominant colour in the Kilogrid is blue (Fig. 1B). Instead, heterogeneous swarms seem to cope with environmental changes better, to adapt quicker, and have a higher performance Λ than homogeneous swarms in any of the tested conditions. We also conducted a few preliminary real-robot experiments on a swarm of 40 robots, for $\omega = 0.2$ and $\eta = 0.2$, showing promising results with environmental inference improving in heterogeneous swarms (videos in supplementary material [59]).

Surprisingly, the swarm robotics results more closely align with those of a multiagent system on a fixed structured network, where the benefits of heterogeneity are clearer, rather than on a random time-varying network, where heterogeneous systems are optimal in fewer conditions. In fact, the interactions among robots should be better described by a network that changes over time through random encounters. While we do not have a definitive explanation yet, we believe that one of the causes could be the neighbourhood correlation (i.e., there are cliques where several neighbours of my neighbours are also my neighbours), but further investigation is needed for confirmation.

5 Discussion and Conclusion

In this paper, we have studied swarm robotics algorithms for the best-of-n problem in dynamic environments. The robots continuously attempt to infer the correct environmental state which changes over time. To alleviate the effect of potentially large errors in making personal observations of the environment, robots exchange information on what they believe to be the true environmental state: thanks to these social exchanges, they make collective decisions that are more accurate than what they could do if operating alone. Both our and prior analyses suggest that heavy reliance on social information can lead to group opinion stagnation, hindering the swarm from acquiring new environmental information and locking the collective into former beliefs [31,34,42,49,51]. We differentiate from previous work that investigated adaptability to environmental changes in collective best-of-n decision-making of robot swarms by considering a behaviourally heterogeneous swarm. In our study, some of the robots use the majority rule while others use the voter rule (these are two simple voting rules frequently studied in the opinion dynamics and swarm robotics literature [4,11,16,18,56]).

Even though the majority rule is more sophisticated and allows the robots to quantify better the option of the rest of the group (using larger neighbourhood sampling), the introduction of a proportion of robots using the simpler voter rule, which is based on one random social sample, can lead to collective benefits. Our analysis shows that there are indeed conditions where having a heterogeneous swarm composed of two groups of robots, one using the majority rule and the other using the voter rule, can lead to a quicker response to environmental changes. Relevant conditions where we find swarm heterogeneity useful are when there are high levels of errors in personal environmental observations (high noise parameter η) and when the robots have low confidence in their observations and give more importance to social information rather than personal information (low weight parameter ω). In addition, our analysis shows that heterogeneous groups can also be quicker to respond to an environmental change (e.g., see insets of Fig. 4). Future research may investigate whether the idea of having a mix of simpler and more sophisticated robots could also be beneficial in other cases, such as when robots are heterogeneous due to manufacturing differences (e.g., robots with different sensor noise levels) [36].

Our analysis based on multiagent and swarm robotics simulations shows the potential benefits of heterogeneity, expanding the theoretical analysis conducted in [42]. Further analysis should better characterise in which conditions homogeneity or heterogeneity are the best strategy. Our intuition is that information noise and social network correlations may be critical factors. We believe a fruitful research direction is considering heterogeneity for cost-efficiency, as shown in our previous work [1], where combining simple and complex robot behaviours can reduce the average cost of running the swarm algorithm.

Acknowledgements. R.Z and M.D. acknowledge support from the Belgian F.R.S.-FNRS. A.R. acknowledges support from DFG under Germany's Excellence Strategy - EXC 2117 - 422037984.

References

1. Antonic, N., Zakir, R., Dorigo, M., Reina, A.: Collective robustness of heterogeneous decision-makers against stubborn individuals. In: Proceedings of the 23rd International Conference on Autonomous Agents and Multiagent Systems (AAMAS 2024), pp. 68—77. IFAAMAS, Richland, SC (2024)
2. Aust, T., Talamali, M., Dorigo, M., Hamann, H., Reina, A.: The hidden benefits of limited communication and slow sensing in collective monitoring of dynamic environments. In: Dorigo, M., et al. (eds.) ANTS 2022. LNCS, vol. 13491, pp. 234–247. Springer, Cham (2022). https://doi.org/10.1007/978-3-031-20176-9_19
3. Bartashevich, P., Mostaghim, S.: Multi-featured collective perception with evidence theory: tackling spatial correlations. Swarm Intell. **15**(1–2), 83–110 (2021)
4. Canciani, F., Talamali, M.S., Marshall, J.A.R., Bose, T., Reina, A.: Keep calm and vote on: Swarm resiliency in collective decision making. In: Proceedings of Workshop Resilient Robot Teams of the 2019 IEEE International Conference on Robotics and Automation (ICRA 2019), p. 4 (2019)
5. Castellano, C., Fortunato, S., Loreto, V.: Statistical physics of social dynamics. Rev. Mod. Phys. **81**(2), 591–646 (2009)
6. Chin, K.Y., Khaluf, Y., Pinciroli, C.: Minimalistic collective perception with imperfect sensors. In: 2023 IEEE/RSJ International Conference on Intelligent Robots and Systems (IROS), Piscataway, NJ, pp. 8862–8868 (2023)
7. Clifford, P., Sudberry, A.: A model for spatial conflict. Biometrika **60**(3), 581–588 (1973)
8. Correll, N., Martinoli, A.: Modeling and designing self-organized aggregation in a swarm of miniature robots. Int. J. Robot. Res. **30**(5), 615–626 (2011)
9. Couzin, I., Krause, J., Franks, N., Levin, S.: Effective leadership and decision making in animal groups on the move. Nature **433**, 513–516 (2005)
10. Crosscombe, M., Lawry, J.: Collective preference learning in the best-of-n problem. Swarm Intell. **15**, 1–26 (2021)
11. Cruciani, E., Mimun, H.A., Quattropani, M., Rizzo, S.: Phase transition of the k-majority dynamics in biased communication models. Distrib. Comput. **36**(2), 107–135 (2023)
12. Dall, S.R., Giraldeau, L.A., Olsson, O., McNamara, J.M., Stephens, D.W.: Information and its use by animals in evolutionary ecology. Trends Ecol. Evol. **20**(4), 187–193 (2005)
13. Dorigo, M., Theraulaz, G., Trianni, V.: Swarm robotics: past, present, and future [point of view]. Proc. IEEE **109**(7), 1152–1165 (2021)
14. Ebert, J.T., Gauci, M., Mallmann-Trenn, F., Nagpal, R.: Bayes bots: collective Bayesian decision-making in decentralized robot swarms. In: 2020 IEEE International Conference on Robotics and Automation (ICRA), pp. 7186–7192. IEEE, Piscataway, NJ (2020)

15. Franks, N.R., Dornhaus, A., Best, C.S., Jones, E.L.: Decision making by small and large house-hunting ant colonies: one size fits all. Anim. Behav. **72**(3), 611–616 (2006)
16. Galam, S.: Majority rule, hierarchical structures, and democratic totalitarianism: a statistical approach. J. Math. Psychol. **30**(4), 426–434 (1986)
17. Goss, S., Aron, S., Deneubourg, J.L., Pasteels, J.: Self-organized shortcuts in the argentine ant. Naturwissenschaften **76**, 579–581 (1989)
18. Holley, R.A., Liggett, T.M.: Ergodic theorems for weakly interacting infinite systems and the voter model. Ann. Probab. **3**(4), 643–663 (1975)
19. Hunt, E.R., Cullen, C.B., Hauert, S.: Value at risk strategies for robot swarms in hazardous environments. In: Unmanned Systems Technology XXIII, p. 117580M. SPIE (2021)
20. Karamched, B., Stickler, M., Ott, W., Lindner, B., Kilpatrick, Z.P., Josić, K.: Heterogeneity improves speed and accuracy in social networks. Phys. Rev. Lett. **125**(21), 218302 (2020)
21. Kengyel, D., Hamann, H., Zahadat, P., Radspieler, G., Wotawa, F., Schmickl, T.: Potential of heterogeneity in collective behaviors: a case study on heterogeneous swarms. In: Chen, Q., Torroni, P., Villata, S., Hsu, J., Omicini, A. (eds.) PRIMA 2015. LNCS, vol. 9387, pp. 201–217. Springer, Cham (2015). https://doi.org/10.1007/978-3-319-25524-8_13
22. Khaluf, Y., Simoens, P., Hamann, H.: The neglected pieces of designing collective decision-making processes. Front. Robot. AI **6**, 16 (2019)
23. Krapivsky, P.L., Redner, S.: Dynamics of majority rule in two-state interacting spin systems. Phys. Rev. Lett. **90**, 238701 (2003)
24. Krause, J., Ruxton, G., Krause, S.: Swarm intelligence in animals and humans. Trends Ecol. Evol. **25**, 28–34 (2009)
25. Lee, C., Lawry, J., Winfield, A.F.T.: Negative updating applied to the best-of-n problem with noisy qualities. Swarm Intell. **15**(1–2), 111–143 (2021)
26. Marshall, J.A.R., Kurvers, R.H.J.M., Krause, J., Wolf, M.: Quorums enable optimal pooling of independent judgements in biological systems. eLife **8**, e40368 (2019)
27. Mobilia, M.: Nonlinear q-voter model with inflexible zealots. Phys. Rev. E **92**(1) (2015)
28. Montes de Oca, M.A., Ferrante, E., Scheidler, A., Pinciroli, C., Birattari, M., Dorigo, M.: Majority-rule opinion dynamics with differential latency: a mechanism for self-organized collective decision-making. Swarm Intell. **5**, 305–327 (2011)
29. Önür, G., Turgut, A.E., Şahin, E.: Predictive search model of flocking for quadcopter swarm in the presence of static and dynamic obstacles. Swarm Intell. 1935–3820 (2024)
30. O'Shea-Wheller, T.A., Hunt, E.R., Sasaki, T.: Functional heterogeneity in superorganisms: emerging trends and concepts. Ann. Entomol. Soc. Am. **114**(5), 562–574 (2020)
31. Pfister, K., Hamann, H.: Collective decision-making and change detection with Bayesian robots in dynamic environments. In: 2023 IEEE/RSJ International Conference on Intelligent Robots and Systems (IROS), pp. 8814–8819. IEEE, Piscataway, NJ (2023)
32. Pinciroli, C., Talamali, M.S., Reina, A., Marshall, J.A.R., Trianni, V.: Simulating Kilobots within ARGoS: models and experimental validation. In: Dorigo, M., Birattari, M., Blum, C., Christensen, A.L., Reina, A., Trianni, V. (eds.) ANTS 2018. LNCS, vol. 11172, pp. 176–187. Springer, Cham (2018). https://doi.org/10.1007/978-3-030-00533-7_14

33. Pinciroli, C., et al.: ARGoS: a modular, parallel, multi-engine simulator for multi-robot systems. Swarm Intell. **6**(4), 271–295 (2012)
34. Prasetyo, J., De Masi, G., Ferrante, E.: Collective decision making in dynamic environments. Swarm Intell. **13**(3), 217–243 (2019)
35. Prasetyo, J., De Masi, G., Ranjan, P., Ferrante, E.: The best-of-n problem with dynamic site qualities: achieving adaptability with stubborn individuals. In: Dorigo, M., Birattari, M., Blum, C., Christensen, A.L., Reina, A., Trianni, V. (eds.) ANTS 2018. LNCS, vol. 11172, pp. 239–251. Springer, Cham (2018). https://doi.org/10.1007/978-3-030-00533-7_19
36. Raoufi, M., Romanczuk, P., Hamann, H.: Individuality in swarm robots with the case study of Kilobots: noise, bug, or feature? In: ALIFE 2023: Proceedings of the 2023 Artificial Life Conference, pp. 35–44. MIT Press, Cambridge (2023)
37. Reina, A., Bose, T., Trianni, V., Marshall, J.A.R.: Effects of spatiality on value-sensitive decisions made by robot swarms. In: Groß, R., et al. (eds.) Distributed Autonomous Robotic Systems. SPAR, vol. 6, pp. 461–473. Springer, Cham (2018). https://doi.org/10.1007/978-3-319-73008-0_32
38. Reina, A., Bose, T., Trianni, V., Marshall, J.A.R.: Psychophysical laws and the superorganism. Sci. Rep. **8**(4387) (2018)
39. Reina, A., Ferrante, E., Valentini, G.: Collective decision-making in living and artificial systems: editorial. Swarm Intell. **15**(1–2), 1–6 (2021)
40. Reina, A., Njougouo, T., Tuci, E., Carletti, T.: Speed-accuracy trade-offs in best-of-n collective decision making through heterogeneous mean-field modeling. Phys. Rev. E **109**, 054307 (2024)
41. Rubenstein, M., Ahler, C., Nagpal, R.: Kilobot: a low cost scalable robot system for collective behaviors. In: 2012 IEEE International Conference on Robotics and Automation (ICRA). IEEE, Piscataway, NJ (2012)
42. Salahshour, M.: Phase diagram and optimal information use in a collective sensing system. Phys. Rev. Lett. **123**(6), 068101 (2019)
43. Salahshour, M., Rouhani, S., Roudi, Y.: Phase transitions and asymmetry between signal comprehension and production in biological communication. Sci. Rep. **9**(1), 3428 (2019)
44. Scheidler, A., Brutschy, A., Ferrante, E., Dorigo, M.: The k-unanimity rule for self-organized decision making in swarms of robots. IEEE Trans. Cybern. **46**, 1175 (2016)
45. Seeley, T.D., Buhrman, S.C.: Nest-site selection in honey bees: how well do swarms implement the "best-of-n" decision rule? Behav. Ecol. Sociobiol. **49**(5), 416–427 (2001)
46. Seeley, T.D., Visscher, P.K., Schlegel, T., Hogan, P.M., Franks, N.R., Marshall, J.A.R.: Stop signals provide cross inhibition in collective decision-making by honeybee swarms. Science **335**(6064), 108–111 (2012)
47. Shan, Q., Mostaghim, S.: Discrete collective estimation in swarm robotics with distributed Bayesian belief sharing. Swarm Intell. **15**(4), 377–402 (2021)
48. Sion, A., Reina, A., Birattari, M., Tuci, E.: Controlling robot swarm aggregation through a minority of informed robots. In: Dorigo, M., et al. (eds.) ANTS 2022. LNCS, vol. 13491, pp. 91–103. Springer, Cham (2022). https://doi.org/10.1007/978-3-031-20176-9_8
49. Soorati, M.D., Krome, M., Mora-Mendoza, M., Ghofrani, J., Hamann, H.: Plasticity in collective decision-making for robots: creating global reference frames, detecting dynamic environments, and preventing lock-ins. In: 2019 IEEE/RSJ International Conference on Intelligent Robots and Systems (IROS), pp. 4100–4105. IEEE, Piscataway, NJ (2019)

50. Talamali, M.S., Marshall, J.A., Bose, T., Reina, A.: Improving collective decision accuracy via time-varying cross-inhibition. In: 2019 International conference on robotics and automation (ICRA), pp. 9652–9659. IEEE, Piscataway, NJ (2019)
51. Talamali, M.S., Saha, A., Marshall, J.A.R., Reina, A.: When less is more: robot swarms adapt better to changes with constrained communication. Sci. Robot. **6**(56), eabf1416 (2021)
52. Valentini, G., et al.: Kilogrid: a novel experimental environment for the Kilobot robot. Swarm Intell. **12**(3), 245–266 (2018)
53. Valentini, G.: Self-organized collective decision-making in swarms of autonomous robots. In: Proceedings of the 13th International Conference on Autonomous Agents and Multiagent Systems (AAMAS 2014), pp. 1703–1704. IFAAMAS, Richland, SC (2014)
54. Valentini, G., Brambilla, D., Hamann, H., Dorigo, M.: Collective perception of environmental features in a robot swarm. In: Dorigo, M., et al. (eds.) ANTS 2016. LNCS, vol. 9882, pp. 65–76. Springer, Cham (2016). https://doi.org/10.1007/978-3-319-44427-7_6
55. Valentini, G., Ferrante, E., Dorigo, M.: The best-of-n problem in robot swarms: formalization, state of the art, and novel perspectives. Front. Robot. AI **4**, 9 (2017)
56. Valentini, G., Ferrante, E., Hamann, H., Dorigo, M.: Collective decision with 100 kilobots: speed versus accuracy in binary discrimination problems. Auton. Agent. Multi-Agent Syst. **30**(3), 553–580 (2016)
57. York, C., Madin, Z.R., O'Dowd, P., Hunt, E.R.: Shaping multi-robot patrol performance with heterogeneity in individual learning behavior. arXiv preprint arXiv:2403.01181 (2024)
58. Zakir, R., Dorigo, M., Reina, A.: Robot swarms break decision deadlocks in collective perception through cross-inhibition. In: Dorigo, M., et al. (eds.) ANTS 2022. LNCS, vol. 13491, pp. 209–221. Springer, Cham (2022). https://doi.org/10.1007/978-3-031-20176-9_17
59. Zakir, R., Salahshour, M., Dorigo, M., Reina, A.: Supplementary material for "Heterogeneity can Enhance the Adaptivity of Robot Swarms to Dynamic Environments (2024). https://iridia.ulb.ac.be/supp/IridiaSupp2024-003/

Impact of Individual Defection on Collective Motion

Swadhin Agrawal[1], Jitesh Jhawar[2], Andreagiovanni Reina[3,4,5], Sujit P. Baliyarasimhuni[1], Heiko Hamann[3,4], and Liang Li[4,5,6(✉)]

[1] Department of Electrical Engineering and Computer Science, Indian Institute of Science Education and Research Bhopal, Bhopal, India
{swadhin20,sujit}@iiserb.ac.in

[2] School of Arts and Sciences, Ahmedabad University, Ahmedabad, Gujarat, India
jitesh.jhawar@ahduni.edu.in

[3] Department of Computer and Information Science, University of Konstanz, Konstanz, Germany

[4] Centre for the Advanced Study of Collective Behaviour, University of Konstanz, Konstanz, Germany

[5] Department of Collective Behaviour, Max Planck Institute of Animal Behavior, Konstanz, Germany
lli@ab.mpg.de

[6] Department of Biology, University of Konstanz, Konstanz, Germany

Abstract. Collective motion modeling has attracted significant attention for gaining insights into the mechanisms of collective behavior and its potential to inspire control strategies for swarm robotics. Most of the existing models assume that individuals within a group strictly adhere to the interaction rules. However, individuals in artificial and natural collectives could occasionally fail to follow the interaction rules, which is distinct from noisy actions. In this study, we analyze how the presence of individuals, who occasionally defect, affects the ordered phase of the group during collective motion. Using Monte Carlo simulations, we study two collective motion models, a non-spatial (pairwise interaction) and a spatial (Couzin) model. In the non-spatial model, when individuals defect with higher probability, both the time required by the agents to reach directional consensus (polarized group motion) as well as the average energy cost of the group to maintain such directional consensus (average rotational energy consumption per individual in highly polarized groups) increases. In the spatial model, there are conditions where the presence of defecting agents can simultaneously reduce the time required by the collective to get highly polarized and the average energy cost in the polarized state. These findings not only enhance our understanding of probabilistic defective behavior in biological systems but can also inspire innovative, efficient, and controllable approaches in swarm robotics.

1 Introduction

Collective motion [46] is a universal phenomenon observed in nature, ranging from macro-molecules [39], to the simplest multicellular organisms [14], insects [7,37], fish [43], birds [4], herds [19], and even humans [34]. Over the past decades, mathematical models [13,24,38,45] have provided insights into how individuals, by interacting with their local neighbors, can collectively move in a coordinated manner. Most of these models assume that every individual adheres to the rules defined in these models with a certain degree of variability which is normally included as random noise. While individuals in natural systems, such as fish shoals or bird flocks, usually adhere to the interaction rules [4,25], they may also occasionally deviate from that behavior in ways that are different from noisy actions. For instance, individuals may choose not to interact and instead take selfish actions to exploit collaboration efforts by their peers without paying relative social coordination costs [11,35]. Another reason why certain individuals may occasionally not follow the rules of collective motion is the fact that individuals have complex behaviors where collective motion is only one of its components. It is reasonable to assume that animals may concurrently forage, look for mates, or avoid dangers and predators; therefore, other individual goals than motion coordination could prevail in determining animals' actions [10,12,22,27,47]. Competition can also play a role in causing a temporary interruption in following collective motion rules [30,33]. We refer to the behavior where individuals deviate from the collective motion rules as *defection*.

There are previous works that modeled intermittent, or occasional, defective behavior in a variety of ways. Such studies considered variable frequencies of interaction, *i.e.*, agents always follow the cooperative rules, however, each of them has a personal frequency of interaction [6,16]. In such an asynchronous interaction system, certain agents update their behaviors (or states) more frequently than others. Slower agents (with lower interaction frequency) may appear as defectors to quicker agents that make (unilateral) updates of their behavior without a direct response. Other studies considered individuals that selectively interact only with a subset of neighbors [23]. This research has provided interesting insights into the evolution of cooperation in collective systems [28,29]. Investigating the impact of occasional defection on collective motion remains an open question [2,48]. In this paper, we study whether having individuals with a given probability of defection (in our case, defection means ignoring the collective motion rule) can be beneficial for the group, and, if so, what level of defection maximizes the collective benefits. Different from previous studies that used frequency-based or selective-interaction approaches to model defection at approximately regular time intervals [24,41], in our models, by introducing a probability for every agent to defect, we break the periodic patterns of defection events across the number of time steps.

Considering the impact of defecting agents in decentralized groups can also be relevant for the design of robotic systems, such as robot swarms [21]. While robots would be programmed to not defect and always follow the collaborative rules, their defective behavior could be the consequence of malfunctioning or

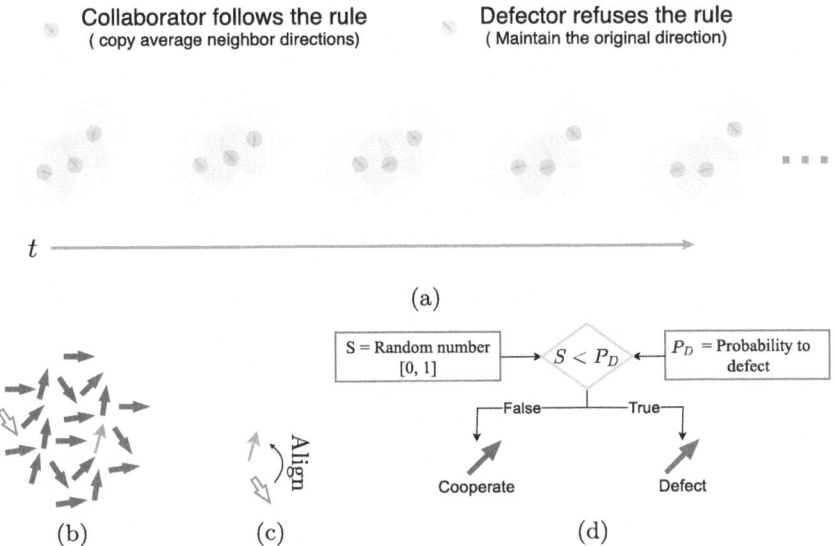

Fig. 1. (a) Illustration of the collective behavior of individuals that probabilistically cooperate by following the collective motion rules or defect by maintaining their original directions. The pairwise model: (b) A pair of interacting agents in a group of fully connected agents, (c) Focal agents' (hollow pink) interaction rule, (d) Mechanism to update the agent's strategy (cooperate/defect) via the defection probability P_D.

noise. For instance, due to communication failures [17,31], noisy sensory inputs, and partial observability of the environment, robots may not always follow the prescribed algorithm [36]. Adversarial tampering is another example where the robot temporarily under the control of a hacker may behave erratically [3,26]. It is thus important to design algorithms that are robust to occasional defection by part of the robots, or that would even exploit defective behavior to improve certain aspects of the group dynamics. This paper makes an initial step towards this research direction by analyzing in what conditions probabilistic defection can improve or harm the group dynamics.

To do so, we analyze two prominent collective motion models, the pairwise model [24] (a non-spatial model) and the Couzin model [13] (a spatially-explicit model). We extend these two models to include the possibility that agents occasionally defect to investigate how this impacts the group dynamics. At each time step, an agent defects with probability P_D and otherwise cooperates by following the interaction rules. Figure 1(a) presents a schematic illustration of individuals defecting in the alignment interaction rule with probability $P_D = 1/3$. That is, an agent cooperates by aligning its orientation with its neighbors in two-thirds of the time steps and, instead, in the remaining one-third of time steps, it defects and maintains its original orientation, ignoring its neighbors. In this study, the agent's decision to cooperate or defect is independent of its neighbors' strategy and solely based on P_D, as illustrated in Fig. 1(d).

We measure the impact of defecting agents in terms of two metrics, polarization time, T, and average energy cost, E, (defined in Sect. 2). Polarization time is the time the group needs to agree on the direction of motion, i.e., to have most of the agents aligned in the same direction. The average energy cost is the amount of energy the agents use to maintain a high level of polarization (i.e., how much rotational energy each agent, on average, consumes to remain aligned in the same direction). For high levels of defection, the group may be unable to achieve a sufficient level of motion alignment, or the group may become polarized only for a short time before returning to a disordered state (unpolarized group).

In Sect. 3, we analyze the non-spatial pairwise model and observe that increasing defection probability P_D causes an increase in both polarization time and average energy cost. In Sect. 4, we analyze the spatially-explicit Couzin model [13]. While the pairwise model only includes an alignment rule, the Couzin model is more complex as it comprises three rules. We find that when individuals defect in certain collective motion rules, as the defection probability increases (up to a certain point), both the polarization time and average energy cost decrease. Note that, in this study, we do not compare the dynamics of the pairwise with those of the Couzin model since they are too different for a fair comparison, which is out of the scope of this study.

2 Methods

In the simulation, the agents move at a constant speed v. The kinematic motion of the agents is implemented using the Dubin's car model [15] described by

$$\dot{x} = v\ \cos(\psi),$$
$$\dot{y} = v\ \sin(\psi),$$
$$\dot{\psi} = \omega. \qquad (1)$$

Here, $\dot{x}$ and $\dot{y}$ are the velocities of the agent along each coordinate axis in the 2D plane, and ψ is the yaw (heading) angle of the agent. The turning rate ω, is constrained by a maximum of $\omega_{\max}$, i.e., $|\omega| < \omega_{\max}$. The desired turning rate for an individual is obtained from the interaction model (Sects. 3 and 4).

Given N agents in a group, we define the group polarization $p(t)$ and the group energy cost $e(t)$ at a given time t. Group polarization $p(t)$ is the average of the unit velocity vectors $\hat{\boldsymbol{v}}_i$ of individual agents i at time t, given by

$$p(t) = \frac{|\sum_{i=1}^{N} \hat{\boldsymbol{v}}_i(t)|}{N}, \qquad (2)$$

and group energy cost $e(t)$ at time t is measured as the sum of the rotational energy of each individual, given by

$$e(t) = -\sum_{i=1}^{N}(\omega_i(t))^2. \qquad (3)$$

We perform Monte Carlo simulations [1,18], 100 independent runs for each condition (set of parameters). Each run lasts $t_{\max} = 200$ s, with a time step $dt = 0.1$ sec. We measure the polarization time T required to achieve a group polarization $p(T)$ beyond a threshold value $p^* = 0.8$, i.e. $p(T) > p^*$, and the average energy cost E per individual after the group gets polarized:

$$E = \frac{1}{N(t_{\max} - T)} \sum_{t=T}^{t_{\max}} e(t). \quad (4)$$

The agents move in an unbounded region, and at the start of each run, they are distributed uniformly at random inside a circular region of radius R_h with random heading angles. The initial states differed across the 100 different runs and across each model. In the case of the Couzin model, the same initial states were used across each type of defection.

3 Defection in the Pairwise Interaction Model

In the pairwise interaction model, the interaction network of the group is fully connected; hence, every agent can interact with any other agent. We use the same simulation algorithm of Jhawar et al. [24]. At each time step t, we chose at random $N/2$ agents as the focal agents that we pair with the rest of the group (non-focal agents). Each focal agent i (pink and hollow arrow in Fig. 1(b)) pairs with another unpaired agent j (thin and green quiver in Fig. 1(b)) chosen at random, irrespective of their Euclidean distance in space to copy j's yaw angle, as shown in Fig. 1(c). Hence, only half of the population, the focal agent group, changes its motion direction (i.e., the rate of change of direction is 0.5, as in [24]). At every simulation step ($dt = 0.1$ s), the focal agents are re-sampled and paired with other random non-focal agents. Mathematically, the desired yaw for a focal agent is described by:

$$\psi_i(t+1) = \begin{cases} \psi_j(t), & \text{if cooperating} \\ \psi_i(t), & \text{if defecting} \end{cases}. \quad (5)$$

Hence, the turning rate of a focal agent is described by

$$\omega_i = \begin{cases} \min\left(\frac{\psi_i(t+1) - \psi_i(t)}{dt}, \omega_{\max}\right), & \text{if cooperating} \\ 0, & \text{if defecting} \end{cases}. \quad (6)$$

Under the above setup, we perform Monte Carlo simulations with $N = 50$ agents. We find that regardless of the defection probability value P_D, the group ultimately gets polarized (i.e., $p(T) > p^*$), except when individuals always defect (hence, for $P_D = 1$ we do not compute polarization time nor energy cost). Both polarization time (Fig. 2(a)) and average energy cost per individual (Fig. 2(b)) increase with increasing P_D. Such an increase is much more pronounced for high defection probabilities, i.e., $P_D > 0.7$.

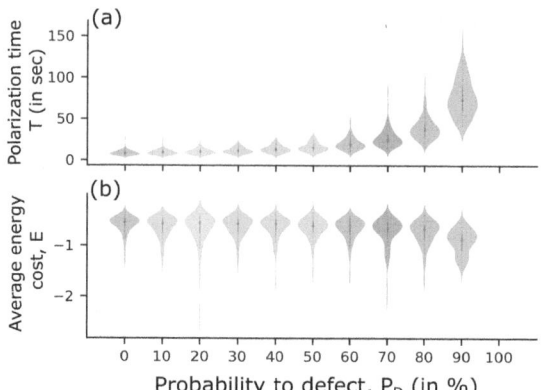

Fig. 2. (a) The polarization time T and (b) the average energy cost E as a function of the probability to defect (P_D) in the pairwise model for $N = 50$ agents. The simulations were repeated 100 times per condition. We only report the data when at least 80% of the 100 simulation runs reached a polarization greater than $p^* = 0.8$. Hence, data for $P_D = 1$ are omitted as the population does not reach polarization.

4 Defection in the Couzin Model

In the Couzin model [13], an agent interacts with its neighbors within interaction range, therefore the interaction network results from Euclidean distances between agents. The interaction range is categorized into three zones: the zone of repulsion (Z_r), the zone of orientation (Z_o), and the zone of attraction (Z_a). Different physical distances and orientations from a focal individual define each of its zones, as shown in Fig. 3(a). When neighbors are present in the interaction zone, the focal agent interacts by getting attracted, repelled, or orienting towards its neighbors as long as they are not in the blind zone (α). In this model, at each simulation step ($dt = 0.1$ s), every agent acts as a focal agent and updates its yaw angle as a combination of the interactions in the three zones as:

$$\psi_i(t+1) = \begin{cases} \tan^{-1}\left(\frac{1}{N_r}\mathbf{d}_r^i(t)\right) & \text{if } Z_r^i \neq \phi \\ \tan^{-1}\left(\frac{1}{N_o}\mathbf{d}_o^i(t)\right) & \text{if only } Z_o^i \neq \phi \\ \tan^{-1}\left(\frac{1}{N_a}\mathbf{d}_a^i(t)\right) & \text{if only } Z_a^i \neq \phi \\ \tan^{-1}\left(\frac{1}{2}(\frac{1}{N_o}\mathbf{d}_o^i(t) + \frac{1}{N_a}\mathbf{d}_a^i(t))\right) & \text{if both } (Z_o^i, Z_a^i) \neq \phi \\ \psi_i(t) & \text{otherwise} \end{cases} \quad (7)$$

Here, $\mathbf{d}_r^i = -\sum_{j \in Z_r} \hat{\beta}_{ij}$, $\mathbf{d}_o^i = \sum_{j \in Z_o} \hat{v}_j$, and $\mathbf{d}_a^i = \sum_{j \in Z_a} \hat{\beta}_{ij}$, where the symbol $\hat{\beta}_{ij}$ represents a unit vector pointing from the position of focal agent i towards the position of the neighbor j (in the coordinate system of agent i). The symbols N_r, N_o, and N_a represent the numbers of neighbors present in each zone (Z_r,

(a) Couzins model   (b) Cooperator   (c) $D - Z_r$ (d) $D - Z_o$ (e) $D - Z_a$ (f) $D - Z_{\text{all}}$

Fig. 3. (a) Focal agents' interaction zones in the Couzin model depicting the zone of orientation (Z_o), zone of attraction (Z_a), zone of repulsion (Z_r), and the blind region (α). Types of defection in the Couzin model: (b) standard Couzin model with repulsion, orientation, and attraction zones, (c–f) types of defection, white regions indicate no interaction (i.e., defective interaction) for each interaction zone.

Z_o, and Z_a, respectively). And $\hat{v}$ is a unit velocity vector. Hence, the desired ω_i of a focal agent at a given simulation step is obtained by:

$$\omega_i = \begin{cases} \min\left(\frac{\psi_i(t+1) - \psi_i(t)}{dt}, \omega_{\max}\right), & \text{if } i \text{ is cooperating} \\ 0, & \text{if } i \text{ is defecting} \end{cases} \quad (8)$$

Given the modified Couzin model defined in Eqs. (7) and (8), we perform Monte Carlo simulations with $N = 20$ and $N = 50$ agents using similar parameter values to the ones used in the seminal work introducing the Couzin model [13], as indicated in Table 1 in detail. With more agents ($N \gg 50$) and the considered set of parameters (Table 1), the group takes longer than our simulation time of 200 seconds [9] to transit to the ordered phase, see supplementary [1] for the case of $N = 80$. In contrast to the pairwise model, we conduct Monte Carlo simulations by defining defection in four different ways based on the zones of behavioral interaction: (1) defection only in the zone of repulsion (D-Z_r, Fig. 3(c)), (2) defection only in the zone of orientation (D-Z_o, Fig. 3(d)), (3) defection only in the zone of attraction (D-Z_a, Fig. 3(e)), and (4) defection in all the three zones at the same time (D-Z_{all}, Fig. 3(f)).

Table 1. Parameters used in the Couzin model for highly parallel formation

Parameter name	Symbol	Value	Unit
Radius of repulsion	R_r	1	units
Radius of orientation	R_o	18	units
Radius of attraction	R_a	20	units
Linear speed	v	5	units/s
Maximum turning rate	$\omega_{\max}$	$\pi/3$	rad/s
Field of invisibility	α	0	rad
Radius of home	R_h	$\sqrt{N}$	units

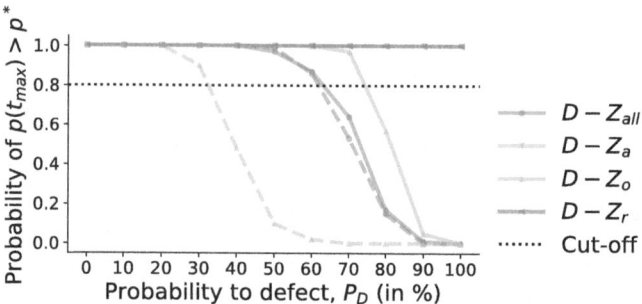

Fig. 4. Probability of having a polarized group (i.e., $p(t_{max}) > p^*$ with $p^* = 0.8$) as a function of the agent's probability P_D to defect. This probability is computed as the proportion of 100 simulation runs where, at time t_{max}, the group had polarization $p(t_{max})$ above $p^* = 0.8$. The solid lines are for $N = 20$, and the dashed lines are for $N = 50$. In the results of Figs. 5 and 6, we do not include data for conditions where the probability of $p(t_{max}) > p^*$ is below our cut-off line of 0.8 (dotted horizontal black line).

First, for each case of N ($N = 20$ and $N = 50$), we examine and report the probability of the group becoming highly polarized by the end of the experiment, i.e., $p(t_{max}) > p^*$. This probability is computed by computing the proportion of the 100 simulation runs (per condition) that are in a polarized state at time $t_{max} = 200$ s. The results in Fig. 4 show that in the case of defection of type $D-Z_r$ and $D-Z_a$, both groups of $N = 20$ and $N = 50$ agents consistently reach high polarization levels irrespectively of the defection probability P_D, including the case of $P_D = 1$. However, in the case of $D-Z_{all}$, the probability of $p(t_{max}) > p^*$ stays below our cut-off line set at 0.8 (horizontal dotted grey line) when defection probabilities are higher than $P_D > 0.6$ for both group sizes. In the case of $D-Z_o$, the system does not reliably reach a polarized state ($p(t_{max}) < p^*$) in 200 s for $P_D > 0.7$ when $N = 20$ and $P_D > 0.3$ when $N = 50$. It means the collective ability to reach a polarized state with defection in zone Z_o decreases with an increasing number of agents.

Figures 5 and 6 show the results of the polarization time T and the average energy cost E for simulations of groups of $N = 20$ and $N = 50$ agents, respectively. The analysis of these two metrics shows that there are large differences depending on the zone where agents probabilistically defect. For both group sizes, defection in the orientation zone Z_o leads to the negative effects of an increase in the polarization time T and higher average energy costs E in maintaining a polarized state (see panels (e–f) of Figs. 5 and 6). We observe the opposite trend for defection in the repulsion zone Z_r (see panels (g–h) of Figs. 5 and 6, and Table 2) where, for both group sizes, the polarization time T and average energy costs E decrease as the defection probability P_D increases. This trend is monotonic in all cases except for E with $N = 50$, where the minimum energy cost is for $P_D = 0.6$, see Fig. 6(h) and Table 2. For defection in the attraction zone Z_a, we find different results for the two considered swarm

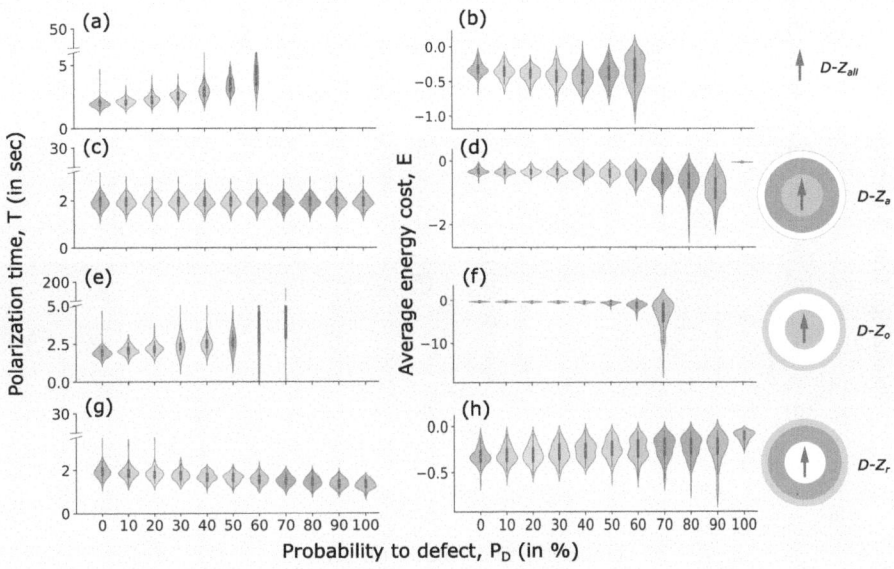

Fig. 5. The polarization time T and the average energy cost E across four types of defection for a group of $N = 20$ agents. The violin plot shows the distribution of values from 100 simulation runs.

Table 2. Mean μ_e and standard deviation σ_e of the average energy cost E for $D - Z_r$, same data as in Fig. 6(h).

	0	10	20	30	40	50	60	70	80	90	100
μ_e	−3.109	−2.648	−2.234	−2.116	−1.877	−1.776	−1.697	−1.775	−2.140	−2.482	−0.595
σ_e	1.944	1.522	0.82	0.875	0.580	0.536	0.562	0.624	1.028	0.939	0.167

sizes. In small swarms, $N = 20$ in Figs. 5(c–d), the inclusion of defection probability does not impact T but worsens average energy cost E. Instead, in larger swarms, $N = 50$ in Figs. 6(c–d), polarization time T decreases for moderate values of defection probability $P_D \leq 0.3$ after plateauing at a constant T. Average energy cost E improves for $P_D < 0.3$, is minimum at $P_D = 0.3$, and increases for higher P_D (see Table 3). Finally, when agents defect in all three zones Z_{all}, the dynamics are quite complex as they are a combination of the trends we observed in the three areas. Polarization time increases for $N = 20$ and has a minimum at $P_D = 0.3$ for $N = 50$ (see Fig. 6(a) and Table 4). Average energy cost E has opposite trends for $N = 20$ and $N = 50$. In the former case, Fig. 5(b) shows a u-shape trend, with a maximum energy cost for $P_D = 0.3$, and in the latter case, Fig. 6(b) shows a bell-shape trend with minimum energy cost for $P_D = 0.3$.

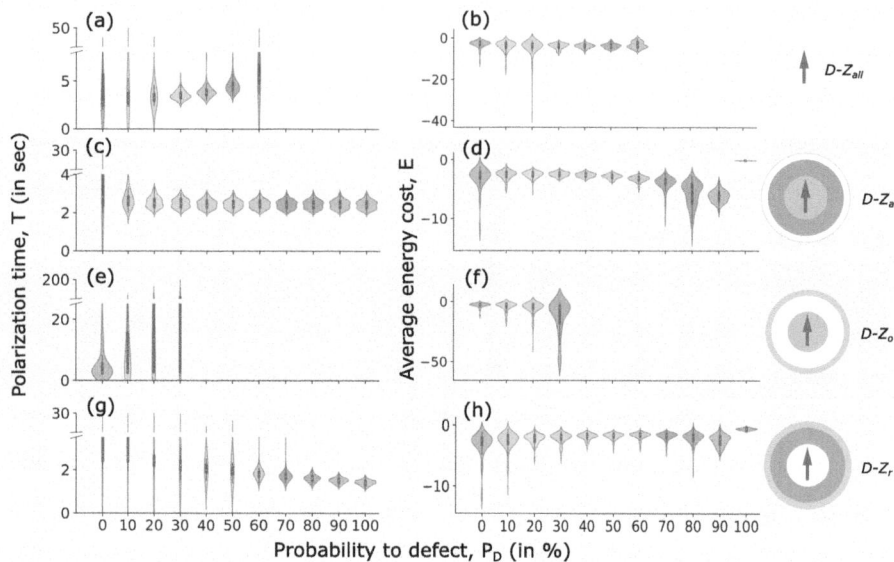

Fig. 6. The polarization time T and the average energy cost E across four types of defection for a group of $N = 50$ agents. The violin plot shows the distribution of values from 100 simulation runs.

Table 3. Mean μ_e and standard deviation σ_e of the average energy cost E for $D - Z_a$, same data as in Fig. 6(d).

	0	10	20	30	40	50	60	70	80	90
μ_e	−3.109	−2.586	−2.538	−2.459	−2.595	−2.841	−3.235	−4.021	−5.891	−6.199
σ_e	1.944	0.769	0.59	0.391	0.371	0.404	0.51	1.294	2.129	0.993

Table 4. Mean μ_t and standard deviation σ_t of the polarization time T for $D - Z_{all}$, same data as in Fig. 6(a).

	0	10	20	30	40	50	60
μ_t	5.501	4.773	3.665	3.522	3.877	4.564	7.408
σ_t	5.713	6.671	3.044	0.482	0.592	0.851	7.167

5 Conclusion

While most of the existing studies on collective motion models assume that individuals always obey the interaction rules without fail, in real-world natural and robotic systems agents may defect and not follow the interaction rules for various reasons, such as selfish behavior, intra-group competitions, and adversarial attacks. In this paper, we introduce probabilistically defecting agents in two—non-spatial and spatial—models of collective motion: the pairwise model [24] and the Couzin model [13]. We analyze the impact of probabilistically defecting agents on the collective motion dynamics by measuring the polarization time T (time needed to reach polarization $p(t)$ above $p^* = 0.8$, $p(T) > p^*$) and the average energy cost E per individual (energy that each agent spends on average to make rotational movements once the group has reached a highly polarized state).

Our analysis of the pairwise interaction model shows that both the polarization time and the average group energy cost increase with increasing defection probability. The Couzin model is a spatial model thus we analyzed four different types of defection depending on the region of space where agents occasionally defect. If agents defect in either the zone of attraction or repulsion, the polarization time reduces with increasing defection probability P_D. With defection in these two zones, the average group energy cost E also has a minimum for defection probability $P_D > 0$. Differently, for defection in the zone of orientation, any value of $P_D > 0$ reduces the performance in terms of both polarization time and average energy cost. In summary, in the considered spatial model (the Couzin model), there are conditions (*e.g.*, defection in the repulsion zone only) that can provide group-level benefits and conditions (*e.g.*, defection in the orientation zone only) that are harmful to the group. The system dynamics and their relationship with the investigated control parameters (defection probability, swarm size, defection zones) are relatively complex. Additional complexity is introduced in the spatial model by the possibility of agents splitting into subgroups moving in different directions with local polarization in each subgroup.

Future research is needed to characterize better the causal relationship between defecting in certain zones and the resulting group dynamics, as well as how this potentially changes with the group size. For example, we hypothesize that measuring and controlling the density of defecting agents inside each interaction zone could give more insights into the observed dynamics or help explain the size-dependent dynamics for defectors in the zone of orientation (Fig. 4). Future studies could also exploit the idea of using zone-selective probabilistic defection to improve the efficiency (polarisation speed and energy cost) of collective motion algorithms for robot swarms, or make the swarm's motion more robust against adversarial attacks [32,44]. Our long-term plans are to use the insights from this study to combine collective motion models with evolutionary games to understand the conditions where certain behavioral traits are likely to manifest, both at the group and individual levels (*e.g.*, splitting and merging of collectives [8], leader-follower [5,20,42] behavior of individuals, and emergent leadership in groups [40]).

Acknowledgements. This work has been financially supported by the Indian Institute of Science Education and Research, Bhopal, India (S.A.,S.P.B.), the Max-Planck Society (L.L.), DFG under Germany's Excellence Strategy - EXC 2117 - 422037984 (S.A.,A.R.,H.H.,L.L.), the Sino-German Centre in Beijing for generous funding of the Sino-German mobility grant M-0541 (L.L.), and Messmer Foundation Research Award (L.L.).

References

1. Agrawal, S., Jhawar, J., Reina, A., Baliyarasimhuni, S.P., Hamann, H., Li, L.: Supplementary materials of the article "impact of individual defection on collective motion". https://github.com/swadhinagrawal/defectorsInCollectiveMotion.git
2. Antonioni, A., Cardillo, A.: Coevolution of synchronization and cooperation in costly networked interactions. Phys. Rev. Lett. **118**(23), 238301 (2017). https://doi.org/10.1103/physrevlett.118.238301
3. Aswale, A., López, A., Ammartayakun, A., Pinciroli, C.: Hacking the colony: on the disruptive effect of misleading pheromone and how to defend against it. In: Proceedings of the 21st International Conference on Autonomous Agents and Multiagent Systems, AAMAS 2022, pp. 27–34. International Foundation for Autonomous Agents and Multiagent Systems, Richland, SC (2022). https://dl.acm.org/doi/abs/10.5555/3535850.3535855
4. Ballerini, M., et al.: Interaction ruling animal collective behavior depends on topological rather than metric distance: evidence from a field study. Proc. Nat. Acad. Sci. **105**(4), 1232–1237 (2008). https://doi.org/10.1073/pnas.0711437105
5. Bernardi, S., Eftimie, R., Painter, K.J.: Leadership through influence: what mechanisms allow leaders to steer a swarm? Bull. Math. Biol. **83**(6), 69 (2021). https://doi.org/10.1007/s11538-021-00901-8
6. Bode, N.W.F., Faria, J.J., Franks, D.W., Krause, J., Wood, A.J.: How perceived threat increases synchronization in collectively moving animal groups. Proc. R. Soc. B: Biol. Sci. **277**(1697), 3065–3070 (2010). https://doi.org/10.1098/rspb.2010.0855
7. Buhl, J., et al.: From disorder to order in marching locusts. Science **312**(5778), 1402–1406 (2006). https://doi.org/10.1126/science.1125142
8. Cardona, G.A., Leahy, K., Vasile, C.I.: Temporal logic swarm control with splitting and merging. In: 2023 IEEE International Conference on Robotics and Automation (ICRA), Piscataway, NJ, pp. 12423–12429 (2023). https://doi.org/10.1109/ICRA48891.2023.10160335
9. Chazelle, B.: The convergence of bird flocking. J. Assoc. Comput. Mach. **61**(4) (2014). https://doi.org/10.1145/2629613
10. Collignon, B., Séguret, A., Chemtob, Y., Cazenille, L., Halloy, J.: Collective departures and leadership in zebrafish. PLoS ONE **14**(5), e0216798 (2019). https://doi.org/10.1371/journal.pone.0216798
11. Couzin, I.D.: Collective animal migration. Curr. Biol. **28**(17), R976–R980 (2018). https://doi.org/10.1016/j.cub.2018.04.044
12. Couzin, I.D., Krause, J., Franks, N.R., Levin, S.A.: Effective leadership and decision-making in animal groups on the move. Nature **433**(7025), 513–516 (2005). https://doi.org/10.1038/nature03236
13. Couzin, I.D., Krause, J., James, R., Ruxton, G.D., Franks, N.R.: Collective memory and spatial sorting in animal groups. J. Theor. Biol. (2002). https://doi.org/10.1006/jtbi.2002.3065
14. Davidescu, M.R., Romanczuk, P., Gregor, T., Couzin, I.D.: Growth produces coordination trade-offs in trichoplax adhaerens, an animal lacking a central nervous system. Proc. Nat. Acad. Sci. **120**(11), e2206163120 (2023). https://doi.org/10.1073/pnas.2206163120
15. Dubins, L.E.: On curves of minimal length with a constraint on average curvature, and with prescribed initial and terminal positions and tangents. Am. J. Math. **79**(3), 497–516 (1957). http://www.jstor.org/stable/2372560

16. Fatès, N.: A tutorial on elementary cellular automata with fully asynchronous updating. Nat. Comput. **19**(1), 179–197 (2020). https://doi.org/10.1007/s11047-020-09782-7
17. Gielis, J., Shankar, A., Prorok, A.: A critical review of communications in multi-robot systems. Curr. Robot. Rep. **3**(4), 213–225 (2022). https://doi.org/10.1007/s43154-022-00090-9
18. Gillespie, D.T.: A general method for numerically simulating the stochastic time evolution of coupled chemical reactions. J. Comput. Phys. **22**(4), 403–434 (1976). https://doi.org/10.1016/0021-9991(76)90041-3
19. Gómez-Nava, L., Bon, R., Peruani, F.: Intermittent collective motion in sheep results from alternating the role of leader and follower. Nat. Phys. **18**(12), 1494–1501 (2022). https://doi.org/10.1038/s41567-022-01769-8
20. Goodrich, M.A., Pendleton, B., Baliyarasimhuni, S.P., Pinto, J.: Toward human interaction with bio-inspired robot teams. In: 2011 IEEE International Conference on Systems, Man, and Cybernetics, pp. 2859–2864. IEEE (2011). https://doi.org/10.1109/ICSMC.2011.6084115
21. Hamann, H.: Swarm Robotics: A Formal Approach. Springer, Cham (2018). https://doi.org/10.1007/978-3-319-74528-2
22. Hensor, E., Godin, J.G., Hoare, D., Krause, J.: Effects of nutritional state on the shoaling tendency of banded killifish, fundulus diaphanus, in the field. Anim. Behav. **65**(4), 663–669 (2003). https://doi.org/10.1006/anbe.2003.2075
23. Jadhav, V., Guttal, V., Masila, D.R.: Randomness in the choice of neighbours promotes cohesion in mobile animal groups. R. Soc. Open Sci. **9**(3), 220124 (2022). https://doi.org/10.1098/rsos.220124
24. Jhawar, J., et al.: Noise-induced schooling of fish. Nat. Phys. **16**(4), 488–493 (2020). https://doi.org/10.1038/s41567-020-0787-y
25. Katz, Y., Tunstrøm, K., Ioannou, C.C., Huepe, C., Couzin, I.D.: Inferring the structure and dynamics of interactions in schooling fish. Proc. Nat. Acad. Sci. **108**(46), 18720–18725 (2011). https://doi.org/10.1073/pnas.1107583108
26. Kumar, Y., Paranjape, A.A., Ghosh, S., Baliyarasimhuni, S.P.: Adversarial fragmentation of robotic teams operating under Reynolds' rules with bounded communication radius. In: 2023 62nd IEEE Conference on Decision and Control (CDC), pp. 2809–2814. IEEE (2023). https://doi.org/10.1109/CDC49753.2023.10383845
27. Leigh, E.G.: How does selection reconcile individual advantage with the good of the group? Proc. Nat. Acad. Sci. **74**(10), 4542–4546 (1977). https://doi.org/10.1073/pnas.74.10.4542
28. Li, A., et al.: Evolution of cooperation on temporal networks. Nat. Commun. **11**(1), 2259 (2020). https://doi.org/10.1038/s41467-020-16088-w
29. Li, L., Chen, C., Li, A.: Autonomy promotes the evolution of cooperation in prisoner's dilemma. Phys. Rev. E **102** (2020). https://doi.org/10.1103/physreve.102.042402
30. MacGregor, H.E.A., Ioannou, C.C.: Collective motion diminishes, but variation between groups emerges, through time in fish shoals. R. Soc. Open Sci. **8**(10), 210655 (2021). https://doi.org/10.1098/rsos.210655
31. Mayya, S., Egerstedt, M.: Safe open-loop strategies for handling intermittent communications in multi-robot systems. In: 2017 IEEE International Conference on Robotics and Automation (ICRA), Piscataway, NJ, pp. 5818–5823 (2017). https://doi.org/10.1109/ICRA.2017.7989683
32. Mendívez Vásquez, B.L., Barca, J.C.: Adversarial scenarios for herding UAVs and counter-swarm techniques. Robotica **41**(5), 1436–1451 (2023). https://doi.org/10.1017/S0263574722001801

33. Metcalfe, N.B., Thomson, B.C.: Fish recognize and prefer to shoal with poor competitors. Proc. R. Soc. Lond. Ser. B: Biol. Sci. **259**(1355), 207–210 (1995). https://doi.org/10.1098/rspb.1995.0030
34. Miller, N., Garnier, S., Hartnett, A.T., Couzin, I.D.: Both information and social cohesion determine collective decisions in animal groups. Proc. Nat. Acad. Sci. **110**(13), 5263–5268 (2013). https://doi.org/10.1073/pnas.1217513110
35. Moreira, J.A., Pacheco, J.M., Santos, F.C.: Evolution of collective action in adaptive social structures. Sci. Rep. **3**, 1521 (2013). https://doi.org/10.1038/srep01521
36. van Otterlo, M.: The logic of adaptive behavior: knowledge representation and algorithms for adaptive sequential decision making under uncertainty in first-order and relational domains. Frontiers in Artificial Intelligence and Applications. IOS Press, Amsterdam, The Netherlands (2009)
37. Ramdya, P., et al.: Mechanosensory interactions drive collective behaviour in drosophila. Nature **519**(7542), 233–236 (2015). https://doi.org/10.1038/nature14024
38. Reynolds, C.W.: Flocks, herds and schools: a distributed behavioral model. Comput. Graph. **21**(4), 25–34 (1987). https://doi.org/10.1145/37402.37406
39. Schaller, V., Weber, C., Semmrich, C., Frey, E., Bausch, A.R.: Polar patterns of driven filaments. Nature **467**(7311), 73–77 (2010). https://doi.org/10.1038/nature09312
40. Strandburg-Peshkin, A., Papageorgiou, D., Crofoot, M.C., Farine, D.R.: Inferring influence and leadership in moving animal groups. Philos. Trans. R. Soc. B: Biol. Sci. **373**(1746), 20170006 (2018). https://doi.org/10.1098/rstb.2017.0006
41. Strömbom, D., Hassan, T., Hunter Greis, W., Antia, A.: Asynchrony induces polarization in attraction-based models of collective motion. R. Soc. Open Sci. **6**(4), 190381 (2019). https://doi.org/10.1098/rsos.190381
42. Tiwari, R., Jain, P., Butail, S., Baliyarasimhuni, S.P., Goodrich, M.A.: Effect of leader placement on robotic swarm control. In: Proceedings of the 16th Conference on Autonomous Agents and MultiAgent Systems, AAMAS 2017, pp. 1387–1394. International Foundation for Autonomous Agents and Multiagent Systems, Richland, SC (2017). https://doi.org/10.5555/3091125.3091316
43. Tunstrøm, K., Katz, Y., Ioannou, C.C., Huepe, C., Lutz, M.J., Couzin, I.D.: Collective states, multistability and transitional behavior in schooling fish. PLoS Comput. Biol. **9**(2), 1–11 (2013). https://doi.org/10.1371/journal.pcbi.1002915
44. Van Calck, L., Pacheco, A., Strobel, V., Dorigo, M., Reina, A.: A blockchain-based information market to incentivise cooperation in swarms of self-interested robots. Sci. Rep. **13**, 20417 (2023). https://doi.org/10.1038/s41598-023-46238-1
45. Vicsek, T., Czirók, A., Ben-Jacob, E., Cohen, I., Shochet, O.: Novel type of phase transition in a system of self-driven particles. Phys. Rev. Lett. **75**(6), 1226–1229 (1995). https://doi.org/10.1103/physrevlett.75.1226
46. Vicsek, T., Zafeiris, A.: Collective motion. Phys. Rep. **517**(3), 71–140 (2012). https://doi.org/10.1016/j.physrep.2012.03.004
47. Yang, W.C., Schmickl, T.: Collective motion as an ultimate effect in crowded selfish herds. Sci. Rep. **9**, 6618 (2019). https://doi.org/10.1038/s41598-019-43179-6
48. You, F., Yang, H.X., Li, Y., Du, W., Wang, G.: A modified Vicsek model based on the evolutionary game. Appl. Math. Comput. **438**, 127565 (2023). https://doi.org/10.1016/j.amc.2022.127565

Minimalist Protocols for Quorum Sensing in Robot Swarms

Fabio Oddi[1,2], Andreagiovanni Reina[3,4], and Vito Trianni[2]

[1] DIAG, Sapienza University of Rome, Rome, Italy
oddi@diag.uniroma1.it
[2] ISTC, National Research Council, Rome, Italy
vito.trianni@istc.cnr.it
[3] CASCB, Universität Konstanz, Konstanz, Germany
andreagiovanni.reina@uni-konstanz.de
[4] Department of Collective Behaviour, Max Planck Institute of Animal Behavior, Konstanz, Germany

Abstract. Quorum sensing is a key mechanism enabling coordinated behaviour in populations of autonomous agents, and is extensively studied in biological systems spanning from bacteria populations to social insects. In swarm robotics too, quorum sensing aroused much interest, but remained mostly constrained to the implementation of collective decisions. Here, we propose protocols to estimate the quorum level that are suitable for resource-constrained robots, and evaluate the precision and speed of the quorum assessment across a large spectrum of swarm state conditions. Through systematic experimentation, we evaluate the proposed protocols for different swarm densities and working area sizes. Our findings shed light on the trade-off between computational requirements and expected performance, aiding in the selection of appropriate quorum sensing protocols for future swarm robotics research.

1 Introduction

Quorum sensing (QS) is a widespread phenomenon in both biological and artificial systems, playing a pivotal role in enabling coordinated system behaviour [2, 3]. It can be defined as a "consistent population-dependent modulation of discrete modes of behaviour" [21], meaning that a population of agents can coordinately switch to a different behaviour when some population-level feature (e.g., density) surpasses a given threshold (i.e., a quorum). QS is largely studied in biology and especially in bacteria [17,36], in which a density-dependent concentration threshold of signalling molecules can trigger a behaviour change (e.g., bioluminescence in *Vibrio fischeri* [13]). In social insects, QS has been studied mainly as a decentralised mechanism necessary to implement a collective decision (e.g., migrating to a new nest site [7,24,31]). Indeed, collective decision making requires two processes: first, building consensus, so that a sufficient number of individuals in the population selects the same alternative (possibly among many); second, decision implementation, which requires that individuals change their behaviour according to the selected alternative. In nest site selection, for instance, first a new nest location must be identified, and then the colony can relocate to the new nest. To minimise colony splitting, a robust QS mechanism must be in place, enabling to implement the decision only when sufficient support is available. This also leads to a speed-accuracy trade-off, because a fast-and-frugal estimation of the quorum can lead to frequent errors in the decision implementation [10].

In swarm robotics too, QS has been conceived as a mechanism for the implementation of collective decisions [6,8,16,20,21]. Indeed, to orchestrate swarm-level activities across multiple functional tasks, the robot swarm must be capable of collective decision making and task switching, whereby in face of multiple alternatives the swarm selects the most profitable or urgent task and coordinately executes it. Minimalist approaches for best-of-N decision problems have been proposed in the past [35]. Recently, it has been shown that a hierarchy of collective decisions can simplify best-of-N problems, improving both speed and accuracy [18]. Such approaches can be generalised to coordinated task switching, whereby the swarm is required to coordinately move from one task to another maintaining coherence and prioritising the most relevant task or suitably distributing among many [1,15]. In both task switching and decision sequences, QS strategies are fundamental to establish when the swarm is ready to move on. This must be optimised for speed to avoid unnecessary delays and energy consumption. Additionally, recovery mechanisms need to be developed for robots that wrongly recognise the quorum or that engage in a different task, by means of systematic coherence check within the swarm.

In this paper, we move beyond previous research on QS for robot swarms [6,20,21] by studying the system dynamics in isolation from the collective decision-making process and by comparing three alternative protocols in terms of their computational and memory requirements. We make a theoretical analysis through an urn model that informs the swarm robotics simulations. We test the ability and speed of the robot swarm to detect a quorum for a wide range of quorum levels. Moreover, we discuss how much QS is impacted by swarm density and size of the working area, which contribute to determining the effectiveness of the sampling strategy. Our results show that effective QS protocols can be implemented on resource-constrained robotic platforms.

2 Background

As mentioned above, QS finds application in swarm robotics, particularly in association with collective decision making, as a mechanism to determine when one alternative gathered sufficient support. Over the years, a few approaches have been developed to estimate the swarm state and react accordingly with a system-level response. Bacteria-inspired QS did not find much application in robotics as it requires that some chemical product is produced and diffused in the environment, and its concentration measured [25]. Much as with pheromone communication, dealing with chemicals imposes technical challenges difficult to overcome [11], or that require some special infrastructure to be simulated effectively [30,32]. Inspired by QS in social insects, approaches based on sampling the state of neighbouring individuals received more attention. Notably, two main approaches can be followed employing either anonymous or identity-aware interactions, each offering distinct advantages and disadvantages in different scenarios.

Anonymous protocols are mostly inspired by studies performed in honeybees and ants during nest site selection [24,31]. In these studies, individuals base QS on the encounter rate with conspecifics, as the estimation takes place at the candidate site: the higher the encounter rate, the stronger the support for the site. Models and algorithms replicate QS via encounter rates, either through a leaky integrator [21] or by maintaining an anonymous buffer of messages [20]. These approaches clearly minimise memory requirements and computational overhead, and can be implemented with very simple logic.

In contrast, identity-aware protocols leverage unique identification mechanisms to recognise each agent within the swarm. These protocols integrate individual identities alongside state information, enabling more precise decision making at the cost of higher memory requirements and larger computational and communication complexity, posing challenges in resource-constrained settings. A class of approaches closely related to QS is decentralised node counting in networks of autonomous agents [12,29], as QS can also be seen as the problem of counting how many agents are in a given state. Closely related to approaches inspired by social insects, population sampling methods allow obtaining estimates of the population state [5,6,8]. In these studies, agents share their state upon encounter, and each agent stores/updates other agents' states to evaluate if the quorum has been reached using a sufficiently large sample of the population. Both majority and k-unanimity voting rules are tested [6], the former presenting faster dynamics.

The choice between anonymous and identity-aware protocols hinges on the specific requirements and constraints of the swarm robotics application at hand. While anonymous protocols offer simplicity and efficiency in memory usage, they may struggle to maintain robustness in scenarios necessitating fine-grained coordination or prolonged interactions among a subset of agents. Conversely, identity-aware protocols provide enhanced precision and adaptability but demand greater computational resources and may introduce complexities in implementation and maintenance. Thus, understanding the trade-offs between these approaches is essential in designing effective QS protocols for swarm robotics applications, also considering performance in terms of precision and speed.

3 Problem Description

We consider a QS problem in which a group of N agents must collectively recognise if a given portion of the group agrees about an opinion. To focus on the QS dynamics aside from the consensus-building process, we consider agents with two possible states, *committed* and *uncommitted*. Agents are randomly initialised in one or the other state, and never change it. As a consequence, the swarm has a constant fraction of *committed* agents, hereafter referred to as the *ground truth* $G \in [0, 1]$. We consider that the quorum is reached when a minimum fraction $Q \in [0, 1]$ of the swarm recognises that the percentage of teammates in the *committed* state is larger than a given *threshold* $\tau \in [0, 1]$. Ideally, one would expect that $Q = 1 \iff G \geq \tau$, and conversely $Q = 0 \iff G < \tau$.

As mentioned in Sect. 2, agents can estimate whether or not the quorum is reached by sampling the state of other agents within the population either anonymously or by being aware of the identity of the interacting individuals. Anonymous protocols minimise assumptions about interactions among agents and reduce the memory requirements, but can be less effective in case of multiple encounters among the same individuals because these individuals will be double counted. This double-counting problem can be resolved through identity-aware protocols, which however require that identity recognition is possible and are more demanding in terms of memory. In swarm robotics, robots are often assigned a unique identification number (ID), which can be communicated to neighbours together with information about the agent's state. Identity-aware protocols can therefore be easily implemented, provided that the memory requirements are considered. In the following, we first discuss theoretical implications related to choosing anonymous or identity-aware protocols, and then we detail the minimal implementations we propose for robot swarms.

3.1 Urn Models of Quorum Sensing

We assume a well-mixed population, where the probability that two agents interact is the same for any couple of agents. Hence, the sampling process carried out by agents to estimate the quorum state can be easily modelled using simple urn models. Despite their simplicity, urn models are often exploited to provide guidance for the understanding of stochastic processes in complex systems such as robotic swarms [14]. Urn models describe probabilistic events as extractions of balls from an urn. In our case, balls represent agents and the ball's colour represents their state (i.e., committed or uncommitted). Therefore the probability of obtaining information from a neighbour agent can be modelled by the extraction of a ball from an urn. Anonymous protocols can be represented by urn sampling with replacement, because the same agent (ball) can be sampled multiple times, therefore the ball is replaced in the urn after being sampled. Instead, identity-aware protocols can be represented by urn sampling without replacement as double-counting is prevented by keeping track of the agent identities. For both cases, we provide models to compute the probability of detecting a quorum as a function of the number of samples n.

Urn Sampling with Replacement. Consider an urn containing N balls with W white balls and $B = N - W$ black balls (i.e., $G = \frac{W}{N}$). The probability of extracting a white ball is $P_w = G$ and of extracting a black ball is $P_b = 1 - G$. Hence, the probability of having k white balls extracted with replacement in n trials is

$$P_r(n,k) = \binom{n}{k} P_w^k P_b^{n-k}, \qquad k \leq n. \tag{1}$$

So, the probability of having at least τ percent (e.g., 80% in our experiments) of the balls extracted within n trials to be white can be written as

$$\mathcal{P}(n,\tau) = \sum_{k=\lceil n\tau \rceil}^{n} P_r(n,k), \tag{2}$$

where $\lceil . \rceil$ is the ceil operator. Note that here $\mathcal{P}(n,\tau)$ does not depend on the population size N, as a consequence of the replacement.

Urn Sampling Without Replacement. Consider an urn containing N balls with W white balls and $B = N - W$ black balls (i.e., $G = \frac{W}{N}$). If there is no replacement, the probability of having $k \leq W$ white balls within n extractions can be written as

$$P_n(n,k) = \frac{\binom{W}{k}\binom{N-W}{n-k}}{\binom{N}{n}}, \qquad k \leq n. \tag{3}$$

Here, the probability of having at least τ percent of the balls being white is

$$\mathcal{P}(n,\tau) = \sum_{k=\lceil n\tau \rceil}^{n} P_n(n,k). \tag{4}$$

3.2 Implementation of Quorum Sensing in Robot Swarms

We consider a swarm of N robots randomly moving in a square arena (side length L). A total of $\lceil GN \rceil$ robots are randomly chosen and initialised in the committed state, the rest are set to the uncommitted state. Robots can communicate with neighbours within a radius r, sharing their unique ID and a bit b_c indicating their state ($b_c = 1$ for committed, $b_c = 0$ for uncommitted). Messages are broadcast every t_c seconds, and upon reception they are stored and processed to evaluate the existence of the quorum. Owing to communication, sampling of the swarm state can be performed.

To estimate the quorum level, each robot maintains a buffer $\mathcal{B}$ of received messages and, at each buffer update, computes the proportion of neighbours in each state and compares it to the threshold τ. To have a large enough sample over which to compute the qualified majority, we impose a minimum buffer dimension B_m before considering the quorum assessment:

$$b_q = |\mathcal{B}| \geq B_m \wedge \sum_{m \in \mathcal{B}} b_c(m) \geq \tau|\mathcal{B}|. \tag{5}$$

Here, b_q is a bit representing the quorum detection state of the agent, and $b_c(m)$ is the commitment bit stored in message m. Note that the quorum detection state can transition to 0 if the conditions on the buffer $\mathcal{B}$ do not hold anymore.

We propose three different approaches over the same problem formulation to implement both anonymous and identity-aware protocols.

Anonymous Protocol. To implement an anonymous protocol, the robot ID contained in a received message m is ignored and only $b_c(m)$ is stored in $\mathcal{B}$. The buffer $\mathcal{B}$ implements a FIFO method for memory management with fixed size $B_M \leq N$, closely following the implementation in [20]. Hence, the maximum memory requirement is B_M bits. Recall that the commitment state of the same robot can appear multiple times within the buffer, especially if robots remain in mutual proximity for a sufficient time. This may bias the estimation of the quorum level and ultimately the accuracy of the algorithm.

Identity-Aware Protocol with Message Broadcasting (ID+B). In this case, a qualified majority is computed only with information coming from different robots, similarly to what is proposed in [6]. The buffer $\mathcal{B}$ stores any received message m in a list of tuples $\langle k(m), b_c(m), t \rangle$, where $k(m)$ is the robot ID, $b_c(m)$ the corresponding commitment state, and t is a timeout for storing a message drawn from an exponential distribution with average T_m. Through this timeout, old messages are removed from the buffer, forcing the robot to make QS estimates on fresh information. If a new message m' is received from robot $k(m')$ and there is already a tuple from the same robot in the buffer, then the new message is discarded only if the status bit is unchanged, otherwise the old message is replaced with the new one and a new timeout T'_m is computed. This approach allows mitigating the effects of repeated encounters among the same robots, while keeping track of changes in the neighbour status. Overall, if we use B_k bits to store the robots' IDs and B_t bits for timeouts, this protocol requires at most $N(B_k + B_t + 1)$ bits to store the message buffer. As we will see in Sect. 4, the buffer size can be limited by reducing the average timeout T_m.

Identity-Aware Protocol with Message Re-broadcasting (ID+R). One limitation of the above protocols is that messages are received only within the communication range r. To address this limitation, we propose a simple rebroadcasting protocol, which allows to diffuse information widely within the swarm by forwarding the same message multiple times. Here, the buffer $\mathcal{B}$ stores any received message m in a list of tuples $\langle k(m), b_c(m), t, b_r \rangle$, where the additional bit b_r indicates whether or not the information has been rebroadcast. The rebroadcast approach implements a FIFO strategy: every time the robot can communicate (i.e., every t_c seconds), it rebroadcasts the oldest message that has not been rebroadcast yet and the corresponding bit is set in $\mathcal{B}$. When there is no message to rebroadcast, the robot shares its own state. Note that the choice of a FIFO may delay the broadcast of one agent's state, as the latter is shared only when the FIFO is empty. However, preliminary tests without a FIFO revealed that such a delay has a negligible effect in practice for the studied settings (data not shown). Concerning memory requirements, if we do not consider a rebroadcasting FIFO buffer (which can be implemented within $\mathcal{B}$ at the expense of some additional computation), this protocol requires at most $N(B_k + B_t + 2)$ bits.

3.3 Implementation with Kilobots

In this study, QS has been tested with Kilobot robots [27,28] simulated through ARGoS [22,23]. To support the experimentation, we use the ARK system that significantly enhances the experimentation possibilities with Kilobots [9,26]. The motion pattern implemented with the Kilobots is a random waypoint model [4], which corresponds to a directed movement towards a position randomly chosen within the working area. This motion pattern keeps the swarm constrained within the predefined working area without the need for collision avoidance—which cannot be performed by Kilobots—and can be easily implemented employing ARK as a global positioning system [33].

Kilobots communicate at a maximum rate of about 2 Hz, that is, $t_c = 0.5$ s. Messages can be effectively received within a radius of $r = 0.1$ m. We have implemented the buffer $\mathcal{B}$ as a doubly linked list to optimise traversal, insertion and deletion of messages. A static vector of indices is also used to quickly check if messages are already present in the buffer, and access them. These structures come with a negligible overhead in memory requirements and can be easily implemented on the Kilobots (see the open-source code [19]).

4 Results

Starting from the probabilistic urn models, we first show how the probability of sampling a qualified majority of white balls changes with and without replacement, mirroring the usage of anonymous and identity-aware protocols in a well-mixed population. To this end, we compute the maximum threshold $\tau_M \leq G$ that ensures a high probability—higher or equal than 80%—of detecting a quorum. In other words, we determine how precise the threshold τ should be (i.e., how close it should be to the ground truth G) to detect the quorum with high probability $\mathcal{P}(n, \tau) \geq 0.8$. Figure 1 shows the value of τ_M normalised on G, for

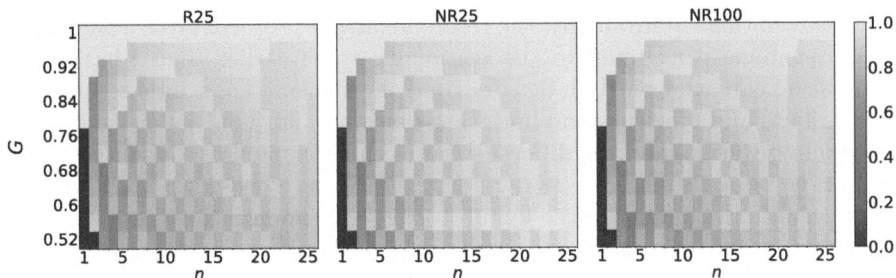

Fig. 1. Prediction of the urn models for sampling with and without replacement. The heatmap shows—for each pair n, G—the normalised maximum threshold τ_M/G that ensures a probability $\mathcal{P}(n, \tau_M) \geq 0.8$. Left: results with replacement and $N = 25$ balls (R25). Centre: results without replacement and $N = 25$ balls (NR25). Right: results without replacement and $N = 100$ balls (NR100).

varying sample size $n \in [1, 25]$ and ground truth $G = i/N, i \in \{\lceil N/2 \rceil, \ldots, N\}$. When $N = 25$, the urn sampling without replacement (NR25) predicts higher precision in detecting the quorum than the case with replacement (R25). On the other hand, when the population size is significantly larger than the number of samples ($N = 100$ in Fig. 1), differences fade away, as it is possible to notice comparing NR100 with R25.[1] Indeed, if the well-mixed assumption holds, an anonymous protocol should be as good as the identity-aware protocol when $N \gg n$. As an additional result, we can determine what is the minimum sampling size B_m to obtain a good precision. Looking at Fig. 1, we can see that for $n < 5$ the precision is not very good for all sampling strategies, but for larger values of n the value of τ_M need not be too distant from G. We therefore fix $B_m = 5$ for all swarm robotics experiments, considering that this is a minimum requirement and that the size of the buffer $\mathcal{B}$ can grow larger than B_m.

Swarm robotics experiments are performed in simulation, testing different swarm densities and arena sizes. Specifically, we consider a low density case (LD25) with $N = 25$ robots in a large square arena ($L = 1\,\mathrm{m}$), a high density case (HD25) with $N = 25$ robots in a small square arena ($L = 0.5\,\mathrm{m}$), and a high density case (HD100) with $N = 100$ robots in a large square arena ($L = 1\,\mathrm{m}$). To give an idea of the consequences of the robot density, consider a random geometric network induced by the robot interactions with average degree $\langle k \rangle = \pi N r^2 / L^2$ [34]. Hence, for LD25 we have less than one neighbour per robot on average ($\langle k \rangle = \pi/4$), while for HD25 and HD100 we have more than three neighbours on average ($\langle k \rangle = \pi$), and a value closer to the percolation threshold $\langle k_c \rangle \approx 4.51$. We study the anonymous and the two identity-aware protocols (ID+B and ID+R) varying the memory requirements, which are determined by the maximum buffer length B_M for the former and by the average timeout T_m for the latter two. Figure 2 shows how different settings influence the buffer length. Given that the anonymous protocol has a fixed buffer length B_M, we plot the number M of unique messages in $\mathcal{B}$, hence excluding double counting. It is possible to notice that the anonymous protocol accumulates a lower amount of information about the swarm than the identity-aware protocols, mainly due to double counting. Additionally, the re-broadcasting protocol gives a significant speed advantage, converging to the stationary value earlier than what simple broadcasting can achieve. High-density conditions ensure good interaction rates among robots and efficient sampling. Low-density conditions suffer from inefficient communication, and the difference between anonymous and identity-aware protocols is reduced.

To evaluate the quality of the QS protocols, we compute the fraction $Q(t)$ of robots that at time t recognise the quorum given a ground truth $G = i/N, i \in \{\lceil N/2 \rceil, \ldots, N\}$ and the threshold $\tau \in [0.5, 1]$ in all the mentioned experimental conditions. We then compute the average quorum detection $\hat{Q}(G, \tau)$ as the average value of $Q(t)$ over $T = 900\,\mathrm{s}$ and 100 independent runs. Figure 3 shows the isolines for $\hat{Q} = 0.8$ and $\hat{Q} = 0.2$, which represent boundaries of regions in which the swarm consistently recognises a quorum ($\hat{Q} \geq 0.8$) or reject it ($\hat{Q} < 0.2$).

---

[1] Recall that the urn model with replacement does not depend on system size N.

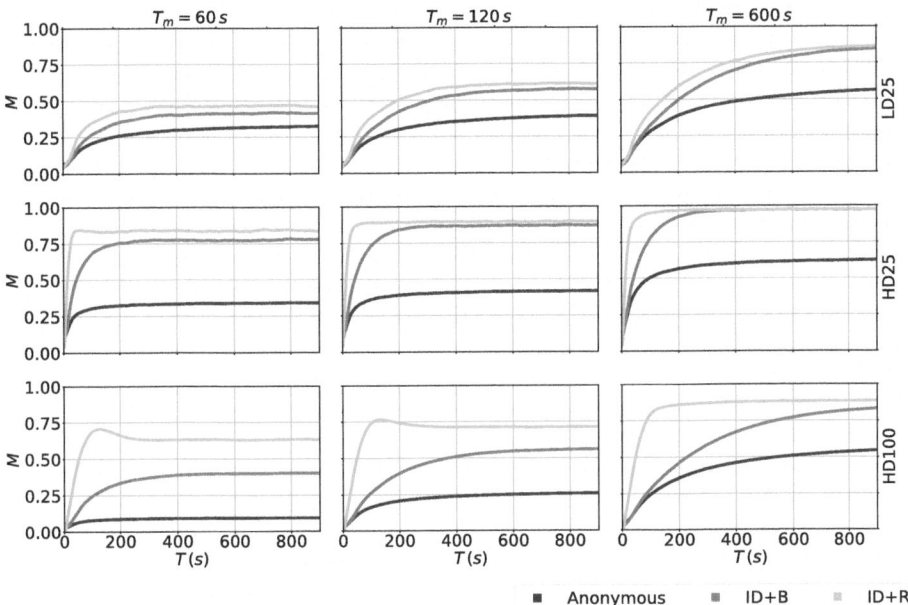

Fig. 2. Evolution over time of the average number M of unique messages in the buffer normalised over the maximum allowable size (i.e., $N-1$). From top to bottom, the density-size scenario is changed, respectively LD25, HD25 and HD100. From left to right, the average timeout T_m and the maximum buffer length B_M values increase. We set $B_m \in \{10, 13, 24\}$ for $N = 25$, and $B_m \in \{10, 32, 99\}$ for $N = 100$. Data are averaged over all robots across $R = 100$ simulation runs.

As mentioned in Sect. 3, ideally the region $\mathcal{R}$ between these isolines should be minimised, as this region corresponds to an undecided state, possibly leading to QS errors. It is possible to notice that in all settings the anonymous protocol leads to a wider region $\mathcal{R}$. Among the two identity-aware protocols, ID+R has a slight advantage consistently across experimental setups. When increasing the timeout T_m and maximum buffer size B_M, the region $\mathcal{R}$ always gets smaller as the qualified majority is evaluated on a larger sample. Similarly, moving from low to high densities and larger numbers of robots increases the sampling size and in turn the quality of the estimation.

Finally, we evaluate how fast the different protocols can lead to a reliable recognition of the quorum. To this end, for each value of $\tau \in [0.5, 1]$, we consider the smallest value $\hat{G}$ that results in a reliable estimation:

$$\hat{G} = \arg\min_{G} \left(\hat{Q}(G, \tau) \geq 0.8 \right), G = i/N, i \in \{\lceil N/2 \rceil, \ldots, N\}. \qquad (6)$$

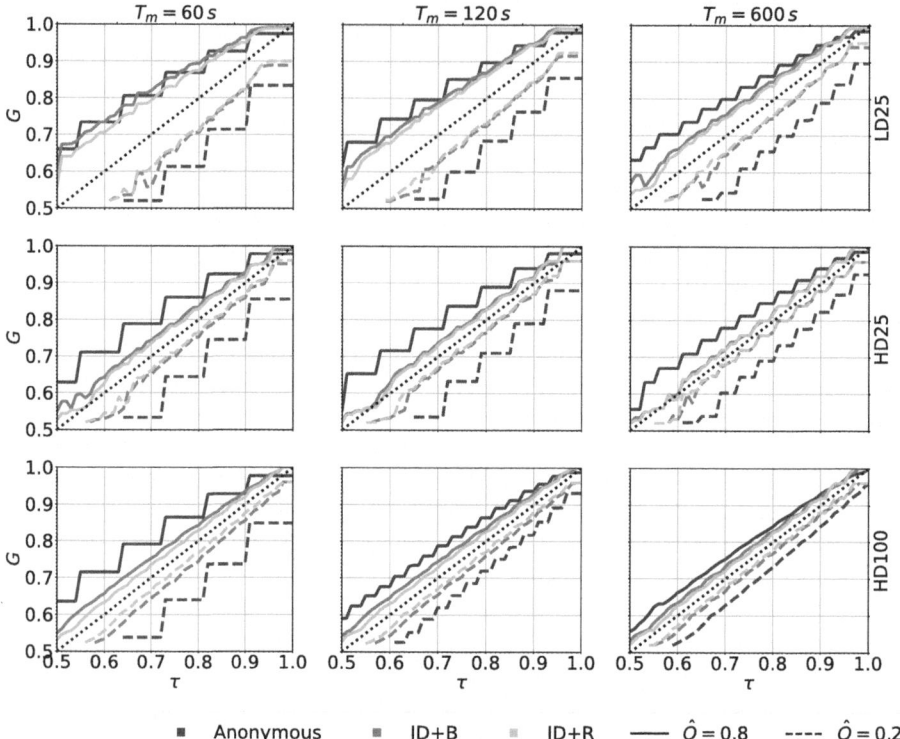

Fig. 3. Isolines of the average quorum detection for $\hat{Q} = 0.8$ (solid lines) and $\hat{Q} = 0.2$ (dashed lines). The isolines are computed through linear interpolation between the grid points used for computing the average quorum detection for all values of G and τ.

For such values of τ and $\hat{G}$, we record the time at which $Q(t)$ exceeds the 0.8 threshold. Figure 4 shows the median quorum recognition time T_c across 100 runs. It is possible to notice that the anonymous protocol is rather fast, but as we have seen in Fig. 3 it trades off precision for speed. Conversely, the identity-aware protocols present similar behaviour with low density. However, ID+R is much faster than the other approaches with high density, because rebroadcasting leads to faster diffusion of information and an improved sampling (as also seen with the growth rate of the message buffer $\mathcal{B}$ in Fig. 2).

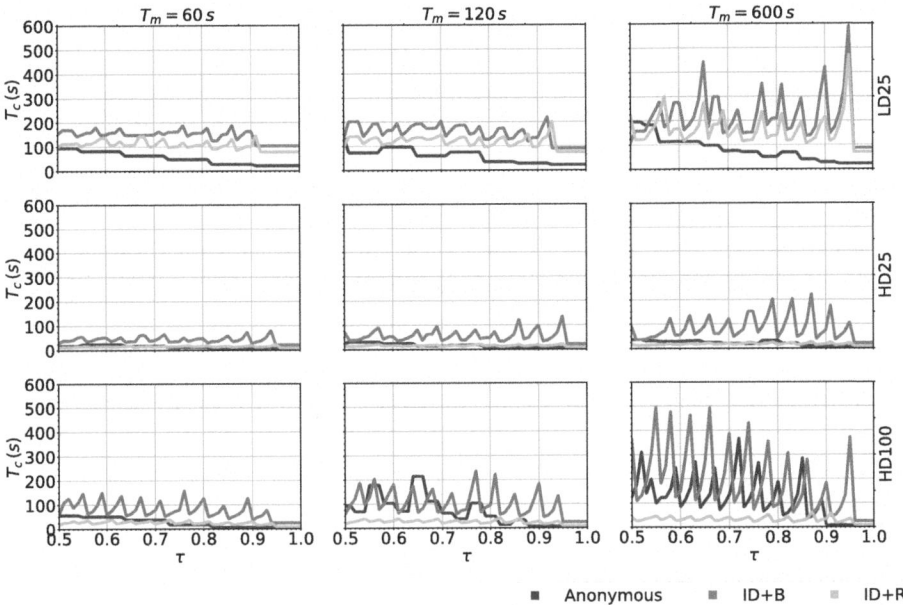

Fig. 4. Median times for the swarm to reliably detect the quorum across 100 different runs. We can interpret the oscillations as an artefact determined by the way in which we select $\hat{G}$, which results in more or less close values of $P(\hat{G}, \tau)$ to the 0.8 threshold: the closer it is, the slower is the quorum detection process.

5 Conclusions and Future Work

In this paper, we explored minimalist approaches to quorum sensing (QS) based on anonymous or identity-aware sampling protocols. Our results demonstrate that anonymous protocols suffer from the double counting problem, which can be mitigated only if the robot swarm population is sufficiently well-mixed. However, robotics settings typically have strong spatial and communication correlations, hence the system is often far from being well-mixed. Identity-aware approaches are effective in both low and high densities, and message rebroadcasting can boost the QS process both in accuracy and speed. We hypothesise that, whenever the density is sufficiently high to enable information sharing, ID+R can be an effective solution to QS in particularly challenging scenarios characterised by large spatial heterogeneities in the robot distribution. Finally, ID+R should also adapt well to dynamic conditions where the ground truth changes over time, owing to the speed of diffusing new information. These conditions will also be the subject of future investigation.

The systematic experimentation we have conducted should provide valuable information to swarm robotics practitioners who need to define how to estimate the state of the swarm in a fast and reliable way. Further characterisation of the mechanisms presented here will improve the ability to make informed choices about the QS protocol, especially in relation to dynamic settings, different motion patterns and varying task requirements. For instance, when robot swarms perform specific tasks (e.g., foraging), their interaction topology stops approximating a well-mixed system, and the performance of the different QS protocols can be affected. Finally, recovery from errors should be explicitly accounted for. While the presented protocols can correct from errors as long as additional samples are collected, the speed and reliability of error recovery needs to be explicitly assessed and possibly adaptive mechanisms should be put in place to enable quick reaction both locally and globally.

Acknowledgements. A.R. acknowledges support from DFG under Germany's Excellence Strategy - EXC 2117 - 422037984. V.T. acknowledges support from the project TAILOR (H2020-ICT-48 GA: 952215) and from the PNRR MUR project PE0000013-FAIR.

References

1. Albani, D., Hönig, W., Nardi, D., Ayanian, N., Trianni, V.: Hierarchical task assignment and path finding with limited communication for robot swarms. Appl. Sci. **11**(7), 3115 (2021)
2. Amir, Y., Abu-Horowitz, A., Werfel, J., Bachelet, I.: Nanoscale robots exhibiting quorum sensing. Artif. Life **25**(3), 227–231 (2019)
3. Atkinson, S., Williams, P.: Quorum sensing and social networking in the microbial world. J. R. Soc. Interface **6**(40), 959–978 (2009)
4. Bettstetter, C., Hartenstein, H., Pérez-Costa, X.: Stochastic properties of the random waypoint mobility model. Wirel. Netw. **10**(5), 555–567 (2004)
5. Cai, G., Sofge, D.: An urgency-dependent quorum sensing algorithm for n-site selection in autonomous swarms. In: AAMAS 2019: Proceedings of the 18th International Conference on Autonomous Agents and Multiagent Systems, pp. 1853–1855. IFAAMAS, Richland, SC (2019)
6. Cody, J.R., Adams, J.A.: An evaluation of quorum sensing mechanisms in collective value-sensitive site selection. In: 2017 International Symposium on Multi-Robot and Multi-Agent Systems (MRS), pp. 40–47. IEEE (2017)
7. Cronin, A.L.: Ratio-dependent quantity discrimination in quorum sensing ants. Anim. Cogn. **17**(6), 1261–1268 (2014)
8. Ebert, J.T., Gauci, M., Nagpal, R.: Multi-feature collective decision making in robot swarms. In: AAMAS 2018: Proceedings of the 17th International Conference on Autonomous Agents and Multiagent Systems, pp. 1711—1719. IFAAMAS, Richland, SC (2018)
9. Feola, L., Reina, A., Talamali, M.S., Trianni, V.: Multi-swarm interaction through augmented reality for Kilobots. IEEE Robot. Autom. Lett. **8**(11), 6907–6914 (2023)
10. Franks, N.R., Dechaume-Moncharmont, F.X., Hanmore, E., Reynolds, J.K.: Speed versus accuracy in decision-making ants: expediting politics and policy implementation. Philos. Trans. R. Soc. B: Biol. Sci. **364**(1518), 845–852 (2009)
11. Fujisawa, R., Dobata, S., Sugawara, K., Matsuno, F.: Designing pheromone communication in swarm robotics: group foraging behavior mediated by chemical substance. Swarm Intell. **8**(3), 227–246 (2014)

12. Ganesh, A.J., Kermarrec, A.M., Merrer, E.L., Massoulié, L.: Peer counting and sampling in overlay networks based on random walks. Distrib. Comput. **20**(4), 267–278 (2007)
13. Girard, L.: Quorum sensing in *Vibrio* spp.: the complexity of multiple signalling molecules in marine and aquatic environments. Crit. Rev. Microbiol. **45**(4), 451–471 (2019)
14. Hamann, H.: Towards swarm calculus: urn models of collective decisions and universal properties of swarm performance. Swarm Intell. **7**(2), 145–172 (2013)
15. Hsieh, M.A., Halász, Á., Berman, S., Kumar, V.: Biologically inspired redistribution of a swarm of robots among multiple sites. Swarm Intell. **2**(2), 121–141 (2008)
16. Leaf, J., Adams, J.: The effect of uneven and obstructed site layouts in best-of-N. Swarm Intell. (2024)
17. Miller, M.B., Bassler, B.L.: Quorum sensing in bacteria. Annu. Rev. Microbiol. **55**(1), 165–199 (2001)
18. Oddi, F., Cristofaro, A., Trianni, V.: Best-of-N collective decisions on a hierarchy. In: Dorigo, M., et al. (eds.) ANTS 2022. LNCS, vol. 13491, pp. 66–78. Springer, Cham (2022). https://doi.org/10.1007/978-3-031-20176-9_6
19. Oddi, F., Reina, A., Trianni, V.: Code associated to "Minimalist Protocols for Quorum Sensing in Robot Swarms" (2024). https://github.com/Fabio930/argos3-kilobot.git
20. Parker, C.A., Zhang, H.: Collective unary decision-making by decentralized multiple-robot systems applied to the task-sequencing problem. Swarm Intell. **4**, 199–220 (2010)
21. Pavlic, T.P., Hanson, J., Valentini, G., Walker, S.I., Pratt, S.C.: Quorum sensing without deliberation: biological inspiration for externalizing computation to physical spaces in multi-robot systems. Swarm Intell. **15**(1–2), 171–203 (2021)
22. Pinciroli, C., Talamali, M.S., Reina, A., Marshall, J.A.R., Trianni, V.: Simulating Kilobots within ARGoS: models and experimental validation. In: Dorigo, M., Birattari, M., Blum, C., Christensen, A.L., Reina, A., Trianni, V. (eds.) ANTS 2018. LNCS, vol. 11172, pp. 176–187. Springer, Cham (2018). https://doi.org/10.1007/978-3-030-00533-7_14
23. Pinciroli, C., et al.: ARGoS: a modular, parallel, multi-engine simulator for multi-robot systems. Swarm Intell. **6**(4), 271–295 (2012)
24. Pratt, S.C.: Quorum sensing by encounter rates in the ant Temnothorax albipennis. Behav. Ecol. **16**(2), 488–496 (2005)
25. Pérez-Velázquez, J., Gölgeli, M., García-Contreras, R.: Mathematical modelling of bacterial quorum sensing: a review. Bull. Math. Biol. **78**(8), 1585–1639 (2016)
26. Reina, A., Cope, A.J., Nikolaidis, E., Marshall, J.A.R., Sabo, C.: ARK: augmented reality for kilobots. IEEE Robot. Autom. Lett. **2**(3), 1755–1761 (2017)
27. Rubenstein, M., Ahler, C., Hoff, N., Cabrera, A., Nagpal, R.: Kilobot: a low cost robot with scalable operations designed for collective behaviors. Robot. Auton. Syst. **62**(7), 966–975 (2014)
28. Rubenstein, M., Cornejo, A., Nagpal, R.: Programmable self-assembly in a thousand-robot swarm. Science **345**(6198), 795–799 (2014)
29. Saha, A., Marshall, J.A.R., Reina, A.: Memory and communication efficient algorithm for decentralized counting of nodes in networks. PLoS ONE **16**(11), e0259736 (2021)
30. Salman, M., Ramos, D.G., Hasselmann, K., Birattari, M.: Phormica: photochromic pheromone release and detection system for stigmergic coordination in robot swarms. Front. Robot. AI **7**, 591402 (2020)

31. Seeley, T., Visscher, P.: Quorum sensing during nest-site selection by honeybee swarms. Behav. Ecol. Sociobiol. **56**, 594–601 (2004)
32. Talamali, M.S., Bose, T., Haire, M., Xu, X., Marshall, J.A.R., Reina, A.: Sophisticated collective foraging with minimalist agents: a swarm robotics test. Swarm Intell. **14**(1), 25–56 (2020)
33. Talamali, M.S., Saha, A., Marshall, J.A.R., Reina, A.: When less is more: robot swarms adapt better to changes with constrained communication. Sci. Robot. **6**(56) (2021)
34. Trianni, V., De Simone, D., Reina, A., Baronchelli, A.: Emergence of consensus in a multi-robot network: from abstract models to empirical validation. IEEE Robot. Autom. Lett. **1**(1), 348–353 (2016)
35. Valentini, G., Ferrante, E., Dorigo, M.: The best-of-n problem in robot swarms: formalization, state of the art, and novel perspectives. Front. Robot. AI **4**, 1–43 (2017)
36. Waters, C.M., Bassler, B.L.: Quorum sensing: communication in bacteria. Ann. Rev. Cell Dev. Biol. **21**(1), 319–346 (2005)

Self-organized Flocking in Three Dimensions

Tugay Alperen Karagüzel[✉], Fuda van Diggelen, Andres Garcia Rincon, and Eliseo Ferrante

Computer Science Department, Vrije Universiteit, Amsterdam, The Netherlands
{t.a.karaguzel,fuda.van.diggelen,e.ferrante}@vu.nl,
a.a.garciarincon@student.vu.nl

Abstract. This paper introduces a novel methodology for achieving self-organized flocking behavior entirely in three dimensions, building upon the active elastic model. Through comprehensive evaluations made with kinematic and dynamic simulations, as well as through a real-world implementation with Crazyflie nano drones, we demonstrate the feasibility and robustness of our approach under various operational constraints. Our findings show that the system scalability to larger swarm sizes has a stronger dependence to parameters, in particular to the strength of the formation, compared to the two dimensional counterpart. The successful deployment of this methodology in an indoor environment, where the system navigates physical boundaries and demonstrates cohesive and ordered motion, showcases its potential applications in surveillance, search and rescue, and environmental monitoring. This work paves the way for future research in advanced swarm behaviors and underscores the practical applicability of three-dimensional flocking in robotic swarms.

1 Introduction

Collective motion [21] is a widespread natural phenomenon where individuals in a group coordinate their movements coherently without a central infrastructure, through simple local interactions. This behavior, observed in various animals, offers numerous benefits, such as enhanced foraging [8], predator evasion [16], efficient navigation [8], and improved hydrodynamic and aerodynamic performance [5]. The term "flocking", closely related to collective motion, typically describes the flight dynamics of birds, which often involve three-dimensional movements. Despite its prevalence, the modeling of three-dimensional swarming behavior in physics and biology has been relatively limited [7].

Collective motion and flocking have been modeled in artificial systems with three kinds of individual-based rules, to ensure cohesive, ordered and coordinated collective motion: *cohesion, separation* and *alignment* [18]. Achieving flocking through distributed principles in artificial platforms offers several advantages, including efficiency, robustness to member loss and the possibility for collective and emergent level problem solving through local interactions [15,19]. These advantages have motivated studying flocking within the swarm robotics field [20].

Applications of collective motion in robotics span various platforms and media. Mobile ground robots, equipped with sensors for detecting peers' positions and orientations, have showcased self-organized motion without a shared goal direction [20]. This concept extends to enable collective motion without the need for orientation perception or communication among ground robots [9], and allows informed robots to lead a swarm with conflicting information when communication is available [10]. In the water media, a robotic fish swarm that exhibits behaviors like synchronization, aggregation, circle formation, and search-capture is introduced in [6], where control and peer sensing occur onboard. Meanwhile in the air, [22] introduced a drone swarm maintaining stable flocking at a fixed altitude, achieving high speeds while navigating clear of boundary zones and obstacles. Flocking in other media other than ground represent a practical reason that motivates achieving flocking in three dimensions.

Unmanned Aerial Vehicles (UAVs) offer application scenarios in three dimensional spaces that can be indoor [14] or large spaces outdoor [12]. Outdoor flocking in the planar dimension has been demonstrated with notably large velocities in a detailed report [22]. In [12], a two-dimensional outdoor flock of quadrotors was designed to operate in search and rescue missions. Agents carry specific sensors and computers for this task and are able to communicate with others over radio or be informed about the specific search locations. The authors of [4] explored the third dimension for UAVs flocking, by proposing a modification of the self-organized flocking rules developed in [20] on a quadrotor swarm. This swarm can perceive the relative position of peers nearby and achieve a formation in three dimension, with the ability to sense and follow the shape of the ground. Nevertheless, collective motion does not occur in three dimension but on the plane parallel to the ground, despite the title being very similar to the current paper presented. In addition, non-holonomic constraints on agents are also removed in this study in order to achieve a more flexible flocking. Finally, [13] presents a reinforcement learning framework that can take system uncertainties, platform unknowns and dynamic environment conditions. Although this controller is designed for UAV swarms operating in 3D as showcased in the extensive simulation experiments, the real flight experiments are limited with fixed-altitude experiments.

In this paper, we present an approach to perform flocking entirely in three dimensions. To the best of our knowledge, this is the first work that achieves flocking entirely in 3D, namely not only the swarm formation is in 3D, but also the self-organized direction of motion is a line in three dimension rather than along the plane parallel to the ground. One important feature of our method is that it does not introduce any additional parameter compared to a two dimensional implementation: it is based on the active elastic model [9,11] and requires the same number of control parameters, namely we one linear speed and one angular speed. The model we present here is comprehensively evaluated in simplified kinematic simulations, validated in dynamics simulations, and demonstrated in a real-world deployment with Crazyflie drones. As per the original active elastic model [14], the method is fully deployed onboard the drones, and the relative positions of other drones are also calculated onboard the drones, necessitating only a shared radio communication channel for interaction among close peers.

The rest of the paper is organized as follows: Sect. 2 introduces the proposed methodology in a platform-agnostic manner. Section 3 details the experimental setup, including kinematic simulations, physics-based simulations, and autonomous drone flights. This section also introduces metrics for assessing flocking performance and analyzes and discusses results for all settings. Finally, Sect. 4 concludes our findings.

2 Methodology

The control logic explained here is applied to each agent in the swarm independently. Each agent can move freely in three-dimensional space in their heading direction, denoted as $\hat{h} = (h_x, h_y, h_z)$. After each control step, a virtual force vector $\boldsymbol{f}_i$ is computed. From $\boldsymbol{f}_i$, a linear speed u_i is determined, directing the agent to move along its current heading. Movement in the local frame of reference is marked as v_x, v_y, and v_z and is based on the heading. Furthermore, an adjustment in the heading direction, denoted by w_i (angular speed), is calculated for the next control step.

At each step, the focal agent i collects only the relative positions (and optionally, the heading directions) of peers within a limited sensing range. The virtual force vector $\boldsymbol{f}_i$ is then calculated using the collected information as follows:

$$\boldsymbol{f}_i = \alpha \boldsymbol{p}_i + \hat{a} \tag{1}$$

In the above equation, $\boldsymbol{p}_i$ represents the proximal vector, $\hat{a}$ represents the alignment vector and α controls the relative weights among them. $\boldsymbol{p}_i$ acts as a virtual spring between peers: it produces repulsion if a peer is closer than the desired distance and is attractive if the peer is further away [9]. $\boldsymbol{p}_i$ is calculated as:

$$\boldsymbol{p}_i = \sum_{m \in N} -\epsilon \left[2\frac{\sigma^4}{(d_i^m)^5} - \frac{\sigma^2}{(d_i^m)^3} \right] \hat{r}_i^m \tag{2}$$

$\boldsymbol{p}_i$ is produced in the direction of relative position vector of neighbour m. This direction is shown by $\hat{r}_i^m$ as a unit vector and magnitude is shown by d_i^m. In the above equation, ϵ changes the strength of $\boldsymbol{p}_i$ and σ changes the desired distance with the peers ($d_{des} = \sigma\sqrt{2}$).

In order to calculate the alignment vector $\hat{a}$, the focal agent collects the heading directions of perceived neighbors ($\hat{h}_m$) and its own ($\hat{h}_i$), sums them all to obtain ($\boldsymbol{h}_{sum}$), and normalizes the sum.

$$\hat{a} = \frac{\boldsymbol{h}_{sum}}{||\boldsymbol{h}_{sum}||} \quad \text{where} \quad \boldsymbol{h}_{sum} = \left(\sum_{m \in N} \hat{h}_m \right) + \hat{h}_i \tag{3}$$

The alignment vector $\hat{a}$ helps the focal agent modify its heading direction to align with the average direction of motion of the swarm, and to achieve improved performance. Importantly however, as in the original active elastic model, spontaneous alignment in 3D can be achieved also when the alignment vector is not included in the control. This is shown in the results section.

When agents are required to operate within a bounded space, a combined boundary repulsion vector $\boldsymbol{b_{ci}}$ is incorporated into the calculation of $\boldsymbol{f_i}$ as follows:

$$\boldsymbol{f}_i = \alpha \boldsymbol{p}_i + \hat{a} + \boldsymbol{b}_{ci}, \quad \boldsymbol{b}_{ci} = \sum_{d \in D} \boldsymbol{b}_{di}, \quad D = [+x, -x, +y, -y, +z, -z] \quad (4)$$

$\boldsymbol{b}_{ci}$ combines specific boundary repulsion $\boldsymbol{b}_{di}$, determined by an agent's distance (di^b) to the closest maximum or minimum boundary values. The direction of $\boldsymbol{b}_{di}$, indicated by the unit vector $\hat{b}_{di}$, is always aimed towards one of the elements in D, creating a repulsive effect. The magnitude of $\boldsymbol{b}_{di}$ ($||\boldsymbol{b}_{di}||$) is calculated similarly to Eq. 5 but with greater strength and a smaller σ value (σ_b). The equation exclusively accounts for repulsion, applicable only when an agent's proximity to a boundary is under $\sigma_b\sqrt{2}$; attractive forces are not considered. In other words, boundaries can only produce repulsion away from them.

$$||\boldsymbol{b}_{di}|| = -5\epsilon \left[2 \frac{\sigma_b^4}{(d_i^b)^5} - \frac{\sigma_b^2}{(d_i^b)^3} \right], \quad \boldsymbol{b}_{di} = ||\boldsymbol{b}_{di}||\hat{b}_{di} \quad (5)$$

Next, to calculate linear speed u_i and the angular speed w_i, the projection of $\boldsymbol{f}_i$ on the focal agent's heading vector $\hat{h}_i$ is found. To calculate the component of $\boldsymbol{f}_i$ along $\hat{h}_i$, the angle (θ) between these 3D vectors is found:

$$\theta = cos^{-1}\left(\frac{\boldsymbol{f}_i \cdot \hat{h}_i}{||\boldsymbol{f}_i||\ ||\hat{h}_i||} \right) \quad (6)$$

After finding (θ), the projection of $\boldsymbol{f}_i$ on $\hat{h}_i$ determines the linear speed u_i after multiplying with linear speed gain $K1$ and adding additional constant speed u_{add}. The term u_{add} ensures minor self-propulsion, promoting a dynamic and moving swarm rather than a stationary formation. To calculate w_i, the orthogonal component of the projection is used after the multiplication with angular speed gain $K2$.

$$u_i = K_1\, ||\boldsymbol{f}_i||\, cos(\theta) + u_{add}, \quad \omega_i = K_2\, ||\boldsymbol{f}_i||\, sin(\theta) \quad (7)$$

Finally, u_i and w_i is clipped between $[0, u_{max}]$ and $[-w_{max}, w_{max}]$ respectively. The linear speeds to be applied (v_x, v_y and v_z) are calculated as follows. These values remain unchanged until they are updated in the next calculation step.

$$v_x = u_i h_x, \quad v_y = u_i h_y, \quad v_z = u_i h_z \quad (8)$$

The magnitude of the heading direction change, w_i, is multiplied by the time between two consecutive calculation steps (dt) to determine the new heading direction $\hat{h}_i^{new}$. Given that $\hat{h}_i$ is defined in 3D space, rotation can occur in an infinite number of directions. To accurately represent the axis of rotation, we propose a vector $\hat{e}_r$, defined as a unit vector orthogonal to both $\hat{h}_i$ and $\boldsymbol{f}_i$. The calculation of $\hat{e}_r$ is as follows:

$$\hat{e}_r = \frac{\hat{h}_i \times \boldsymbol{f}_i}{||\hat{h}_i \times \boldsymbol{f}_i||} \qquad \hat{e}_\perp = \frac{\hat{h}_i \times \hat{e}_r}{||\hat{h}_i \times \hat{e}_r||} \qquad (9)$$

Next, $\hat{h}_i$ is rotated around $\hat{e}_r$ in the amount of $w_i\, dt$. According to Rodrigues' rotation formula, new heading $\hat{h}_i^{new}$ is calculated as follows:

$$\hat{h}_i^{new} = \hat{h}_i\, cos(w_i dt) + \hat{e}_\perp\, sin(w_i dt) = (h_x^{new}, h_y^{new}, h_z^{new}) \qquad (10)$$

3 Experimental Setup and Results

We test our method in three different setups: 1) *Kinematic agent simulator*, where agents lack inertial properties and thus dynamics; 2) *physics-based simulator*, featuring realistic dynamic models, physical constraints, accurate control algorithms, and process noises, and where we control the linear velocity of Crazyflie models; 3) *Real UAV platform*, Crazyflie 2.1 nano drones are programmed to utilize the methodology through their onboard controllers, allowing for fully autonomous operation without the need for an external computer. Source material of the experimental setup can be found in our supplementary repository. Sample experiment videos can be found on the playlist[1]. Details of these three setups, along with the results of each, are given next.

3.1 Kinematic Agent Simulations

Initially, we conduct a thorough analysis using an efficient kinematic multi-agent simulator. This phase aims to explore the impact of key parameters mentioned in Sect. 2 on performance. Despite its simplicity, the simulator enables comprehensive analysis through the systematic variation of several key parameters.

In kinematic agent simulations, we simulate swarms of 5, 100, and 1000 agents. Their states are updated using discrete integration at a fixed time step of $dt = 0.05$, with updates and perceived peer positions incorporating a small uniform noise. To assess their impact on flocking performance, we systematically vary swarm size and key parameters: the relative proximal control weight α, linear speed gain $K1$, and angular speed gain $K2$.

The environment in which agents operate is a boundless and infinite 3D space. To align the capabilities of the kinematic agents with the Crazyflie 2.1 nano drone, used in autonomous drone experiments and its dynamical model in physics-based simulations, we establish some limits. The maximum linear speed u_i is set to $0.3\,\text{m/s}$, the maximum angular speed w_i is set to π rad/s, and the peer sensing range is set at $3.0\,\text{m}$. Finally, desired distance is chosen to be 0.7 m which makes σ equal to 0.5. ϵ is set to 12.0.

Agents are initialized at random positions to prevent isolation or disconnected groups and given random heading directions. Kinematic experiments run for 600

---

[1] https://bit.ly/3D_AE_flocking.

simulated seconds, with states recorded every 10 time steps. We explore the impact of varying parameters on performance by setting three value levels for each: Low [L], Moderate [M], and High [H], keeping others at a moderate level to isolate effects. The values for α, $K1$, and $K2$ are set as $\alpha = [0.01, 0.1, 1.0]$, $K1 = [0.05, 0.5, 5.0]$, and $K2 = [0.01, 0.1, 1.0]$ respectively. Each configuration is tested 50 times. Additionally, we conduct experiments with all parameters at moderate levels but omit the alignment vector $\boldsymbol{h}$, focusing on proximal control.

To evaluate the swarm's alignment agreement, we utilize the order metric (Ψ), which quantifies the consensus on heading direction among all agents. The order reaches its peak of 1 with full alignment and nears zero as consensus wanes. Ψ is calculated as follows, where $\hat{h}_i$ denotes the unit vector of an agent's heading in a swarm of N agents:

$$\Psi = \frac{|| \sum_{i \in N} h_{i,x}, \sum_{i \in N} h_{i,y}, \sum_{i \in N} h_{i,z} ||}{N} \quad (11)$$

To measure the swarm's coordination, we monitor the center of mass (CoM) speed. If agents keep even distances and move cohesively, the CoM speed should approach the maximum allowed linear speed. Deviations suggest agents are either rotating around each other, diverging from the average heading and creating angular instead of linear speed, or the swarm has lost cohesion, leaving agents without nearby peers to interact with.

Lastly, we display heat maps illustrating the distribution of all swarm agents around the CoM throughout the experiment, as three 2D images for XY, YZ, and XZ projections. The color intensities, not on a uniform scale, are adjusted to improve color contrast for clearer communication. In Fig. 1, we discuss metrics for 5 agents, including swarm order, CoM speed, and distance distributions around the CoM across different settings. These results, like others in this section, are based on 50 experiment aggregations. In (a), moderate parameter settings lead to the swarm rapidly achieving high order and CoM speed, demonstrating successful flocking and consistent movement, even with a sparser agent distribution. (b) shows a denser distribution with high α, which does not alter the swarm order or CoM speed. Meanwhile (c) illustrates the effects of low $K2$, resulting in a delayed rise to comparable order and speed but with a disordered agent distribution around the CoM.

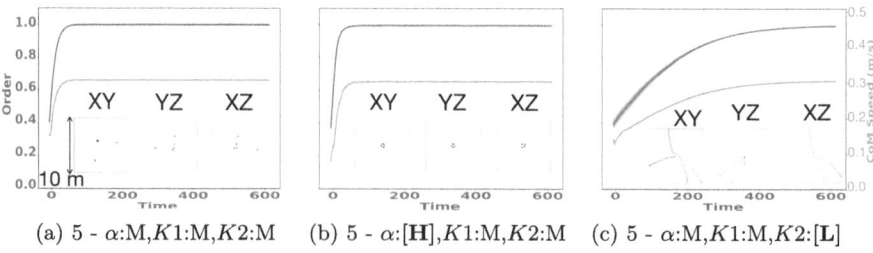

Fig. 1. Flocking metrics for 5 agents, kinematic simulations. (a) Moderate α, K_1, and K_2; (b) High α; (c) Low K_2. Metrics show order, CoM speed and distribution.

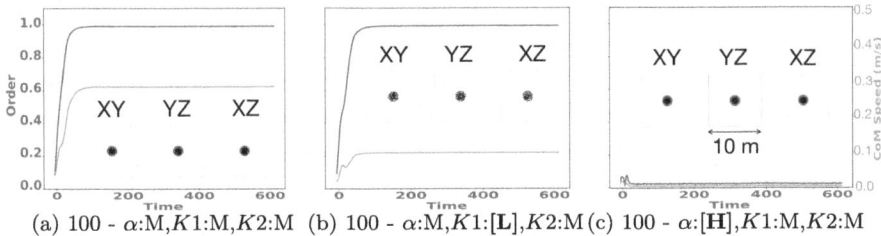

(a) 100 - α:M,$K1$:M,$K2$:M (b) 100 - α:M,$K1$:[**L**],$K2$:M (c) 100 - α:[**H**],$K1$:M,$K2$:M

Fig. 2. Flocking metrics for 100 agents. (a) Moderate α, K_1, and K_2; (b) Low K_1; (c) High α. Shows impact of parameters on order, CoM speed, and agent distribution.

In Fig. 2, we display flocking metrics for 100 agents. Under moderate parameters in (a), the swarm's order and CoM speed mirror those of 5 agents, yet the distance distribution becomes denser and spherical. Setting $K1$ to low in (b) keeps the order stable but lowers CoM speed, a difference more marked in 100 agents than in 5. A high α in (c) significantly lowers both order and CoM speed, creating a dense, immobile agent cluster. This shows how high α affects 100 agents differently compared to 5. In Fig. 3, we present results for 1000-agents, highlighting distinct patterns from those seen with 5 and 100 agents. With moderate settings in (a), the swarm's order stabilizes at around 0.6, failing to reach its peak, and the CoM speed caps at 0.2 m/s, alongside a uniform and spherical distribution. Setting α to low in (b) allows for maximum order and CoM speed, leading to a stretched distribution due to the lower α, indicating a significant change in α's impact on flocking behavior. Conversely, a low $K2$ in (c) drastically limits performance, with both order and CoM speed dropping considerably, yet the swarm remains cohesive, showing a lack of dynamic utilization of its linear speed capability.

Figure 4 presents results from experiments without the alignment vector, with parameters at moderate levels. For 5 agents in (a), the order increases slowly and peaking at 0.8, which is lower and slower compared to using alignment control. Their CoM speed does not reach the maximum but stays close to u_{add}, with distribution remaining cohesive and uniform. With 100 agents in (b), outcomes

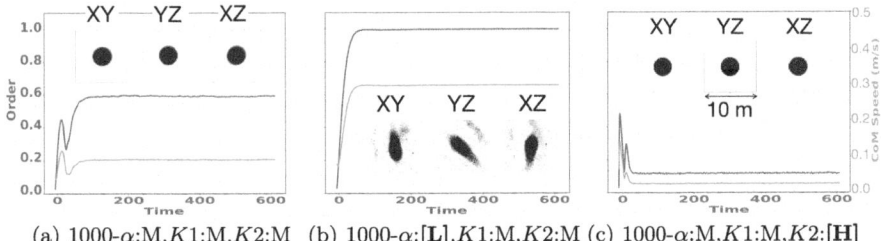

(a) 1000-α:M,$K1$:M,$K2$:M (b) 1000-α:[**L**],$K1$:M,$K2$:M (c) 1000-α:M,$K1$:M,$K2$:[**H**]

Fig. 3. Flocking metrics for 1000 agents. (a) Moderate α, K_1, and K_2; (b) Low α; (c) High K_2. Shows impact on order, CoM speed, and distribution.

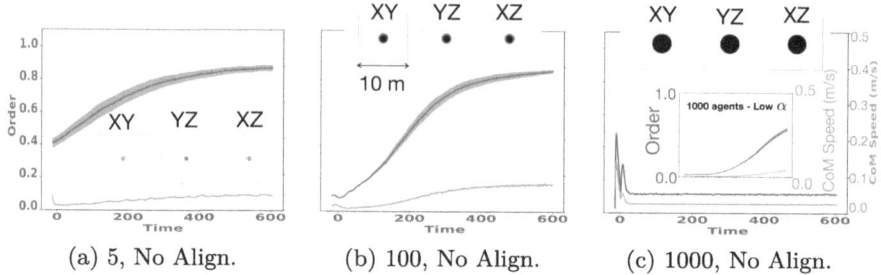

Fig. 4. Metrics without alignment vector for 5, 100, and 1000 agents. Moderate α, K_1, and K_2. Performance with varied α levels for 1000 agents.

are similar but with quicker order convergence and slightly higher CoM speed. For 1000 agents in (c), moderate parameters yield less optimal results: order and CoM speed are low, though distribution stays spherical. A sub-figure shows improved performance with lower α, consistent with findings that lower α benefits larger swarms. This underscores a stable link between performance and low α across large swarms, even without alignment control.

In kinematic agent experiments, we used an efficient simulator to explore how key parameters influence flocking in varying swarm sizes. Analysis of performance metrics reveals moderate parameter levels typically offer optimal outcomes for all swarm sizes. Increasing α enhances distribution for 5 agents, yet moderate levels suffice, suggesting a small swarm doesn't fully utilize multi-agent dynamics. This implies proximal force, generated per agent, might not scale well. In contrast, swarms with 1000 agents excel with lower α, indicating too many neighbors make moderate α overwhelming, so reducing it is beneficial. Swarms of 100 agents, in an intermediary position, perform best at moderate α levels, showcasing a balanced impact of α.

The $K1$ and $K2$ parameters show more uniform performance at moderate levels, especially when compared to the variations in α. Properly setting α often eliminates the need to adjust $K1$ and $K2$ for achieving optimal flocking. $K1$ mainly influences the swarm's speed, helping it approach the maximum linear speed without notably affecting coherence or order. Conversely, $K2$ has a noticeable adverse effect on order, as seen in the 5 and 1000 agent experiments. Low $K2$ values delay achieving order and optimal CoM speed, while high values can disrupt order. Based on these insights, we opted to continue with moderate parameter values for physics based simulation and drone experiments.

3.2 Physics Based Simulations

Simulations with dynamics are conducted using a physics-based simulator that features a realistic dynamic and sensory model of the Crazyflie nano aerial vehicle [17]. Chosen for its alignment with the Crazyflie platform, this simulator allows drones to freely navigate in 3D space without encountering collidable objects. Physics simulations run at 240 Hz, with the Crazyflie's PID controller at 48 Hz, and our flocking algorithm at 20 Hz. Due to the increased computational demands of dynamic simulations, only moderate level parameters are tested, maintaining consistency with kinematic simulations. The study includes setups of 5 and 50 drones, each repeated 20 times.

Drone positions are updated by the simulator following control and physics calculations for the given velocity commands (v_x, v_y, and v_z). The heading directions of the agents are not directly reflected in the drone orientation, as quadrotor aerodynamics do not permit such calculations. Instead, the heading direction components h_x, h_y, and h_z are maintained and updated as internal variables, used solely in the calculation of velocity commands. The simulator lacks a model for a sensor capable of performing relative localization. Thus, we simulate the peer sensing concept in a manner transparent to the proposed algorithm: Drones are still able to sense peers within a specific sensing distance and, if required by the setting, obtain heading direction information.

For each experiment, drones are initialized in the same manner as with kinematic agent simulations. The experiment runs for 300 simulated seconds. Drone states are recorded as previously described, and swarm order, CoM speed, and distance distribution heat maps are generated from the collected data. Snapshots from physics based and kinematic simulations can be seen here[2].

In Fig. 5, we present flocking performance using the same metrics as in the kinematic simulator, aggregating data from 20 runs. Results show high consistency. With moderate parameters, order swiftly maximizes and stabilizes, while CoM speed reaches about half its maximum, then levels off. For both 5 and 50 agents, trends are similar. The 5-agent distance distribution is uniformly spherical; for 50 agents, it spreads wider around the CoM but remains non-spherical. Despite distribution differences, they minimally impact order or CoM speed.

Results from Fig. 5 confirm that outcomes from the kinematic simulator can be effectively replicated in a physics-based simulator. Being able to do so proves our method's robustness to real-world disturbances and its feasibility under physical and dynamic constraints. The Crazyflie nano drones successfully interact within predetermined speed and distance limits. Achieving this with up to 50 drones validates our choice of moderate parameter settings.

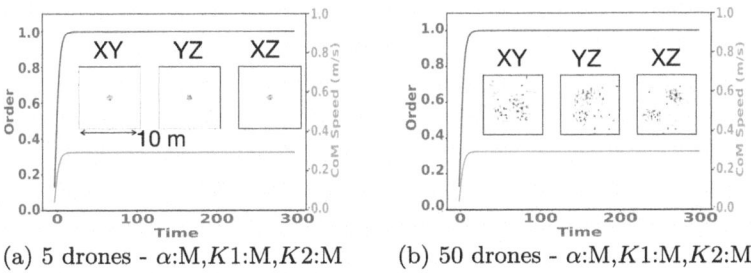

Fig. 5. Metrics for physics-based simulations with (a) 5 drones and (b) 50 drones.

---

[2] https://bit.ly/3dflocking_snapshots.

3.3 Autonomous Drone Experiments

The final phase of the experiments employs the Crazyflie nano drone. Chosen for its compact size (9.2 cm motor-to-motor distance) and suitability for indoor flights, this platform serves as a test bed to assess the proposed method's applicability, taking into account physical limitations, inaccuracies, and delays.

The proposed algorithm is implemented directly on the Crazyflie using the *App-Layer* functionality [1], allowing for autonomous operation without a central computer. Due to the absence of a fully developed solution for accurate on-board peer localization, we establish a communication network between drones via a radio channel to share their positions and, if necessary, heading direction components. This information is utilized according to the proposed methodology to calculate relative positions and, if needed, the average heading direction. For agent communication, we utilize the *Peer to Peer API* [2]. On-board localization is facilitated by the Lighthouse indoor positioning system, which employs specialized receivers on the drones. A central computer logs data by acting as a listener on the common radio channel.

The flight arena measures 6.5 m (x), 4.5 m (y), and 1.5 m (z). With the boundaries being introduced for the first time, we use the boundary repulsion vector from Sect. 2, with σ_b set to 0.05, causing repulsion for agent-boundary distances under 0.07 m. The corresponding σ_b value is chosen to be 0.05, ensuring repulsion for distances between an agent and the boundary closer than 0.07 m. The controller detects boundaries only within 0.5 m. Drone experiments consist of a total of six flights, each lasting 270 s. At the start of each experiment, drones are randomly positioned within a neighborhood. After takeoff, each drone calculates random heading direction components. In Fig. 6, we display flocking metrics from a single run of swarm drone flight experiment due to the unique challenge of boundary interactions in our flight arena. Attempting to synchronize the drones' boundary bounces proved infeasible, and aggregating data from multiple flights would obscure individual experiment dynamics. For comprehensive details and plots across all runs, please refer to our supplementary repository [3].

When observing the swarm order values in Fig. 6(a), it is evident that the swarm can achieve high order levels. However, interactions with a boundary trigger repulsion in the opposite direction, necessitating a collective turn, even

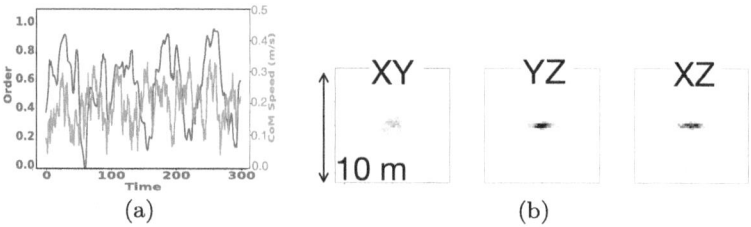

Fig. 6. Autonomous flight metrics for 5 drones. (a) Order and CoM speed, (b) Distance distribution. Shows recovery after boundary interactions.

for those that have not yet interacted with the boundary. This phenomenon is marked by the lows in the order. Nonetheless, these lows are followed by peaks in order, demonstrating the swarm's ability to recover its order after significant disturbances. The CoM speed values in (a) exhibit a pattern similar to the swarm order, with peaks and lows aligning in timing. This correlation is expected: when the swarm encounters a collective disturbance, it temporarily loses order, affecting the CoM speed as well, as discussed previously. However, in the absence of such disturbances, both the order and the CoM speed recover each time.

Figure 6(b) shows the distance distribution of 5 drones around the CoM, indicating cohesion and uniformity. While the distribution appears spherical in the XY plane, it is elliptically compressed in the YZ and XZ planes. This validates our boundary avoidance method, reflecting the arena's wider X and Y dimensions compared to the shorter Z axis. The compression along the Z axis limits the space for a spherical formation, resulting in an elongated appearance from other perspectives.

4 Conclusions

This work introduces a novel method for achieving three-dimensional, self-organized flocking using the active elastic model. Through kinematic and dynamic simulations, as well as real-world experiments with Crazyflie nano drones, we demonstrated the effectiveness and robustness of our approach. Key findings highlight the critical role of parameter selection, particularly the proximal control weight α, in optimizing swarm behavior for cohesion and flexibility across different swarm sizes. Moderate α levels generally yield the best results. However, simulations show that smaller swarms benefit from stronger proximal control, while larger swarms need weaker proximal control for optimal performance. This nuanced relationship between parameter settings and swarm size, not explored in previous 2D active-elastic model studies [9,11], is due to the greater agent capacity within a 3D sensing range sphere compared to a 2D circle.

Implementing this method on Crazyflie drones, despite limited onboard computation and the need for autonomous operation without a central computer, has proven the model's viability in real-world confined environments. This highlights its potential for various UAV applications. Moreover, the ability to navigate in three-dimensional space fully unlocks the potential of flying robot swarms for tasks such as surveillance, search and rescue, and other emergent capabilities like three-dimensional source localization with limited sensing [15]. Overall, this study paves the way to achieving autonomous 3D flocking in robotic swarms, setting the stage for future advancements in swarm robotics.

References

1. Bitcraze - crazyflie firmware - app layer documentation (2024). https://www.bitcraze.io/documentation/repository/crazyflie-firmware/master/userguides/app_layer/
2. Bitcraze - crazyflie firmware - p2p documentation (2024). https://www.bitcraze.io/documentation/repository/crazyflie-firmware/master/functional-areas/p2p_api/
3. Supplementary repository (2024). https://github.com/tugayalperen/3D_AE
4. Albani, D., Manoni, T., Saska, M., Ferrante, E.: Distributed three dimensional flocking of autonomous drones. In: 2022 International Conference on Robotics and Automation (ICRA), pp. 6904–6911. IEEE (2022)
5. Beaver, L.E., Malikopoulos, A.A.: An overview on optimal flocking. Annu. Rev. Control. **51**, 88–99 (2021)
6. Berlinger, F., Gauci, M., Nagpal, R.: Implicit coordination for 3D underwater collective behaviors in a fish-inspired robot swarm. Sci. Robot. **6**(50), eabd8668 (2021)
7. Camperi, M., Cavagna, A., Giardina, I., Parisi, G., Silvestri, E.: Spatially balanced topological interaction grants optimal cohesion in flocking models. Interface Focus **2**(6), 715–725 (2012)
8. Couzin, I.D., Krause, J., Franks, N.R., Levin, S.A.: Effective leadership and decision-making in animal groups on the move. Nature **433**(7025), 513–516 (2005)
9. Ferrante, E., Turgut, A.E., Huepe, C., Stranieri, A., Pinciroli, C., Dorigo, M.: Self-organized flocking with a mobile robot swarm: a novel motion control method. Adapt. Behav. **20**(6), 460–477 (2012)
10. Ferrante, E., Turgut, A.E., Stranieri, A., Pinciroli, C., Birattari, M., Dorigo, M.: A self-adaptive communication strategy for flocking in stationary and non-stationary environments. Nat. Comput. **13**(2), 225–245 (2014)
11. Ferrante, E., Turgut, A.E., Dorigo, M., Huepe, C.: Elasticity-based mechanism for the collective motion of self-propelled particles with springlike interactions: a model system for natural and artificial swarms. Phys. Rev. Lett. **111**(26), 268302 (2013)
12. Horyna, J., et al.: Decentralized swarms of unmanned aerial vehicles for search and rescue operations without explicit communication. Auton. Robot. **47**(1), 77–93 (2023)
13. Jafari, M., Xu, H., Carrillo, L.R.G.: A biologically-inspired reinforcement learning based intelligent distributed flocking control for multi-agent systems in presence of uncertain system and dynamic environment. IFAC J. Syst. Control **13**, 100096 (2020). https://doi.org/10.1016/j.ifacsc.2020.100096
14. Karagüzel, T.A., Retamal, V., Ferrante, E.: Onboard controller design for Nano UAV swarm in operator-guided collective behaviors. In: 2023 IEEE International Conference on Robotics and Automation (ICRA), pp. 3268–3274. IEEE (2023)
15. Karagüzel, T.A., Turgut, A.E., Eiben, A., Ferrante, E.: Collective gradient perception with a flying robot swarm. Swarm Intell. **17**(1), 117–146 (2023)
16. Olson, R.S., Hintze, A., Dyer, F.C., Knoester, D.B., Adami, C.: Predator confusion is sufficient to evolve swarming behaviour. J. R. Soc. Interface **10**(85), 20130305 (2013)
17. Panerati, J., Zheng, H., Zhou, S., Xu, J., Prorok, A., Schoellig, A.P.: Learning to fly-a gym environment with pybullet physics for reinforcement learning of multi-agent quadcopter control. In: 2021 IEEE/RSJ International Conference on Intelligent Robots and Systems (IROS), pp. 7512–7519 (2021). https://doi.org/10.1109/IROS51168.2021.9635857

18. Reynolds, C.W.: Flocks, herds and schools: a distributed behavioral model. In: Proceedings of the 14th Annual Conference on Computer Graphics and Interactive Techniques, pp. 25–34 (1987)
19. Şahin, E.: Swarm robotics: from sources of inspiration to domains of application. In: Şahin, E., Spears, W.M. (eds.) SR 2004. LNCS, vol. 3342, pp. 10–20. Springer, Heidelberg (2005). https://doi.org/10.1007/978-3-540-30552-1_2
20. Turgut, A.E., Çelikkanat, H., Gökçe, F., Şahin, E.: Self-organized flocking in mobile robot swarms. Swarm Intell. **2**(2–4), 97–120 (2008)
21. Vicsek, T., Zafeiris, A.: Collective motion. Phys. Rep. **517**(3–4), 71–140 (2012)
22. Vásárhelyi, G., Virágh, C., Somorjai, G., Nepusz, T., Eiben, A.E., Vicsek, T.: Optimized flocking of autonomous drones in confined environments. Sci. Robot. **3**(20) (2018)

Swarm-Inspired Controller: An Inference-Free Approach to Distributed Manipulation

Nicolas Bessone[(✉)], Kasper Stoy, and Payam Zahadat

Department of Computer Science, IT University of Copenhagen,
Copenhagen, Denmark
{nbes,ksty,paza}@itu.dk

Abstract. This paper proposes an approach to a self-organizing decentralized controller for modular robotic manipulation surfaces. Distributed manipulator surfaces use a grid of independently controlled actuators to induce motion on objects through multiple contact points. These systems are primarily designed to achieve precise positioning and orientation of objects. The proposed approach is introduced in the context of a simulated virtual environment featuring a generic distributed planar manipulator. Due to its high level of abstraction, this approach is well-suited for use with different actuation principles and sensing devices, concentrating its focus on the information and module capabilities required to perform the task. A series of experiments were designed to assess the system's self-organization capabilities for objects of different shapes alongside assessing its performance and fault tolerance. We compared the results of the proposed inference-free method against a technique that relies on information about global object properties, where this information is provided by an external source or inferred by the actuators, finding comparable outcomes. The experiments provide evidence of the effectiveness of the developed control rules, thereby highlighting the approach's potential as a viable solution for achieving efficient and resilient decentralized control for robotic manipulation surfaces.

1 Introduction and Related Works

A distributed manipulation surface (DMS) consists of multiple actuators distributed across a surface, along with a control strategy for coordinating these actuators to manipulate objects positioned on it. A common use of DMS is to carry objects to precise placements, regardless of their initial configurations. Figure 1 shows an example DMS as a grid of nine actuators, each able to exert a planar force on the object resting on its top. In the past years, DMS has attracted considerable attention for single-part manipulation, which entails positioning and orienting a single object with precision [12].

There are various forms of DMS, and their actuation mechanisms have been extensively studied. These include delta-manipulator arrays [17], wheeled conveying system [19], planar airflow and electrostatic fields [13] and single degree

Fig. 1. Example of a Wheeled Distributed Manipulator.

of freedom actuators [3]. The trends in technological development in manipulations have historically moved towards decentralized manipulators with sensors [2]. Distributed manipulators that utilize sensor information for feedback loop strategies have been proven to be advantageous, allowing stable orientations of the object to be achieved [12].

Regardless of their manipulation approach, all these systems are governed by a control system that aims to coordinate the individual movements of their components to achieve the desired planar manipulation of objects, often involving linear translation and rotation. Different methodologies have been employed to attain this behavior, such as selective braking of actuators in contact with a moving object which relies on dynamics-dependent models [4], algorithmic models with velocity fields that ensure object positioning [11,12], and distributed behavior-based control [19]. Many manipulation control systems rely on generating static vector fields, a method often constrained by assumptions such as mirror symmetry in objects [2] or simplicity in object's shapes [13]. Additionally, these systems typically adopt centralized control architectures, assigning independent movement vectors to each actuator [19]. However, this centralized approach limits scalability, as a single computer may struggle to manage numerous actuators [2], and reduces robustness due to heavy reliance on a singular control unit.

Given the distributed complexity of such systems, designing control policies proves challenging. Even robust approaches often fail to generalize to objects of varying shapes and sizes [17]. Early work [8] has demonstrated the feasibility of a shape-agnostic, self-organizing distributed manipulator controller, assuming agents possess information about global object properties like center and orientation. However, current methods to meet these assumptions often rely on centralized feedback mechanisms such as overhead cameras, introducing a centralization point that compromises system resilience, since a failure in this centralized feedback mechanism can lead to complete system breakdown [14,15].

An alternative to externally providing feedback information is to endow the multi-agent system with the ability to infer these global properties of interest through local communication based on available information. Previous approaches for acquiring this information can be found in [5,16] in which the authors describe a methodology for computing the center and orientation of simple shapes and static objects. In [9] we demonstrated that while the system exhibits remarkable precision in inferring properties of objects that were part of the training set, its precision diminishes for objects with geometries differing from those in the training set.

To address these limitations, this paper proposes a shape-agnostic inference-free approach as a decentralized control system for distributed manipulation based on self-organizing systems with emergent behavior. The system allows for local communication among neighboring actuators, enabling the generation of a global attraction force toward a target position. Additionally, the system relies on local excitation signals triggered by the presence of the object, thereby coordinating the actuators mediated by the object's presence, to rotate it until it aligns with the target. By eliminating the need for both a centralized control architecture and inference capabilities of the agents, the proposed approach could lead to more adaptable, and robust systems. In this paper, we will contrast our proposed approach with the method outlined in [8], which necessitates agents to know the object's global properties, i.e., its geometric center and orientation. This will serve as our baseline for performance comparison.

The proposed approach referred to as the Swarm-Inspired Controller draws on models of computation such as cellular automata [18] and amorphous computing [1], as well as models of reaction-diffusion systems [10], and chemo-taxis [6]. The central idea is that each individual captures relevant information from the environment, makes decisions based on predefined behavior rules, and adjusts its state (i.e., applied force vector) accordingly, thereby contributing to the overall collective behavior of the system. The information exchange occurs through both direct communication with neighbors and indirect interaction induced by the object. Despite the simplicity of behavior rules, the multiple interactions between the agents produce self-organization and enhance the system's capabilities, resulting in complexities beyond those achievable by a simple actuator.

2 Method

To assess the proposed methodology, the task selected for examination is the placement of an object, consisting of its positioning and orientation. This task has been chosen due to its prominence as a primary function of distributed manipulators. Each actuator within the multi-agent system, referred to as a tile, possesses the capability to induce a planar force on the object situated on top of it. This force has both magnitude and direction and therefore is referred to as a vector actuation. The combination of individual actuation vectors results in the creation of a force field or vector field within the system, which will guide the object toward a desired target placement.

The proposed approach centers around the notion that each agent within the system will contribute to both disseminating relevant information and contributing to a force field. The tiles will use their knowledge of the local neighborhood to make decisions about how to contribute to the force field, while also sharing information through local communication within their immediate vicinity.

To validate and evaluate this approach, a high-level abstraction model was used, which enabled independence from specific manipulation principles and instead focused on the control rules, the necessary information for goal attainment, the individual agent capabilities, and the emergent system behavior.

2.1 The Environment[1]

The proposed methodology will be evaluated within a two-dimensional simulated model based on previously developed modular manipulators created for experimental purposes [12,14,15] and used in the development and assessment of [8,9]. Two-dimensional simulations for grid computation, although limited in their faithfulness to the real world due to the reality gap, are useful for understanding the control strategies for manipulator surfaces and can help identify potential issues that may arise when dealing with physical systems. The high level of abstraction allows the independence of any particular actuation principle, and the easiness and cost-free scalability of a virtual model allows the exploration of the behavior of the systems with different numbers of individual populations [3].

A representation of the environment and its elements is shown in Fig. 2. This system consists of a square grid of cellular automata-like units, referred to as *tiles*. An object is placed on top of the tiles, and its geometric center and orientation can be determined. The objective of the system is to move the object to a given final position and orientation. From this point forward, this final placement will be referred to as the target.

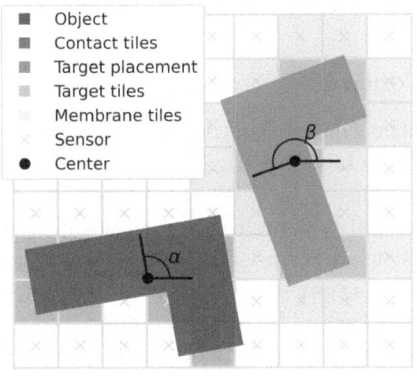

Fig. 2. Environment.

---

[1] Based on [8] and Included for Completeness.

Target. Recalling that the goal of the system is to carry an object to a target placement, i.e., position and orientation, we define a target region as the set of tiles coming into contact with the object when it reaches its target placement. The tiles in the target region are called *target tiles* and those adjacent to the target tiles are called *membrane tiles*.

Neighborhood. Each tile in the system is able to communicate with those in its immediate neighborhood. Here, the immediate neighbors of a tile are the four adjacent ones at its up, down, right, and left, i.e., Manhattan distance of one.

2.2 Tiles

Here the term tile refers to a square-shaped individual agent organized in a grid. All tiles exhibit uniform physical capabilities and are governed by the same logic. Each tile is a single unit or module, and the aggregation of many tiles produces a multi-agent system capable of achieving the task of translating and rotating an object to reach a desired placement. To exhibit such behavior, each tile is capable of interacting with the objects on top and with its neighboring tiles.

Sensing. Each tile is equipped with a binary sensor positioned at its center, enabling it to detect the presence of an object when it comes into contact with the sensor. A tile with its sensor detecting an object is referred to as a *contact tile*.

Actuation. When an object is detected by a tile, the tile can induce a planar linear force on the object. The force is assumed to be constant and exerted from the center of the tile. If multiple tiles are in contact with an object, the object's movement is the result of all the forces induced on various points.

2.3 Behavior and Control

State Variables. All the tiles in this work are assumed to be homogeneous in both capabilities and behavior logic. The logic is a set of behavior rules that act upon state variables to update the tile's actuation vector. The state variables, summarized in Table 1, include information about the tile's status in the environment and its interactions. Their values are determined by different mechanisms, e.g., the `is target` status of a tile is task-specific and constant during the execution of the task. The `neighborhood` and `neighbors' relative position` information are static and dependent on the manipulator system. The `is contact` status is the result of the interaction between a tile and the object, `object center` is externally provided or inferred, and the `vector actuation` is computed based on the information diffusion from neighbors and behavior rules. In this work we compare a proposed approach against the *Information Diffusion* approach presented in [8] and therefore Table 1 shows the state variables used in both approaches.

Table 1. State variables of a tile in both approaches

State variables	Information Diffusion	Swarm-Inspired
`neighborhood`	system-specific	system-specific
`neighbors' rel. pos.`	system-specific	system-specific
`is target`	task-specific	task-specific
`is contact`	object-tile interaction	object-tile interaction
`object center`	provided	-
`object angle` (α)	provided	-
`target center`	provided	-
`target angle` (β)*	diffusion and behavior rules	-
`vector translation` (v_t)	diffusion and behavior rules	-
`vector rotation` (v_r)	diffusion and behavior rules	-
`vector actuation` (v)	diffusion and behavior rules	diffusion and behavior rules
`excitation signal`**	-	behavior rules

*Provided if tile is set as target.
**Used only in membrane tiles.

Diffusion. Some state variables are diffusible. For such variables, the information diffusion is done by each agent averaging the values of the corresponding variables of their neighbors.

Control Loop. In each step of the simulation, all the tiles within the system execute their control loop in an asynchronous arbitrary order. The control loop of a tile consists of updating state variables through information diffusion and interaction with the object (sensing the object), collecting other information from neighbors if needed, and then applying the behavior rules. Finally, the value of the `vector actuation` variable of the tile is utilized to determine the direction in which the actuator will induce a force to the object on top of it.

2.4 Behavior Rules of the Proposed Method

The proposed method relies on generating a force field that attracts the object toward the target and rotates it to fit into the target region. The force field is the result of the tiles' actuation vectors, their directions are determined by the behavior rules, and their magnitude is one unit. The behavior rules depend on the position of the tile with respect to the target: if the tile is a target tile, a membrane tile, or an ordinary tile (neither of the previous two types of tiles). Target tiles are always dormant and do not perform any actuation or diffusion to their neighbors, i.e., the actuation vector, v, for all target tiles is zero: $v = 0$. Hence, if the object is completely within the target region, it does not move anymore.

The membrane tiles play a crucial role in directing the force field, which is propagated across the environment by the ordinary tiles. When an object is not in contact with the membrane tiles, these will adjust their actuation vectors to ensure that the force field attracts the object toward the general direction of the target. When the object comes into contact with the membrane, the attraction might be accompanied by rotation. This rotational movement can be an emergent effect caused by the attraction of opposite sides of the object in opposite directions toward the target region.

The ordinary tiles propagate the vector field via an information diffusion process, as in [8], by setting their actuation vectors to the average of their neighbors, $v = 1/|N| \sum_{v_i \in N} v_i$, where N is the set of neighbors and $|N|$ is the set's size. Although this propagation may be sufficient to position the object in the target, sometimes, depending on the initial position of the object and target, and the geometry of the object, the generated force field reaches a state of equilibrium in which the object, despite not being in its desired target position, remains static due to the cancellation of the forces acting over it. The rotational component of the proposed approach acts by generating a disruption in the force field, thus forcing the object to escape the state of equilibrium.

The behavior of a membrane tile depends on whether it is excited or not. When unexcited, the tile's actuation vector aligns with a unit vector, v_t, pointing toward the centroid of its adjacent target tiles, generating attraction. The vector is calculated using the neighbors' relative positions. When excited, the membrane tile explicitly aims to induce a counter-clockwise rotation around the target, based on the direction of v_t:

$$v = \begin{cases} \nwarrow & \text{if } v_t \in \{\nwarrow, \leftarrow, \swarrow\} \\ \swarrow & \text{if } v_t = \downarrow \\ \searrow & \text{if } v_t \in \{\nearrow, \rightarrow, \searrow\} \\ \nearrow & \text{if } v_t = \uparrow \end{cases} \quad (1)$$

How effective the actuation is in inducing the rotation depends on the number of excited membrane tiles and factors like the object's shape.

Initially, all the membrane tiles are unexcited. For a membrane tile to get excited, it must be in contact with the object for extended periods of time. For that, each membrane tile utilizes an excitation signal, s, which increases in the event of contact with the object, and constantly decays over time. The signal's dynamics can be represented by the following exponential moving average:

$$s(t+1) = \lambda \cdot s(t) + (1 - \lambda) \cdot I(t) \quad (2)$$

where I denotes the binary status of `is contact` variable, λ is a multiplier for the decay, and t is the simulation step. To determine if a membrane tile is excited or not, its s value is compared to a threshold τ: if $s < \tau$, the tile remains unexcited; if $s \geq \tau$, the tile is excited.

In this work, the values of λ and τ were determined through a coarse-grained grid search. Specifically, we conducted a soft optimization across 10 equally spaced values ranging from 0 to 1 for a fixed set of 20 randomly generated manipulation tasks. The loss was computed as the mean of the inverse final coverage (see performance metric at Subsect. 3.1) across the tasks. The results of the grid search are shown in Fig. 3. Despite the minimum loss being found at $\lambda = 0.33$ and $\tau = 0.89$, the adopted values were $\lambda = 0.5$ and $\tau = 0.16$, being an approximate centroid of a more robust local minimum indicated by the larger dark region.

Figure 4 and 5 show the rotation behavior's influence. Figure 4 highlights the absence of symmetry breaking in the force field, resulting in the object reaching a mispositioned equilibrium point. Notably, an early disruption of the force field by excitement-induced rotation in the membrane tiles (Fig. 5, Step 15) prevents the object from settling into such an equilibrium and enables it to reach the target. Implementation of the approach and reproduction kit is available at [7].

3 Experiments

A series of experiments were carried out to validate and study the potential and limitations of the proposed decentralized controller. The experiments assessed the controller's adaptability and robustness under failure conditions, ensuring a methodological evaluation of its performance.

We will compare the performance of the presented approach with the *Information Diffusion* approach [8]. To work with standardized object shapes, this paper will use Tetrominoes - geometric shapes composed of four orthogonally connected squares - widely recognized from the game of Tetris (Fig. 6).

3.1 Performance Comparison

This experiment aims to quantify the performance of the proposed approach across various metrics of interest. It involved positioning an object within the system and tracking its orientation and position relative to a predefined target. The system comprises a 20×20 grid of tiles, with the initial position, angle, and shape of the object being randomly selected. Conversely, the target position is consistently set at the center of the grid, with its angle initialized randomly. A total of 100 manipulation tasks were executed for 1000 asynchronous update

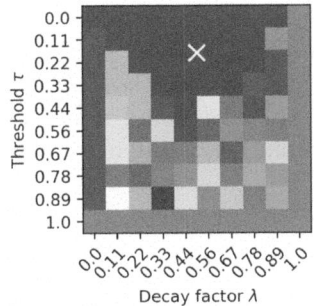

Fig. 3. Grid Search Results: darker colors indicate lower loss values. (Color figure online)

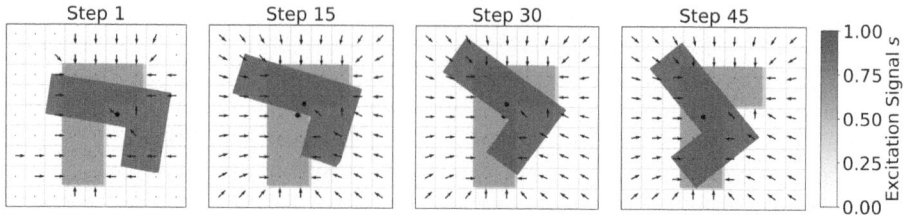

Fig. 4. Behavior without explicit rotation: the object arrived at a mispositioned location due to a state of stable equilibrium within the force field.

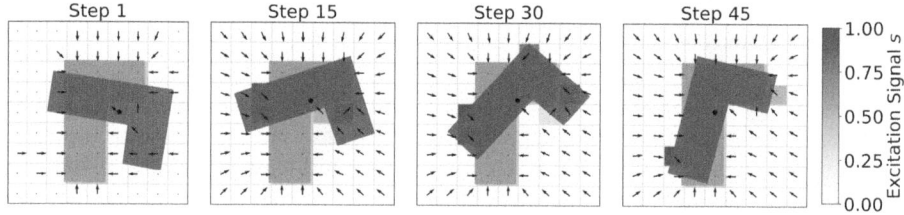

Fig. 5. Behavior with explicit (excitement-induced) rotation: this added force generates a rupture in the symmetry of the force field, preventing the object from falling into such an equilibrium state.

steps. The experiment is replicated for both the proposed approach and the *Information Diffusion* method, facilitating a comparative analysis.

We will assess the performance of both behaviors across four key metrics. The **Position Error** measures the accuracy of the object's final position relative to the length of a tile side, thus position error is in the scale of tile sizes. The **Angle Error** quantifies the deviation in object orientation after the manipulation task, extracted from the last update step. The **Coverage** indicates the proportion of target tiles the object contacts at the end of the task. Finally, the **Operation Period** is represented by the number of update steps required for the system to complete the task.

There are two significant aspects of the Swarm-Inspired Controller behavior that deserve further discussion. Firstly, the assumption made in [8] asserts perfect inference regarding global properties, assuming that all agents possess accurate information about the center and orientation of objects and targets (Table 1). An approach to fulfill these inference capabilities was presented in

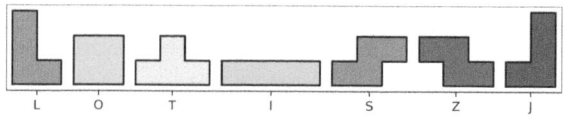

Fig. 6. Tetrominoes.

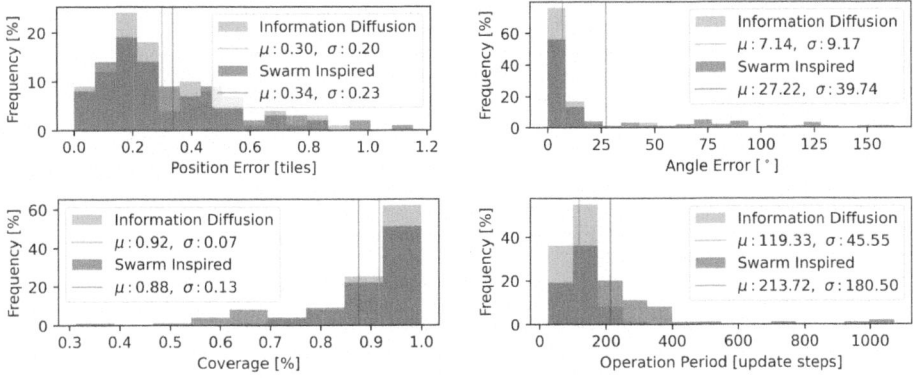

Fig. 7. Performance Metrics.

[9], in which the center estimation was, in the first place, imperfect, and in the second place, presented an inconsistent accuracy on object shapes outside the training set. In contrast, the Swarm-Inspired Controller is both inference-free and shape-agnostic, representing a considerable simplification over prior methods. Secondly, regarding the angle error metric, agents in the proposed controller lack knowledge about object or target orientations. Alignment arises from the explained behavioral rules, primarily influenced by `membrane tiles` and `contact tiles`. Consequently, symmetrical shapes, like $'O'$ (in Fig. 6), yield identical sets of `contact tiles` across all rotational symmetries, such as every 90° for the 'O' shape. In such scenarios, the proposed approach remains agnostic to rotational symmetries. Thus, for applications where rotational symmetries are inconsequential, the percentage coverage serves as a more suitable metric for performance evaluation. The induced rotation is consistently counterclockwise, regardless of the relative angle between the object and the target. Even when the object and target orientations are nearby, this can result in the object rotating until it reaches the next symmetrical position. This behavior may account for the longer operation period observed with the proposed approach. Figure 7 illustrates the distribution of performance metrics from the experimental results. Notably, despite not requiring information about the object position or orientation, the proposed method achieves competitive performance across all metrics when compared to the previous method.

3.2 Fault Tolerance Comparison

One central aspect of a multi-agent system lies in its ability to withstand faults, allowing it to operate effectively even when certain agents are non-operational. This experiment seeks to delve into the fault resistance capabilities of the distributed controllers.

The task remains unchanged from the previous experiment: moving an arbitrary object from its arbitrary initial position to a final one. However, an addi-

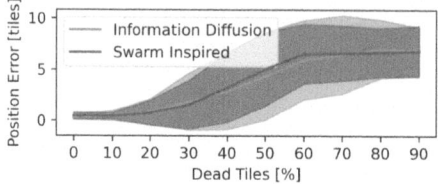 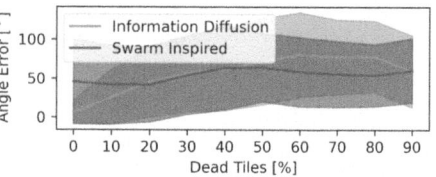

Fig. 8. Fault Tolerance.

tional variable has been introduced: the percentage of faulty tiles. In each trial, a portion of tiles is randomly selected from the system, following a uniform distribution, and designated as faulty. This emulates a common scenario of agent failures. The target tiles are exempt from potential faults to ensure a valid end condition and maintain the experiment's integrity. This precaution safeguards against scenarios where no target exists on the board. A faulty tile is functionally inert, exerting no influence on the system's dynamics and remaining undetected by neighboring tiles, effectively rendering it an absent element within the system.

The experiment was conducted 100 times per percentage of faulty tiles. In Fig. 8, the mean and standard deviation of both distance and angle errors are depicted. It can be noted that although the position error resembles that of the *Information Diffusion* approach, with a slight reduction in the deviation, the angle error of the proposed method demonstrates lower variability across varying percentages of faulty tiles. This observation shows that the Swarm-Inspired approach has a slight robustness advantage over its counterpart.

4 Conclusions

This paper presents a straightforward decentralized approach for managing distributed manipulator surfaces, offering an alternative to established centralized architectures lacking agent interaction. The method showcases self-organization and adaptability traits during planar manipulation tasks, accommodating objects of varying shapes. Moreover, it demonstrates a potential for failure tolerance. The proposed approach maintained a competitive performance level when compared against another controller of a similar nature. Despite not revealing significant superiority in capabilities, its notable advantage lies in its simplicity and independence from complex inference requirements regarding feedback information. Overall, the proposed approach presents a rational and promising methodology for governing the behavior of modular manipulation surfaces. It holds the potential to offer valuable insights for the ongoing development and optimization of such systems.

Acknowledgements. Funded by the European Union. Views and opinions expressed are however those of the author(s) only and do not necessarily reflect those of the European Union or the European Commission. Neither the European Union nor the granting authority can be held responsible for them.

References

1. Abelson, H., et al.: Amorphous computing. Commun. ACM **43**(5), 74–82 (2000). https://doi.org/10.1145/332833.332842
2. Agarwal, P.K., Kavraki, L.E., Mason, M.T.: Velocity field design for the modular distributed manipulator system (MDMS). In: Robotics: The Algorithmic Perspective, pp. 45–58. A K Peters/CRC Press (1998)
3. Barr, D.R., Walsh, D., Dudek, P.: A smart surface simulation environment. In: 2013 IEEE International Conference on Systems, Man, and Cybernetics, pp. 4456–4461. IEEE, Manchester (2013).https://doi.org/10.1109/SMC.2013.758, http://ieeexplore.ieee.org/document/6722513/
4. Bedillion, M., Hoover, R., McGough, J.: A distributed manipulation concept using selective braking. In: 2014 American Control Conference, pp. 3322–3328. IEEE, Portland, OR, USA (2014). https://doi.org/10.1109/ACC.2014.6858624, http://ieeexplore.ieee.org/document/6858624/
5. Bedillion, M.D., Parajuli, D., Hoover, R.C.: Distributed sensing in actuator array manipulation. In: Volume 4A: Dynamics, Vibration and Control, p. V04AT04A002. American Society of Mechanical Engineers, San Diego, California, USA (2013). https://doi.org/10.1115/IMECE2013-62499
6. Berg, H.C., Brown, D.A.: Chemotaxis in Escherichia coli analysed by three-dimensional tracking. Nature **239**(5374), 500–504 (1972). https://doi.org/10.1038/239500a0
7. Bessone, N.: nhbess/Swarm-Inspired-Controller-An-Inference-Free-Approach-to-Distributed-Manipulation (2024). https://github.com/nhbess/Swarm-Inspired-Controller-An-Inference-Free-Approach-to-Distributed-Manipulation
8. Bessone, N., Zahadat, P., Stoy, K.: Decentralized control of distributed manipulators: an information diffusion approach. In: Proceedings of the 23rd International Conference on Autonomous Agents and Multiagent Systems (AAMAS 2024), Auckland, New Zealand, May 6–10, 2024, IFAAMAS, p. 3 (2024). (In press)
9. Bessone, N., Zahadat, P., Stoy, K.: Neural cellular automaton for decentralized inference in distributed manipulation systems. In: Proceedings of the 22nd International Conference on Practical applications of Agents and Multi-Agent Systems (PAAMS 2024), Salamanca, Spain, June 26-28, 2024 (2024). (In press)
10. Coore, D., Nagpal, R.: Implementing reaction-diffusion on an amorphous computer. In: Proceedings of 1998 MIT Student Workshop on High-Performance Computing in Science and Engineering, pp. 189–192. Citeseer (1998)
11. Luntz, J., Messner, W.C., Choset, H.: Velocity field design for the modular distributed manipulator system (MDMS). In: In Proceedings of Workshop on Algorithmic Found. Robot., vol. 3, pp. 35–47 (1998)
12. Luntz, J.E., Messner, W., Choset, H.: Distributed manipulation using discrete actuator arrays. Int. J. Robot. Res. **20**(7), 553–583 (2001). https://doi.org/10.1177/02783640122067543
13. Moon, H., Luntz, J.: Distributed manipulation of flat objects with two airflow sinks. IEEE Trans. Rob. **22**(6), 1189–1201 (2006). https://doi.org/10.1109/tro.2006.882921
14. Murphey, T.D., Burdick, J.W.: Feedback control methods for distributed manipulation systems that involve mechanical contacts. Int. J. Robot. Res. **23**, 763–781 (2004). https://doi.org/10.1177/0278364904045480
15. Murphey, T.D., Burdick, J.W., Burdick, J.: The power dissipation method and kinematic reducibility of multiple-model robotic systems. IEEE Trans. Rob. **22**(4), 694–710 (2006). https://doi.org/10.1109/tro.2006.878971

16. Parajuli, D., Bedillion, M.D., Hoover, R.C.: Actuator array manipulation using low resolution local sensing. In: Volume 4A: Dynamics, Vibration, and Control, p. V04AT04A003. American Society of Mechanical Engineers, Montreal, Quebec, Canada (2014). https://doi.org/10.1115/IMECE2014-37147
17. Thompson, S., Mannam, P., Temel, Z., Kroemer, O.: Towards robust planar translations using delta-manipulator arrays. In: 2021 IEEE International Conference on Robotics and Automation (ICRA), pp. 6563–6569. IEEE, Xi'an, China (2021). https://doi.org/10.1109/ICRA48506.2021.9561003, https://ieeexplore.ieee.org/document/9561003/
18. Wolfram, S.: Cellular automata as models of complexity. Nature **311**(5985), 419–424 (1984). https://doi.org/10.1038/311419a0
19. Yaemglin, T., Charoenseang, S.: Distributive behavior-based control for a flexible conveying system. In: 2002 IEEE International Conference on Industrial Technology, 2002. IEEE ICIT 2002, vol. 1, pp. 24–29 (2002). https://doi.org/10.1109/ICIT.2002.1189853

Swarming Out of the Lab: Comparing Relative Localization Methods for Collective Behavior

Rafael Gomes Braga[1(✉)], Vivek Shankar Varadharajan[2], Giovanni Beltrame[2], and David St-Onge[1]

[1] Department of Mechanical Engineering, École de Technologie Supérieure, Montreal, Canada
`rafael.gomes-braga.1@ens.etsmtl.ca`
[2] Computer and Software Engineering Department, Polytechnique Montréal, Montreal, Canada

Abstract. Efficient collaboration within a robot swarm hinges on the precise localization of swarm members relative to their neighbors. However, in real-world scenarios, such as indoor GPS-denied environments, access to accurate global localization systems is typically limited, and relative localization poses challenges due to the absence of a global reference frame. This paper compares the localization accuracy of three methods: IR-based, visual-inertial, and ultra-wideband localization systems. We evaluate these systems to ascertain the relative localization accuracy of neighboring robots engaged in collective behaviors. We develop a simulation model for the three localization systems and conduct accuracy studies. Furthermore, we deploy two swarms, one consisting of five flying robots and one consisting of five ground robots performing three distinct behaviors to validate the simulation experiments. Through simulation and robot experiments, we present the characteristics of each system, including estimation accuracy, deployment cost, communication overhead, and behavior performance accuracy.

1 Introduction

The collective behaviors of social animals like insects and birds have inspired the design of many robot behaviors, such as flocking [40] inspired by birds, milling [2] inspired by fish, and foraging [23] inspired by insects. Some approaches deploy robots with collective behaviors to perform tasks like search-and-rescue [23] and exploration [45]. An essential requirement for members of a swarm to coordinate their movement and behaviors is to be aware of the relative positions of other swarm members. In nature, organisms like ants, bees, and birds coordinate their movements by directly sensing the positions of their peers through vision and hearing, or indirectly through cues like pheromone gradients. Drawing from these natural coordination mechanisms, researchers have developed both direct [7,22] and indirect techniques [25,41,43] for neighbor localization in robot swarms.

Direct methods involve using embedded sensors to detect neighbors and estimate their range and bearing (RAB). These methods do not require communication between swarm members or a common frame of reference; however, they may suffer from insufficient bearing resolution and sensing noise. In contrast, indirect methods employ techniques such as situated communication, where robots estimate their positions in a global reference frame and share this information along with peer-to-peer messages within the swarm. This approach can be highly effective if a common frame of reference is available, though it comes with the trade-off of increased communication load.

Several techniques have been proposed for estimating the RAB of neighboring robots, including Infrared transceivers [7,33], camera-based approaches that illuminate robots for identification [2], Ultra-wideband (UWB) transceivers [23], Global Positioning System (GPS) receivers [40], visual-inertial simultaneous localization and mapping (SLAM) [42,45], and motion capture systems [29]. Table 1 categorizes the methods available in the literature as direct or indirect techniques. In direct methods, IR-based [7,22] method uses several concentric IR transceivers to measure the range and bearing of neighboring robots. Kilobot [30], tiny tabletop robots, use eight IR-based transceivers to perform range-only localization while operating the robots on a reflective surface. Camera-based [2,31] localization methods use an IR transmitter or a visual marker and detect this marker or IR transmitter on a camera to determine the location of the neighboring robots. Direct UWB-based methods use the time-of-flight measurements from peer to peer to determine the location of the neighboring robots. Some methods have investigated Bluetooth RSSI measurements [9] to neighbor localization.

Indirect methods encompass various strategies for each robot to determine its own position and send the resulting estimate to its neighbors during communication. Approaches that use GPS [35,40] exchange the GPS coordinates among the robots and estimate their relative position. Motion capture systems have been widely used in a laboratory setting to develop many collective behaviors [29,36]. With centralized motion capture systems, a communication ground station is used to exchange the estimates with the robots and estimate neighbor positions. Some approaches use UWB anchors deployed in the operational space to act as static points with known locations [38]; robots then estimate their location with respect to these static anchors using swarm-compliant methods like Time Difference of Arrival (TDOA) [5]. Simultaneous Localization and Mapping (SLAM) has been used with both lidars [46] and visual [25] features to estimate a position for the robots and exchange these positions to compute the neighbor locations. Hybrid techniques [43] combine multiple modalities like UWB and visual-inertial odometry to estimate the robot position and its neighbors.

Direct and indirect camera-based techniques can only be used in well illuminated environments [12]. However, UWB-based techniques can be used in low illumination conditions in both indoor and outdoor settings [44]. The accuracy of the UWB system depends on the number of anchors available for localizing and the type of antenna deployed [20]. Visual methods heavily depend on illumination but also on the availability of features in the environment to track. Direct and

SLAM-based indirect methods are suitable for scenarios where a fixed infrastructure, like deploying cameras or anchors in the environment, cannot be performed. UWB-based direct localization [6,11], where peers use signal strength to localize neighbors, appears promising for infrastructure-less deployment. However, these methods struggle to scale to a large number of robots [18,32].

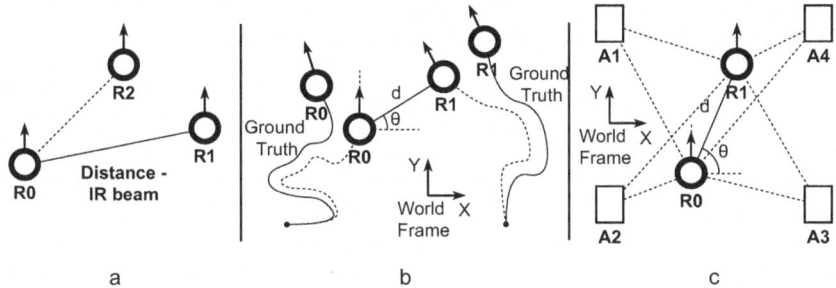

Fig. 1. Three categories of localization methods used in the study for estimating the neighbor positions: a. Direct Measurement: the robots use concentric IR-transceivers to directly measure the distance and bearing of neighbors. b. Indirect Visual-inertial SLAM based: each robot uses a visual SLAM approach onboard and relies on situated communication to localize neighboring robots. c. Indirect Infrastructure-based (UWB/GPS): robots utilize fixed anchors with known locations in the environment for localization and use situated communication.

The properties of the direct and indirect techniques, like behavior performance capability, ranging and bearing inaccuracies, and communication overheads, are not clearly known. In this work, we classify the types of RAB estimation techniques and examine their properties using three simulated sensor models (see Fig. 1) with large-scale simulation experiments. Then we deploy a small team of ten ground and flying robots to perform three different collective behaviors to further evaluate the properties of these techniques on physical systems.

Table 1. A comparison table highlighting the various localization methods for robots swarms in the literature.

RAB Technique		Domain		
		Ground	Air	Underwater
Direct	IR Based	[7,33], Mergeable nervous system [22]	[3]	*
	Camera Based	[21]	[31], FACT [19]	Blueswarm [2]
	UWB	[6,11,18], SGBA [23]	[10,17,24]	*
Indirect	GPS	[35]	[40]	*
	Motion Capture	[36]	[13], Crazyswarm [29]	*
	UWB	[37]	[16]	*
	Overhead camera	[15,28]	[42,45], SwarmLIO [46]	*
	and SLAM	SWARM SLAM [14]	VIOSwarm [25]	
	Hybrid	FG-RCL [41]	Omni-Swarm [43]	*

2 Methodology

2.1 System Model

In this work we focus on the study of 2D swarm behaviors. Even when deploying aerial swarms, we navigate them at fixed altitude, so all the motions and interactions happen in a plane. In this setting, for $i \in 1, ..., k$ where k is the number or robots in the swarm, the state of the robot i is given by:

$$\mathbf{X}_i = (\mathbf{x}_i, \theta) \qquad (1)$$

where $\mathbf{x}_i = (x_i, y_i)$ is the vector of Cartesian coordinates and θ is the orientation of the robot in the global frame. Each robot obtains the data relative to its own state from any one of the localization modalities used in this study - IR, visual-inertial odometry, UWB infrastructure, and motion capture.

For the robots in the swarm to be aware of their neighbors, we employ a form of situated communication by embedding the state information and broadcasting it through communication to all other robots in the swarm. Each robot gathers information, determines the distance and direction to nearby robots, and maintains a list of these neighbors. This list is updated whenever new information is received. The RAB computation is straightforward from the following equations:

$$r_i = \sqrt{(x_i - x)^2 + (y_i - y)^2}, \text{ and} \qquad (2)$$

$$\theta_i = arctan\left(\frac{(y_i - y)}{(x_i - x)}\right), \qquad (3)$$

where r_i and θ_i are the range and bearing to neighbor i with coordinates (x_i, y_i), and (x, y) are the coordinates of the receiving robot.

2.2 Swarm Behaviors

Many decentralized behavior (algorithms) have been developed for robot swarms, designed to produce emergent behaviors from local interactions between robots and their surroundings [34,39,40]. We selected three behaviors for our comparison study: the Lennard-Jones potential, a form of cyclic pursuit, and a third behavior ("fireworks") created from the composition of other common swarm primitives (aggregation, dispersion and state barrier) (Fig. 2).

Software Infrastructure. Implementing even the most basic swarm behaviors can pose challenges due to their decentralized nature. To address this, we utilize Buzz [26], a programming language designed to simplify the implementation of swarm behaviors. Buzz offers constructs that streamline communication, swarm management, and neighbor operations. It operates as a virtual machine on each robot, executing identical scripts, and is platform-agnostic, allowing for seamless deployment across various hardware configurations.

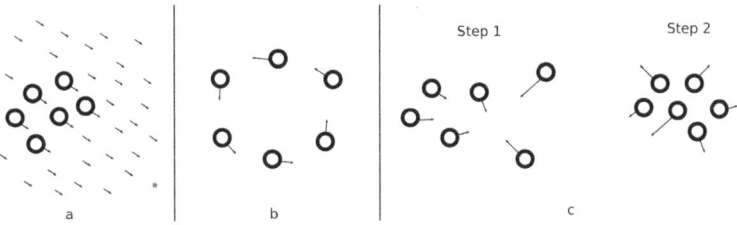

Fig. 2. Schematics of the three swarm behaviors selected for comparison: a. Lennard-Jones potential (attraction point as a red star), b. cyclic pursuit and c. Fireworks. (Color figure online)

Lennard-Jones Potential. A straightforward and efficient method for guiding robots in a swarm involves using potential functions. These functions generate attractive forces that pull robots toward a designated goal and repulsive forces that push them away from obstacles. By utilizing a potential function, each robot is drawn toward its neighbors, fostering group cohesion as they collectively navigate toward a destination point, similar to the fluid motion observed in large flocks of birds. The velocity vector ($\mathbf{u}$) of each robot is governed by the following relation:

$$\mathbf{u} = \sum_{i=1}^{k} f_i(d_i)(\mathbf{p}_i - \mathbf{p}) \qquad (4)$$

where k is the number of detected neighbors, d_i is the distance to neighbor i, $\mathbf{p}_i$ is neighbor i'S position vector and $f(d_i)$ is the potential function. A common choice for that function is the Lennard-Jones potential, given by:

$$f_{LJ} = \frac{\epsilon}{d}\left(\left(\frac{t}{d}\right)^4 - \left(\frac{t}{d}\right)^2\right) \qquad (5)$$

where d is the distance to the neighbor and ϵ is a control gain. This function provides an attractive part and a repulsive one (positive and negative terms in Eq. 5). These two factors cancel each other when the desired distance $d = t$ is reached, attracting the robots to each other but avoiding collisions. Equation 4 is not limited to neighboring swarm members, and can also integrate key points in the environment in its summation.

Cyclic Pursuit. Cyclic pursuit stands out as one of the most captivating swarm behaviors, demonstrating how organized motion can emerge from straightforward rules [34]. In this behavior, robots continuously advance while gently adjusting their course to the right if a neighboring robot is detected ahead, or to the left otherwise. By adhering to these two simple rules, the robots gradually converge into a smooth circling motion. Letting v represent the linear velocity and ω the angular velocity of the robot, it follows that v remains consistently positive, while ω will follow the rule:

$$\begin{cases} \omega > 0, \text{if neighbor is detected} \\ \omega < 0, \text{if neighbor is not detected} \end{cases} \quad (6)$$

Depending on the area to be covered and the desired size of the resulting circle, the left and right angular velocities may not be identical and can be fine-tuned through simulation.

Fireworks. Here, we merge two classic behaviors – aggregation and diffusion – to form a motion we term "Fireworks." In this motion, robots cyclically gather together before dispersing, akin to the particles in an explosion. This behavior serves as an effective stress test for localization systems, particularly against repetitive motion and robots operating in close proximity. The control vector is given by:

$$f = \begin{cases} \frac{1}{k} \sum_{i=1}^{k} \mathbf{p}_i, t < threshold \\ \frac{1}{k} \sum_{i=1}^{k} (\mathbf{p} - \mathbf{p}_i), t >= threshold \end{cases} \quad (7)$$

where t represents a cyclic timer and $threshold$ is the time-step when the behavior changes between aggregation (upper part) and diffusion (lower part).

3 Experimental Setup

We first compare relative localization strategies in a simulation setup detailed in this section and then extend to the physical setup. While simulations enable testing with a large number of robots, real robot experiments provide authentic noise on the network and measurements.

3.1 Simulations

We obtain theoretical results by developing three simulation models (see Fig. 1) for direct localization inspired by IR-based methods, visual-inertial SLAM, and UWB sensor models using the ARGoS3 [27] physics-based simulator. The IR-based sensor model assumes robots are equipped with IR transceivers, featuring eight sensors around each robot with a resolution of $\pi/4$. Range measurements include Gaussian noise with a zero mean and standard deviation σ_{IR}, while bearing measurements maintain a resolution of $\pi/4$.

The visual-inertial SLAM and UWB-based sensor models simulate indirect localization methods utilizing situated communication. In these models, robots exchange position information in a common frame of reference through communication and estimate neighbor positions accordingly. The visual-inertial SLAM model simulates a fixed drift accumulation of ρ with motion, where position estimates are updated every LO meters to simulate SLAM drift and loop closures. For example, a $\rho = 0.1$ signifies a drift of 0.1 m for every meter of motion. The UWB sensor model simulates a fixed jump in localization estimate at each time step using a zero-mean Gaussian distribution. These simulated sensor models closely mirror behaviors observed in actual hardware platforms. For our experimental evaluations, we set the noise standard deviation for the IR and UWB sensors to 0.1 m and $\rho = 0.1$ for visual-inertial SLAM, with loop closure fixes occurring every 5 m of motion.

Using these three simulation models, we conducted large-scale experiments with an increasing number of robots, ranging from five to one hundred, performing cyclic pursuit behavior for evaluation. Each experimental configuration was repeated 30 times. The recorded performance metrics include the time to reach a stable formation, root mean square (RMS) error in RAB measurements. Stability of formation is determined by measuring the centroid of circling robots and computing the distance of each robot to the centroid over time; formation is declared stable if the difference in distance to the centroid remains below a predefined threshold. The pursuit behavior was chosen for the simulations because stability during formation can be used for experiment termination, unlike other behaviors. Average RMS values are computed for each neighbor and averaged over the experimental period. The number of neighbors depends on the communication range of the robots, which is set to 3 m, considering a simulated robot footprint of 0.15 m.

Additionally, we analyze the cost of robots and communication overhead. Robot cost is computed by assigning fixed equipment costs per robot: 20 for IR-based systems, 90 for UWB anchors, and 350 for SLAM sensors like T265. Communication overhead is calculated based on the number of neighbors, with each neighbor's presence resulting in a 12-byte additional data transfer within the system. We assume robots exchange pose and heading information in $\mathbb{R}^2$, totaling three floating point values per exchange.

3.2 Physical Deployment

We conducted hardware experiments in an large room to analyze the performance of UWB-based indirect localization with predeployed anchors and Visual Inertial SLAM localization using Intel T265 sensors compared to a motion capture system reference. In the environment, we deployed eight UWB anchors for robots to localize the robots UWB tags using Time Difference of Arrival. Throughout the behavior execution, we utilized a set of four Optitrack cameras to record ground truth estimates.

Two swarms, one terrestrial and one aerial, each consisting of five robots, were deployed for the experiments. Each swarm performed three runs of the three swarm behaviors, totaling 18 runs in the experiment. The experimental setup is depicted in Fig. 3.

The terrestrial swarm comprises five Clearpath Dingos [8] equipped with Realsense Intel T265 cameras for visual-inertial localization and BitCraze UWB modules [4]. The rover and all sensors are integrated within ROS. For the aerial swarm, we used five custom Cogniflies [1] that were also equipped with BitCraze UWB modules.

Fig. 3. Experimental setup for the swarm behaviors experiment, left showing a team of 5 flying robots and right a team of 5 ground robots.

4 Results

4.1 Simulations

Figure 4 illustrates the simulation outcomes observed during the experiments. The ranging error of the time-of-flight sensor demonstrates significantly lower values compared to those of UWB and SLAM sensors. Conversely, the IR-based method exhibits a larger error in bearing due to the sensor's resolution limitations. Particularly for smaller swarms, the bearing error is pronounced as the number of neighbors decreases.

Regarding convergence time, it increases from ten to thirty robots and stabilizes with larger swarms. Convergence time in pursuit behavior seems to correlate with the number of neighbors; more neighbors lead to quicker convergence. For smaller swarms, the convergence time with the IR-based technique is favorable for pursuit behavior, as it only requires coarse bearing estimates of the robots ahead. As swarm size increases, convergence time appears similar across all localization methods, stabilizing when sufficient neighbors are available during pursuit behavior.

In terms of deployment cost, robots equipped with visual-inertial sensors incur the highest expenses compared to those with UWB and IR sensors. Deploying UWB sensors necessitates additional costs for anchors in the environment. Optimal performance with UWB sensors requires eight anchors, adding a cost of 1360 to the system. Communication overhead correlates with the number of neighbors in the system, linearly increasing as the number of neighbors rises.

4.2 Physical Deployment

Figure 5 depicts the trajectory of the robots using various localization methods compared to measurements from the motion capture system. The trajectory disparity between the two localization methods fits the behaviors observed in the simulation.

Visual-inertial odometry appears to accumulate drift over motion and corrects on loop closures. Conversely, UWB estimates exhibit saw-tooth-like motion,

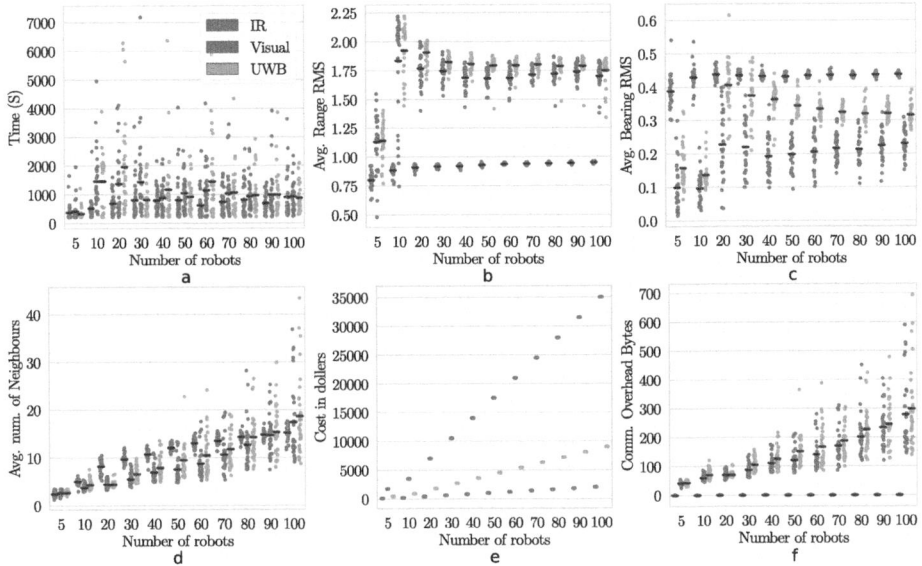

Fig. 4. Simulation properties of the various sensor models: (a) shows the time taken to reach a stable formation with pursuit behavior; (b) the average RMS error in ranging across neighbors; (c) the average RMS error in bearing across neighbors; (d) the average number of neighbors during the behavior; (e) the cost of deploying the various system; (f) the communication overhead created in the whole system by using the type of localization.

characterized by jumps in localization estimates that are subsequently corrected in later estimates.

Table 2 shows the average Root Mean Square error in range and bearing of the neighbors for each swarm and each behavior. The RMS errors in localization were found to be the lowest for the UWB-equipped flying robot team during pursuit, due to their clear line of sight to the anchors. However, the ground robots using UWB exhibited higher ranging errors compared to the flying robots, likely due to poor line-of-sight conditions at the borders of the arena. With visual-inertial odometry, bearing errors were significantly higher as orientation changes induced drifts in visual SLAM. The continuous circular motion in the pursuit behavior, akin to fish milling, resulted in poor bearing estimates with visual-inertial SLAM.

During the fireworks behavior, the flying robot team performed more evasive maneuvers for aggregating and diffusing, which resulted in lower accuracy in UWB range estimates. However, the bearing error was lower for the flying robots performing fireworks. For the ground robots, the linear motion towards and away from the center of the arena in the fireworks behavior resulted in lower range errors compared to the flying robots. The bearing estimate error for ground robots using UWB was higher during pursuit due to non-line-of-sight communication near the arena borders. A similar effect was observed during the Lennard-Jones potential-based behavior.

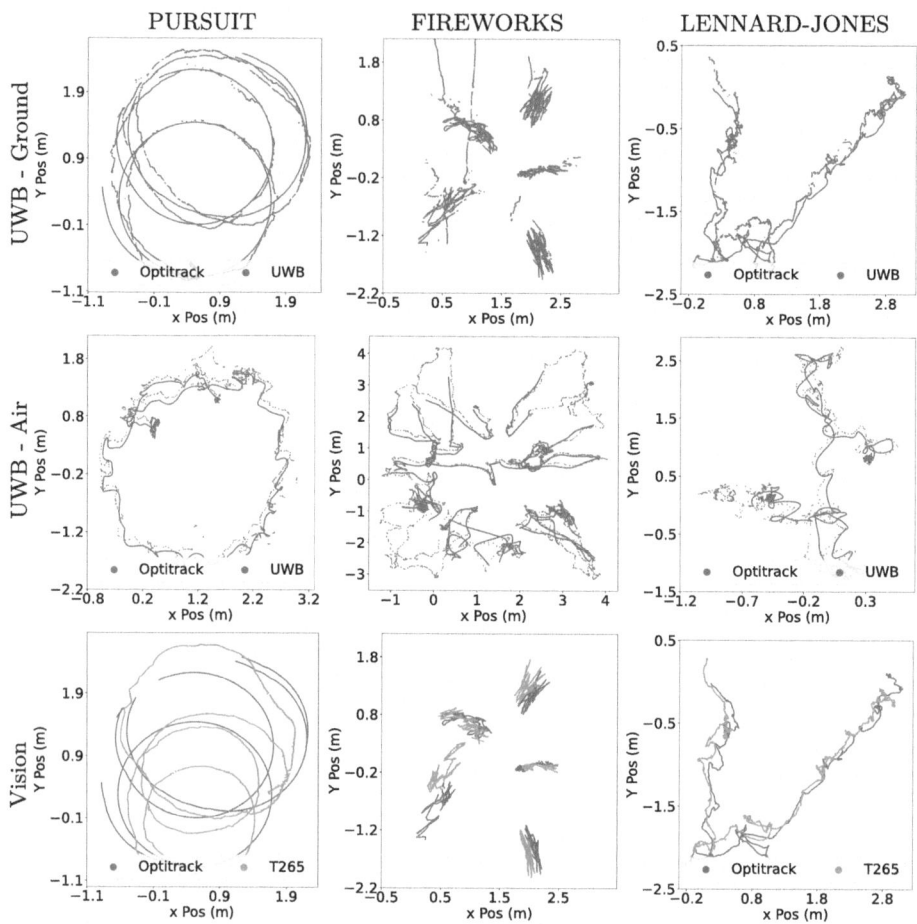

Fig. 5. Visualization of the trajectories of the robots with UWB localization and visual SLAM performing the three swarm behaviors.

With visual SLAM, the fireworks behavior produced very small bearing errors due to minimal orientation changes. In contrast, the Lennard-Jones-based behavior exhibited higher bearing errors due to frequent orientation changes. These observations highlight the impact of various motion patterns on localization accuracy using UWB and visual-inertial odometry. Flying robots benefit from the UWB-based localization as a good line-of-sight could be achieved. On the ground robots, the visual-inertial odometry appears to degrade the RAB estimates when the behavior requires large orientation changes.

Videos showcasing the robots performing the three swarm behaviors studied using UWB infrastructure for state estimation and situated communication for relative localization are available online[1].

---

[1] https://bit.ly/3VKFc36.

Table 2. Average RAB RMS error for each behavior and each localization method

Method	Behavior					
	Pursuit		Fireworks		Lennard-Jones	
	Range	Bearing	Range	Bearing	Range	Bearing
UWB Air	0.071	0.27	0.163	0.15	0.100	0.25
UWB Ground	0.327	0.24	0.124	0.65	0.090	0.98
Vision Ground	0.161	2.23	0.071	0.16	0.068	0.87

5 Conclusion

Most collective algorithms in swarm robotics rely on information about the relative distance and azimuth between robots, typically obtained through situated communication and IR-based RAB sensors. However, current IR-based techniques often exhibit significant bearing estimation errors, which can limit their effectiveness for behaviors requiring precise bearing measurements. In contrast, SLAM and UWB localization methods have shown promise for real-world deployments, as their bearing estimates are notably more accurate.

This study has showcased the characteristics and performance of IR-based localization systems, visual SLAM, and UWB-based RAB estimation. UWB-based systems are particularly viable in environments where the deployment of infrastructure is feasible, providing reliable localization even in challenging conditions. Visual-inertial odometry excels in well-lit, feature-rich environments, offering accurate estimates of position and orientation.

While visual SLAM and UWB-based techniques offer improved accuracy, they also introduce additional communication overheads due to the need for exchanging position estimates and computing RAB estimates. Despite this, these indirect methods demonstrate scalability with increasing numbers of robots, making them suitable for a wide range of swarm robotics applications.

In conclusion, visual SLAM and UWB-based localization methods offer significant advantages over traditional IR-based systems, particularly in terms of bearing estimation accuracy. Their scalability and adaptability to different environments make them promising candidates for broader deployment in various swarm robotics applications. Future research should focus on improving the scalability of direct UWB-based methods and enhancing the robustness of indirect methods for diverse operational scenarios.

References

1. de Azambuja, R., Fouad, H., Bouteiller, Y., Sol, C., Beltrame, G.: When being soft makes you tough: a collision-resilient quadcopter inspired by arthropods' exoskeletons. In: 2022 International Conference on Robotics and Automation (ICRA), pp. 7854–7860. IEEE (2022)
2. Berlinger, F., Gauci, M., Nagpal, R.: Implicit coordination for 3D underwater collective behaviors in a fish-inspired robot swarm. Sci. Robot. **6**(50), eabd8668 (2021)

3. Bilaloğlu, C., Şahin, M., Arvin, F., Şahin, E., Turgut, A.E.: A novel time-of-flight range and bearing sensor system for micro air vehicle swarms. In: Dorigo, M., et al. (eds.) ANTS 2022. LNCS, vol. 13491, pp. 248–256. Springer, Cham (2022). https://doi.org/10.1007/978-3-031-20176-9_20
4. Bitcraze: Crazyflie platform overview (2024). https://www.bitcraze.io/documentation/system/platform/. Accessed 19 Mar 2024
5. Bottigliero, S., Milanesio, D., Saccani, M., Maggiora, R.: A low-cost indoor real-time locating system based on TDOA estimation of UWB pulse sequences. IEEE Trans. Instrum. Meas. **70**, 1–11 (2021)
6. Cao, Y., Chen, C., St-Onge, D., Beltrame, G.: Distributed TDMA for mobile UWB network localization. IEEE Internet Things J. **8**(17), 13449–13464 (2021)
7. Chen, J., Gauci, M., Li, W., Kolling, A., Groß, R.: Occlusion-based cooperative transport with a swarm of miniature mobile robots. IEEE Trans. Rob. **31**(2), 307–321 (2015)
8. Clearpath Robotics: Dingo indoor mobile robot (2024). https://clearpathrobotics.com/dingo-indoor-mobile-robot/. Accessed 19 Mar 2024
9. Coppola, M., McGuire, K.N., Scheper, K.Y., de Croon, G.C.: On-board communication-based relative localization for collision avoidance in micro air vehicle teams. Auton. Robot. **42**, 1787–1805 (2018)
10. Güler, S., Abdelkader, M., Shamma, J.S.: Peer-to-peer relative localization of aerial robots with ultrawideband sensors. IEEE Trans. Control Syst. Technol. **29**(5), 1981–1996 (2020)
11. Guo, K., Qiu, Z., Meng, W., Xie, L., Teo, R.: Ultra-wideband based cooperative relative localization algorithm and experiments for multiple unmanned aerial vehicles in gps denied environments. Int. J. Micro Air Veh. **9**(3), 169–186 (2017)
12. Kasper, M., McGuire, S., Heckman, C.: A benchmark for visual-inertial odometry systems employing onboard illumination. In: 2019 IEEE/RSJ International Conference on Intelligent Robots and Systems (IROS), pp. 5256–5263. IEEE (2019)
13. Kushleyev, A., Mellinger, D., Powers, C., Kumar, V.: Towards a swarm of agile micro quadrotors. Auton. Robot. **35**(4), 287–300 (2013)
14. Lajoie, P.Y., Beltrame, G.: Swarm-slam: sparse decentralized collaborative simultaneous localization and mapping framework for multi-robot systems. IEEE Robot. Autom. Lett. **9**(1), 475–482 (2023)
15. Le Goc, M., Kim, L.H., Parsaei, A., Fekete, J.D., Dragicevic, P., Follmer, S.: Zooids: building blocks for swarm user interfaces. In: Proceedings of the 29th Annual Symposium on User Interface Software and Technology, pp. 97–109 (2016)
16. Li, J., Bi, Y., Li, K., Wang, K., Lin, F., Chen, B.M.: Accurate 3D localization for MAV swarms by UWB and IMU fusion. In: 2018 IEEE 14th International Conference on Control and Automation (ICCA), pp. 100–105. IEEE (2018)
17. Li, M., Liang, G., Luo, H., Qian, H., Lam, T.L.: Robot-to-robot relative pose estimation based on semidefinite relaxation optimization. In: 2020 IEEE/RSJ International Conference on Intelligent Robots and Systems (IROS), pp. 4491–4498. IEEE (2020)
18. Li, S., Coppola, M., De Wagter, C., de Croon, G.C.: An autonomous swarm of micro flying robots with range-based relative localization. arXiv preprint arXiv:2003.05853 (2020)
19. Li, Y., et al.: Fact: Fast and active coordinate initialization for vision-based drone swarms. arXiv preprint arXiv:2403.13455 (2024)
20. Li, Z., Fang, H., Zhao, J., Pang, L.: A multi-node collaborative and iterative UWB localisation algorithm for indoor complex environments. Int. J. Sens. Netw. **44**(3), 133–143 (2024)

21. Liu, S., Yu, J., Ke, Z., Dai, F., Chen, Y.: Aerial-ground collaborative 3D reconstruction for fast pile volume estimation with unexplored surroundings. Int. J. Adv. Rob. Syst. **17**(2), 1729881420919948 (2020)
22. Mathews, N., Christensen, A.L., O'Grady, R., Mondada, F., Dorigo, M.: Mergeable nervous systems for robots. Nat. Commun. **8**(1), 439 (2017)
23. McGuire, K., De Wagter, C., Tuyls, K., Kappen, H., de Croon, G.C.: Minimal navigation solution for a swarm of tiny flying robots to explore an unknown environment. Sci. Robot. **4**(35), eaaw9710 (2019)
24. Nguyen, T.H., Xie, L.: Relative transformation estimation based on fusion of odometry and UWB ranging data. IEEE Trans. Robot. (2023)
25. Nguyen, T., Mohta, K., Taylor, C.J., Kumar, V.: Vision-based multi-MAV localization with anonymous relative measurements using coupled probabilistic data association filter. In: 2020 IEEE International Conference on Robotics and Automation (ICRA), pp. 3349–3355. IEEE (2020)
26. Pinciroli, C., Beltrame, G.: Buzz: an extensible programming language for heterogeneous swarm robotics. In: 2016 IEEE/RSJ International Conference on Intelligent Robots and Systems (IROS), pp. 3794–3800. IEEE (2016)
27. Pinciroli, C., et al.: Argos: a modular, parallel, multi-engine simulator for multi-robot systems. Swarm Intell. **6**, 271–295 (2012)
28. Pires, A.G., Rezeck, P.A., Chaves, R.A., Macharet, D.G., Chaimowicz, L.: Cooperative localization and mapping with robotic swarms. J. Intell. Robot. Syst. **102**(2), 47 (2021)
29. Preiss, J.A., Honig, W., Sukhatme, G.S., Ayanian, N.: Crazyswarm: a large nanoquadcopter swarm. In: 2017 IEEE International Conference on Robotics and Automation (ICRA), pp. 3299–3304. IEEE (2017)
30. Rubenstein, M., Ahler, C., Hoff, N., Cabrera, A., Nagpal, R.: Kilobot: a low cost robot with scalable operations designed for collective behaviors. Robot. Auton. Syst. **62**(7), 966–975 (2014)
31. Saska, M., et al.: System for deployment of groups of unmanned micro aerial vehicles in GPS-denied environments using onboard visual relative localization. Auton. Robot. **41**, 919–944 (2017)
32. Shan, F., Huo, H., Zeng, J., Li, Z., Wu, W., Luo, J.: Ultra-wideband swarm ranging protocol for dynamic and dense networks. IEEE/ACM Trans. Networking **30**(6), 2834–2848 (2022)
33. Slavkov, I., Carrillo-Zapata, D., et al.: Morphogenesis in robot swarms. Sci. Robot.' **3**(25), eaau9178 (2018)
34. St-Onge, D., Pinciroli, C., Beltrame, G.: Circle formation with computation-free robots shows emergent behavioural structure. In: 2018 IEEE/RSJ International Conference on Intelligent Robots and Systems (IROS), pp. 5344–5349. IEEE (2018)
35. St-Onge, D., Varadharajan, V.S., Švogor, I., Beltrame, G.: From design to deployment: decentralized coordination of heterogeneous robotic teams. Front. Robot. AI **7**, 51 (2020)
36. Sun, G., et al.: Mean-shift exploration in shape assembly of robot swarms. Nat. Commun. **14**(1), 3476 (2023)
37. Tiemann, J., Eckermann, F., Wietfeld, C.: Atlas-an open-source TDOA-based ultra-wideband localization system. In: 2016 International Conference on Indoor Positioning and Indoor Navigation (IPIN), pp. 1–6. IEEE (2016)
38. Tiemann, J., Wietfeld, C.: Scalable and precise multi-UAV indoor navigation using TDOA-based UWB localization. In: 2017 International Conference on Indoor Positioning and Indoor Navigation (IPIN), pp. 1–7. IEEE (2017)

39. Trianni, V., Campo, A.: Fundamental collective behaviors in swarm robotics. In: Kacprzyk, J., Pedrycz, W. (eds.) Springer Handbook of Computational Intelligence, pp. 1377–1394. Springer, Heidelberg (2015). https://doi.org/10.1007/978-3-662-43505-2_71
40. Vásárhelyi, G., Virágh, C., Somorjai, G., Nepusz, T., Eiben, A.E., Vicsek, T.: Optimized flocking of autonomous drones in confined environments. Sci. Robot. **3**(20), eaat3536 (2018)
41. Wang, D., Lian, B., Liu, Y., Gao, B., Zhang, S.: Resilient cooperative localization based on factor graphs for multirobot systems. Remote Sens. **16**(5), 832 (2024)
42. Weinstein, A., Cho, A., Loianno, G., Kumar, V.: Visual inertial odometry swarm: an autonomous swarm of vision-based quadrotors. IEEE Robot. Autom. Lett. **3**(3), 1801–1807 (2018)
43. Xu, H., et al.: Omni-swarm: a decentralized omnidirectional visual-inertial-UWB state estimation system for aerial swarms. IEEE Trans. Rob. **38**(6), 3374–3394 (2022)
44. Yang, B., Yang, E., Yu, L., Loeliger, A.: High-precision UWB-based localisation for UAV in extremely confined environments. IEEE Sens. J. **22**(1), 1020–1029 (2021)
45. Zhou, X., et al.: Swarm of micro flying robots in the wild. Sci. Robot. **7**(66), eabm5954 (2022)
46. Zhu, F., et al.: Swarm-lio: decentralized swarm lidar-inertial odometry. In: 2023 IEEE International Conference on Robotics and Automation (ICRA), pp. 3254–3260. IEEE (2023)

Short Papers

BittyBuzz: A Swarm Robotics Runtime for Tiny Systems

Ulrich Dah-Achinanon, Emir Khaled Belhaddad, Guillaume Ricard[(✉)], and Giovanni Beltrame[ORCID]

Department of Computer and Software Engineering, Polytechnique Montreal, Montreal, QC, Canada
{ulrich.dah-achinanon,emir-khaled.belhaddad,guillaume.ricard, giovanni.beltrame}@polymtl.ca

Abstract. Swarm robotics is an emerging field of research which is increasingly attracting attention thanks to the advances in robotics and its potential applications. However, despite the enthusiasm surrounding this area of research, software development for swarm robotics is still a tedious task. That fact is partly due to the lack of dedicated solutions, in particular for low-cost systems to be produced in large numbers and which can have important resource constraints. To address this issue, we introduce BittyBuzz, a novel runtime platform: it allows Buzz, a domain-specific language, to run on microcontrollers while maintaining dynamic memory management. BittyBuzz is designed to fit a flash memory as small as 32 kB (with usable space for scripts) and work with as little as 2 kB of RAM. We introduce the BittyBuzz implementation, its differences from the original Buzz virtual machine, and its advantages for swarm robotics systems. We show that BittyBuzz is successfully integrated with three robotic platforms with minimal memory footprint and conduct experiments to show computation performance of BittyBuzz. Results show that BittyBuzz can be effectively used to implement common swarm behaviors on many microcontroller-based robot systems.

1 Introduction

Swarm robotics studies coordination in groups of relatively simple robots, leveraging local interactions to generate emergent global behaviors. It is inspired by societies of insects, where groups can achieve tasks beyond capabilities of individuals [15]. With properties such as scalability, robustness, autonomy, decentralization [4], swarm robotics allow a wide variety of applications, including nanomedicine, space exploration, search and rescue missions, and generally tasks in dangerous areas [4]. Despite many possible applications, most of swarm robotics deployments are still at the research stage. For reasons related to cost, time, space and complexity, swarm applications require small and low-cost robots. Multiple robotic platforms meeting these specifications are widely used for research (Crazyflie [9], Kilobot [22]). Due to their low-cost, such systems come with a limited amount of computing resources. While programming cooperative behaviours for swarm robotics has always been a challenging task [26], implementing the same strategies on resource-constrained swarm systems renders development harder still. To prevent reinventing the wheel everytime, some previous works proposed domain specific languages (DSLs) to raise the level of abstraction and provide common programming primitives for developers [5,8,17].

Unfortunately, these DSLs do not tackle any resource limitation problems in their implementations and may therefore be unusable in many practical applications. Furthermore, the RAM consumption during application execution can easily exceed that available on some robots (as little as 2 kB for the Kilobot). Facing such issues, we propose a system offering built-in swarm capabilities for multi-agent systems composed of resource-constrained robots. We hence introduce BittyBuzz, an implementation of the Buzz Virtual Machine for microcontrollers. This implementation uses the same interface as Buzz, with very few limitations to address resource constraints. The main contributions of this paper are: 1) the development of a swarm-oriented runtime environment capable of defining the behavior of heterogeneous swarms of robots with as small as 32 kB of flash and 2 kB of RAM; 2) the adaptation and optimization of Buzz's capabilities (generality, mixed bottom-up/top-down logic, etc.) to resource-constrained platforms; 3) the integration of BittyBuzz with three different robot hardware platforms used for research in swarm robotics. BittyBuzz is released as open-source software and can be downloaded from our repository[1]. We evaluated its performance on the Crazyflie [20], Kilobot [21], and Zooid [14] robotic platforms.

The rest of this paper is organized as follows. In Sect. 2, we discuss some of the work related to BittyBuzz. Section 3 presents BittyBuzz's features and design principles while comparing them to Buzz. Section 4 explains the design choices that were made to overcome the resource constraints. Section 5 presents an overview of the BittyBuzz-integrated robotic platforms and evaluates their VM execution performance. We then draw the concluding remarks in Sect. 6.

2 Related Work

Previous works introduced dedicated tools for swarm systems and optimized frameworks for resource-constrained devices (see Table 1 for a short summary). Micropython [1] is a lightweight implementation of the Python language. This 256 kB runtime runs with 16 kB of RAM and was used for IoT development [7] and educative mobile robots [13]; yet it uses more memory and does not offer swarm features. Artoo [10] is another micro-framework designed for robotics. It provides a powerful DSL for robot control and physical computing, borrowing from Ruby and Sinatra [23]. Supporting 15 different platforms, Artoo allows developers to create solutions that incorporate different hardware devices at the same time. However, Artoo is not designed for decentralized swarm application design as it requires a centralized control computer. The Zephyr project [2] lets users build a small, scalable real-time operating system (RTOS) optimized for

---

[1] https://github.com/buzz-lang/BittyBuzz.

resource-constrained devices on multiple architectures. Its OS uses a different approach to the resource constraint problem by adapting user-configured limits.

Table 1. Comparison of existing frameworks for robots applications development.

System/Framework	Swarm support	Heterogeneous	DSL	Embedded	Resource constraints
MicroPython [1]		N/A	N/A	✓	✓
Artoo [10]		N/A	N/A		✓
Zephyr project [2]		N/A	N/A	✓	✓
DRONA [5]	✓		P	✓	
Protelis [17]	✓	✓	Protelis	✓	
Koord [8]	✓	✓	Koord	✓	
Actor-based framework [26]	✓	✓	(unnamed)	✓	
Emsbot [16]	✓	✓	N/A	✓	✓
OpenSwarm [24]	✓	✓	N/A	✓	✓
Buzz [18]	✓	✓	Buzz	✓	
BittyBuzz	✓	✓	Buzz	✓	✓

Other existing frameworks introduce DSLs for swarm applications. For instance, Koord [8] is a swarm-oriented language developed to make platform-independent code portable and verifiable. Koord proposes useful features for coordination in a swarm, such as shared variables between the robots and state recording. However, Koord does not focus on resource constraints. Another example is DRONA [5], a framework for building reliable distributed mobile robotics applications. DRONA provides a state-machine based language (P) for event-driven programming that compiles to C code directly deployable on ROS. Protelis [17] is a Java-based language developed to provide a universal platform for aggregate programming, a multi-device paradigm that may apply to robot swarms. Yi et al. [26] propose an actor-based programming framework for swarm robotic systems: the "actor" control unit abstracts away robot-specific features and capabilities. Some lighter DSL-based tools also address resource constraints: Peng et al. [16] present EmSBoT, a component-based framework targeting embedded devices (using 13 kB of flash memory and 5 KB of RAM). OpenSwarm [24] is an OS designed for severely computationally constrained robots (using 1 kB RAM and 12 kB of flash) enabling the developer to design platform-independent solutions. Both EmSBoT and OpenSwarm only provide low-level programming and work on a small subset of platforms, while Buzz and BittyBuzz are based on a virtual machine (VM), meaning that the Buzz code can run on any system where the virtual machine is supported without recompiling.

Overall, although some of those frameworks are lightweight, they either do not focus on resource-constrained systems or do not meet the key requirements of a successful programming language for swarm robotics (decentralized control, spatial computing, neighbor communication, etc. [18]). Buzz is a swarm-specific language that was developed to address the lack of software development tools in swarm robotics [18]. Buzz offers interesting features for swarm application development, including heterogeneous support, swarm-level abstractions, neighbor operations and a consensus system (the "virtual stigmergy"). The work presented in this paper is an adaptation of Buzz optimized to fit resource-constrained systems such as microcontrollers. We extended the support of BittyBuzz to three hardware platforms: the Bitcraze Crazyflie [20], Zooids [14] and the Kilobot [21].

3 BittyBuzz Structure

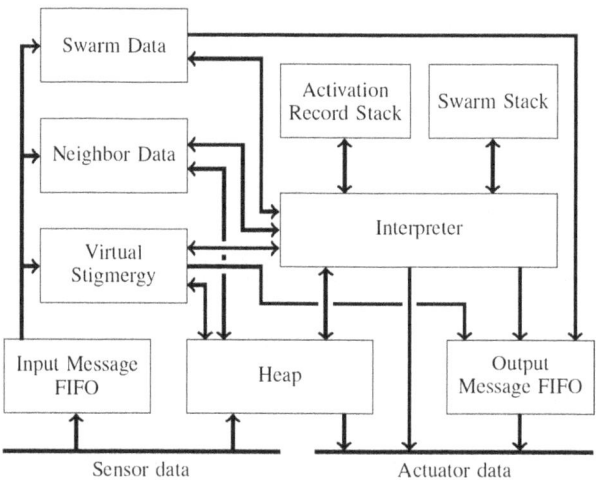

Fig. 1. BittyBuzz virtual machine structure [18]

Our framework is a run-time platform based on a custom virtual machine written in C. The BittyBuzz virtual machine (BBZVM) structure and operation are the same as Buzz: the virtual machine operates in discrete time steps, each of which consists in a sequence of sub-steps: 1) the BBZVM collects the robot's sensor readings; 2) the BBZVM collects incoming messages and updates relevant data structures related to communications ; 3) the BittyBuzz interpreter is called to execute a Buzz script; 4) and finally the BBZVM outputs actuator signals and outgoing messages. The BBZVM has a variable size, and can be as small as 17.1 kB (see Sect. 4.1). The main VM components are presented in Fig. 1.

BBZVM preserves the Buzz capability to handle robot swarms as a first-class language object through the *swarm* primitive type. sub-swarms can thus be easily created and robots can join or leave a swarm based on any conditions.

Buzz also implements the Virtual Stigmergy (VS), a conflict-free replicated data structure that is used to provide consensus on a set of key-value pairs across the entire swarm [19]. The VS is essentially seen by the programmer as a shared table: each *put* operation by a member of the swarm triggers an automatic update in the others, with mechanisms to avoid conflicts. The VS is available in BittyBuzz with its complete features, and it is completely transparent to the user as a key-value data store through the *put* and *get* methods. Buzz and BittyBuzz additionnaly provide a *neighbors* data structure that allows data collection and processing from neighboring robots. This type of neighbor operation is the foundation of spatial computing [27] and widely used in swarm robotics. Ours supports peer-to-peer communication with publish/subscribe and broadcasting.

4 Overcoming Resource Constraints

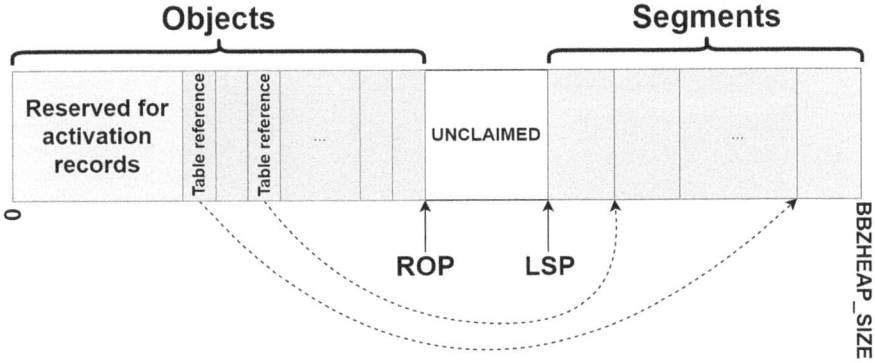

Fig. 2. Illustration of BittyBuzz's heap: the object section containing the activation records section and the allocated objects (non-structured types and table references) in light green, the unclaimed section in white, the segment section in light blue, and the two pointers ROP and LSP. (Color figure online)

Different strategies were used to make BittyBuzz resource consumption more efficient. This section presents the memory layout used and other optimizations:

4.1 Dynamic Memory Management

BittyBuzz has a pre-allocated heap of developer-specified whose size. The heap is a static buffer with three sections: the object section (also containing the activation records of closure calls), the segment section and the unclaimed section. Two pointers are used to access the heap: the rightmost object pointer (ROP) and the leftmost table segment pointer (LSP) as shown in Fig. 2. Each object contains the payload information and an additional metadata byte. Unstructured types such as nil, int, float and string are written from left to right with ROP always pointing to the last added object. For structured types (tables), an object is stored in the objects' section, referring to a data segment stored from right to left in the segments section. Each data segment has a user-defined number of key-value pairs and two metadata bytes containing information about the segment validity and a pointer to the next segment, if any. To save space, arrays are inlined akin to table management in Lua [11]. BittyBuzz also uses a simple *mark-and-sweep* garbage collector algorithm called at each instruction to clear unused objects from heap while preserving global swarm-shared structures.

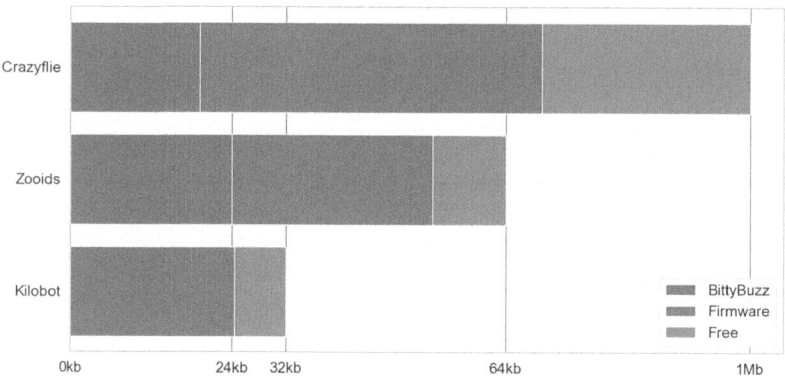

Fig. 3. Sizes of the BBZVM, the robot native firmware, and the available space for user bytecode for the three integrated robotic platforms: Bitcraze Crazyflie, K-Team Kilobot and Zooids

4.2 Optimizations and Limitations

To reduce the time complexity for message queue management, BittyBuzz implements its own ring buffer. Ring buffers seamlessly loop on themselves allowing constant time *push* and *pop* operations for queues [6]. Dynamic memory space requirements are also addressed by placing a user limit on the number of tracked swarms and neighbors. The available static flash memory is also an important factor in resource consumption: to reduce space, bytecode files use 16-bit encoding for small Buzz integer and float values. Strings are replaced by a unique identifier throughout user code to avoid the storage of usual implementations [3].

5 Performance

Figure 3 presents the flash memory usage of the BittyBuzz virtual machine and full-feature framework on the three following robotic platforms used in different swarm robotic projects. In all instances the memory footprint is under 24 kB with some variability between platforms based on available compiler optimizations.

We measured the overhead of the BittyBuzz runtime platform running on the Zooids to derive maximum utilization constraints and compare performance metrics with the original Buzz virtal machine implementation. We consider the total time taken for underlying VM operations, message sharing etc. and maximum execution speed (in VM instructions per second) to determine the amount of work that can be done in the remaining time frame. Buzz measurements were performed on the Khepera IV [12], a wheeled robot running a complete OS on top of a 800 MHz Cortex-A8 processor. Comparing BittyBuzz with a more complex platform lets us assess more benefits of Buzz for embedded swarm development.

Table 2 displays the differences in VM overhead for the two hardware platforms. We experience a slightly higher delay on the Khepera despite the faster CPU due to interaction with the OS e.g. network and I/O buffering. Furthermore, Table 2 details the Buzz virtual machine instruction budget for 100 ms of execution in a timestep The upper bound on the constrained Zooid hardware occupies 459 instructions (i.e. at most 918 bytes) amounting to 13% of the remaining space (see Fig. 3). The instruction budget represents the amount of code that can be executed in a timestep, thus the remaining 87% can be used to store additional behaviours to switch to and from at will during VM executions.

Table 2. BittyBuzz performance evaluation for a 10-Hertz timestep

VM Implementation	Hardware Platform	VM Overhead	Max instructions
Buzz [18]	Khepera IV	4.92 ms	156 000
BittyBuzz	Zooids	4.45 ms	459

Moreover, we carried out swarm tasks with Zooids in the context of hierarchical swarms [25] where a group of guide robots herd a lighter worker swarm and have it move across the environment, thus validating our embedded framework.

6 Conclusion

We presented BittyBuzz, a novel virtual machine for the Buzz language designed for resource-constrained microcontrollers. The contributions of this work include: 1) the development of a virtual machine for a dynamically-typed language with dynamic memory management with as little as 32 kB of flash and 2 kB of RAM; 2) the full inclusion of Buzz's programming model (including neighbor queries, consensus, etc.) to resource-constrained platforms; 3) the integration of BittyBuzz with three different robotic platforms used for research in swarm robotics. In future works, we believe that resource-constrained robotic platforms have an important part to play in the future of IoT. A dedicated language such as BittyBuzz will ease the development and prototyping of swarm applications. Our experiments show that BittyBuzz can be used for established swarm behaviors and that the virtual machine has sufficient performance even on severely constrained devices, enabling user behaviors to perform without noticeable delays.

Acknowledgements. This work was supported by the Fonds de recherche du Québec Nature et technologies (FRQNT) under grant 296737 and by the National Research Council Canada (NRC).

References

1. Micropython - python for microcontrollers. https://micropython.org/. Accessed 1 July 2021
2. Zephyr Project | Applications (2020). https://www.zephyrproject.org/learn-about/applications. Accessed 2 July 2021
3. Busbee, K.L., Braunschweig, D.: String data type. Programming Fundamentals (2018)
4. Cheraghi, A.R., Shahzad, S., Graffi, K.: Past, present, and future of swarm robotics. In: Arai, K. (ed.) IntelliSys 2021. LNNS, vol. 296, pp. 190–233. Springer, Cham (2022). https://doi.org/10.1007/978-3-030-82199-9_13
5. Desai, A., Saha, I., Yang, J., Qadeer, S., Seshia, S.A.: Drona: a framework for safe distributed mobile robotics. In: Proceedings of the 8th International Conference on Cyber-Physical Systems, pp. 239–248 (2017)
6. Feldman, S., Dechev, D.: A wait-free multi-producer multi-consumer ring buffer. ACM SIGAPP Appl. Comput. Rev. **15**(3), 59–71 (2015)
7. Gaspar, G., Fabo, P., Kuba, M., Flochova, J., Dudak, J., Florkova, Z.: Development of IoT applications based on the MicroPython platform for industry 4.0 implementation. In: 2020 19th International Conference on Mechatronics-Mechatronika (ME), pp. 1–7. IEEE (2020)
8. Ghosh, R., Hsieh, C., Misailovic, S., Mitra, S.: Koord: a language for programming and verifying distributed robotics application. Proc. ACM Program. Lang. **4**(OOPSLA), 1–30 (2020)
9. Giernacki, W., Skwierczyński, M., Witwicki, W., Wroński, P., Kozierski, P.: Crazyflie 2.0 quadrotor as a platform for research and education in robotics and control engineering. In: 2017 22nd International Conference on Methods and Models in Automation and Robotics (MMAR). pp. 37–42. IEEE (2017)
10. Group, T.H.: Artoo platforms. http://artoo.io/documentation/guides/what_is_artoo/. Accessed 1 July 2021
11. Ierusalimschy, R., de Figueiredo, L., Celes, W.: The implementation of Lua 5.0. J. Univ. Comput. Sci. **11**(7), 1159–1176 (2005)
12. K-Team: Khepera IV (2016). https://www.k-team.com/khepera-iv
13. Khamphroo, M., Kwankeo, N., Kaemarungsi, K., Fukawa, K.: Integrating micropython-based educational mobile robot with wireless network. In: 2017 9th International Conference on Information Technology and Electrical Engineering (ICITEE), pp. 1–6. IEEE (2017)
14. Le Goc, M., Kim, L.H., Parsaei, A., Fekete, J.D., Dragicevic, P., Follmer, S.: Zooids: building blocks for swarm user interfaces. In: Proceedings of the 29th Annual Symposium on User Interface Software and Technology, pp. 97–109 (2016)
15. Navarro, I., Matía, F.: An introduction to swarm robotics. Int. Sch. Res. Not. **2013** (2013)
16. Peng, L., Guan, F., Perneel, L., Timmerman, M.: EmSBot: a modular framework supporting the development of swarm robotics applications. Int. J. Adv. Rob. Syst. **13**(6), 1729881416663662 (2016)

17. Pianini, D., Viroli, M., Beal, J.: Protelis: practical aggregate programming. In: Proceedings of the 30th Annual ACM Symposium on Applied Computing, pp. 1846–1853 (2015)
18. Pinciroli, C., Beltrame, G.: Buzz: an extensible programming language for heterogeneous swarm robotics. In: 2016 IEEE/RSJ International Conference on Intelligent Robots and Systems (IROS), pp. 3794–3800. IEEE (2016)
19. Pinciroli, C., Lee-Brown, A., Beltrame, G.: A tuple space for data sharing in robot swarms. In: Proceedings of the 9th EAI International Conference on Bio-inspired Information and Communications Technologies (formerly BIONETICS), pp. 287–294 (2016)
20. Preiss, J.A., Honig, W., Sukhatme, G.S., Ayanian, N.: Crazyswarm: a large nanoquadcopter swarm. In: 2017 IEEE International Conference on Robotics and Automation (ICRA), pp. 3299–3304. IEEE (2017)
21. Rubenstein, M., Ahler, C., Hoff, N., Cabrera, A., Nagpal, R.: Kilobot: a low cost robot with scalable operations designed for collective behaviors. Robot. Auton. Syst. **62**(7), 966–975 (2014)
22. Rubenstein, M., Ahler, C., Nagpal, R.: Kilobot: a low cost scalable robot system for collective behaviors. In: 2012 IEEE International Conference on Robotics and Automation, pp. 3293–3298. IEEE (2012)
23. Sinatra: Github - sinatra/sinatra: Classy web-development dressed in a dsl. https://github.com/sinatra/sinatra. Accessed 1 July 2021
24. Trenkwalder, S.M., Lopes, Y.K., Kolling, A., Christensen, A.L., Prodan, R., Groß, R.: Openswarm: an event-driven embedded operating system for miniature robots. In: 2016 IEEE/RSJ International Conference on Intelligent Robots and Systems (IROS), pp. 4483–4490. IEEE (2016)
25. Varadharajan, V.S., Dyanatkar, S., Beltrame, G.: Hierarchical control of smart particle swarms (2022)
26. Yi, W., et al.: An actor-based programming framework for swarm robotic systems. In: 2020 IEEE/RSJ International Conference on Intelligent Robots and Systems (IROS), pp. 8012–8019. IEEE (2020)
27. Zambonelli, F., Mamei, M.: Spatial computing: an emerging paradigm for autonomic computing and communication. In: Smirnov, M. (ed.) WAC 2004. LNCS, vol. 3457, pp. 44–57. Springer, Heidelberg (2005). https://doi.org/10.1007/11520184_4

Collective Random Walks of Flocking Agents Through Emergent Implicit Leadership

Andres Garcia Rincon[✉], Tugay Alperen Karagüzel, Fuda van Diggelen, and Eliseo Ferrante

Computer Science, Vrije Universiteit Amsterdam, Amsterdam, The Netherlands
a.a.garciarincon@student.vu.nl,
{t.a.karaguzel,fuda.van.diggelen,e.ferrante}@vu.nl

Abstract. This paper presents a novel method to achieve collective exploration with a mobile robot swarm. The method enables collective random walks: Swarms of simple individuals using only local interactions flock together while navigating in a random direction. In this paper, we introduce swarm leaders that emerge through a mechanism based on local sensing. Upon emerging, leaders combine social interactions with a goal direction, that is generated using Lévy or Correlated Random Walks processes. Key findings highlight that the aggregated swarm behaviours are random-walk-like, and tend towards Lévy-like dynamics increasing swarm size, irrespective of the emergent leader process. Furthermore, we explore how leader distribution affects the trajectory of the swarm, revealing the potential for optimized movement strategies through the manipulation of leadership roles. The introduction of emergent implicit leadership offers a novel perspective on achieving complex navigational behaviors in swarm robotics, presenting significant implications for applications requiring adaptive, decentralized exploration strategies.

1 Introduction

Random walks are frequently utilized to describe search patterns observed in natural systems, engaged in searching for targets whose locations are not known [2]. Random walk, especially Brownian motion, is a concept originating in the examination of pollen particle motion in a fluid by botanist Robert Brown [7]. Although various models have been deployed over time, two of them are representative in modeling how natural living systems explore the environment. Correlated Random Walks (CRWs), described by a degree of correlation among turning angles [19], have been harnessed to explain the locomotion of diverse insects such as white butterflies [14]. Lévy Walks (LWs), instead, are characterized by a power-law distribution of step-lengths and have been instrumental in explaining search patterns across multiple domains, including predatory behaviors exhibited by African Jackals [1] and basking sharks [22], as well as foraging activities undertaken by zooplankton [16] and wandering albatrosses [26].

In swarm robotics, CRWs have been used in area exploration [11] and pollination [4], while LWs in foraging [17], search and rescue [24], and underwater reconnaissance [23]. However, the literature typically considered individual robots performing random walks rather than collective random motion. Swarm robotics relies on limited individual capabilities enhanced in collective behaviors, such as extended sensing ranges [3] and emergent gradient-sensing [13]. Combining collective motion with random walk strategies could improve swarm exploration. Limited studies integrate random walks with collective motion; one noteworthy work uses LWs with Reynolds-rules-based flocking for surveillance [21], though its design was tailored for a specific use case.

In this paper, we propose a method for collective random walks in a group of robots with self-organized flocking, through emergent implicit leaders. Flocking is achieved through local information [12], with random walks of emergent leaders defining goal directions. Implicit leadership, studied in [9], typically involves predefined leaders and is crucial in tasks like predator evasion [18] and accessing food [20]. Emergent leaders [5] have been studied in numerous papers focusing on natural phenomena [6,25]. To the best of our knowledge, this paper proposes the inaugural implementation of a collective random walk algorithm for swarms of robots via emergent implicit leadership and investigates its properties across different swarm sizes and leader distributions. Flocking is achieved through local information [12], with random walks of emergent leaders defining goal directions. Our research question is: *When emergent leaders follow a given random walk pattern, what pattern emerges at the collective level?*

2 Methodology

We consider N agents in an unbounded 2D environment, each sensing the relative range and bearing of neighbors within sensing range. Agents are non-holonomic: they can only move along their current heading with displacement based on linear speed, and control their heading independently via angular speed. Our method uses Active Elastic (AE) to ensure flocking and motion control [12] (see Sect. 2.1). Our work extends AE with a random walk model for emergent leaders (Sect. 2.2) and the leader emergence mechanism (Sect. 2.3).

AE provides distributed control that calculates (interaction) forces between agents in a swarm. At each step, the focal agent i calculates a virtual force vector $\boldsymbol{f}_i$ as follows:

$$\boldsymbol{f}_i = \gamma_1 \boldsymbol{p}_i + \gamma_2 \boldsymbol{h}_i + \gamma_3 \boldsymbol{g}_i, \tag{1}$$

where $\boldsymbol{p}_i$ is the proximal control vector, $\boldsymbol{h}_i$ the heading alignment vector and $\boldsymbol{g}_i$ the goal direction vector. γ_1, γ_2, and γ_3 are used to balance the contribution of the respective vectors. The goal direction vector is used only by emergent leaders and is dependent on the specific random walk.

2.1 Flocking and Motion Control

Proximal Control. The proximal control vector $\boldsymbol{p}_i$ acts in a similar fashion to a spring, applying a repulsive force on the focal agent if its distance is below the desired distance, and attractive if above, while applying no force if they are exactly at the desired distance from each other. The vector is calculated as:

$$\boldsymbol{p}_i = \sum_{m \in N} p_i^m(d_i^m, \sigma_i) \angle e^{j\phi_i^m}, \qquad (2)$$

where $p_i^m(d_i^m, \sigma_i)$ is the magnitude of the vector pointing from the focal agent to each perceived neighbor, and $\angle e^{j\phi_i^m}$ its corresponding angle. The magnitude of the vector is computed through a virtual force obtained from a modified Lennard-Jones potential function as follows:

$$p_i^m(d_i^m, \sigma_i) = -\epsilon \left[2 \frac{\sigma_i^4}{(d_i^m)^5} - \frac{\sigma_i^2}{(d_i^m)^3} \right], \qquad (3)$$

where d_i^m is the relative distance sensed by i, σ is a parameter proportional to the desired distance between individuals (with coefficient $\sqrt{2}$), and ϵ is the gain of the potential function.

Alignment Control. The alignment control vector applies a vector to the focal agent to align its heading direction with the mean heading of its perceived neighbor, including its own heading and is calculated as follows:

$$\boldsymbol{h}_i = \frac{\angle e^{j\theta_i} + \sum_{m \in \mathcal{R}} \angle e^{j\theta_m}}{|| \angle e^{j\theta_i} + \sum_{m \in \mathcal{R}} \angle e^{j\theta_m} ||}, \qquad (4)$$

where $\angle e^{j\theta_i}$ is the heading of the focal agent, and $\angle e^{j\theta_m}$ is the heading of a neighbor m, and headings are expressed in common frame of reference.

Goal Direction Control. The goal direction vector encourages the movement of the leaders in a random direction θ_r sampled as explained in Sect. 2.2. The vector is defined in the agent's local frame as follows:

$$\boldsymbol{g}_i = \angle e^{j(\theta_i + \theta_r)} \qquad (5)$$

Motion Control. After calculating the virtual force vector, motion control is achieved by finding the components of $\boldsymbol{f}_i$ to compute the agent's linear and angular speed at each step. Agents use a right-handed reference frame with the x-axis parallel to the heading, projecting $\boldsymbol{f}_i$ to its orthogonal axes to obtain f_x and f_y. The linear and angular speeds are calculated as follows:

$$U_i = K_1 f_x + U_c, \qquad (6)$$
$$\omega_i = K_2 f_y. \qquad (7)$$

To compute the linear speed U_i, we multiply the virtual force component f_x by the linear speed gain K_1 and add a speed constant U_c. The angular speed ω_i is determined by multiplying f_y by the angular speed gain K_2. U_i and ω_i are bounded by $[0, U_{\max}]$ and $[-\omega_{\max}, \omega_{\max}]$ respectively.

2.2 Random Walk Model

Emergent leaders follow a random direction for a time that depends on a random step length. The following two distributions implement this mechanism.

Step Length Distribution. We sample the step lengths via a power-law distribution: longer lengths are sampled with lower probability than shorter lengths. It can be defined by its Fourier transformation as:

$$F(k) = e^{-\beta |k|^{\alpha}}, \quad 0 < \alpha \leq 2, \tag{8}$$

where β controls the distribution scale, and α the power scaling. The mean is finite when $\alpha \geq 1$, so values between 1 and 2 are used: $\alpha = 2$ yields to a Gaussian distribution, while $\alpha \to 1$ yields to an increasingly heavy-tailed distribution [10].

Turning Angle Distribution. The sampling of the random turning angle θ_r follows a wrapped Cauchy distribution as follows:

$$f_w(\theta; \mu, \rho) = \frac{1}{2\pi} \frac{1 - \rho^2}{1 + \rho^2 - 2\rho \cos(\theta - \mu)}, \quad 0 < \rho < 1, \tag{9}$$

where μ is the mean value of the distribution, which remains at $\mu = 0$ along all experiments and ρ controls the skewness of the distribution, whereas $\rho \to 1$ the angles grow increasingly correlated [10].

2.3 Emergence of Leaders

Leaders emerge through a self-organized mechanism. The focal agent calculates local *stability*, a local measure of order and cohesion. When stability is high, we make the focal agent more likely to become a leader because there are likely fewer other leaders in the swarm. Leaders are temporary and demote themselves after a certain number of walks. This mechanism is formally explained next.

Stability Measure. The stability is locally calculated by each focal agent with respect to its neighbors within their sensing range. It is composed of two elements: the *heading element* and the *distance element*.

The *heading element* Ψ measures how well-aligned the headings of the focal agents sub-swarm is and is defined as:

$$\Psi = \frac{\|\sum_{i \in N} e^{j\theta_i}\|}{N}, \tag{10}$$

The *distance element* Φ measures how well separated are agents of the focal sub-swarm according to the encouraged desired distance, and is defined as:

$$\Phi = 1 - \frac{\frac{1}{N}\sum_{i \in N}(d_i^m - d_{\text{des}})^2}{(d_{\text{des}})^2}, \tag{11}$$

The final stability measure is the weighted sum of both elements:

$$\Omega = \Phi \cdot w_d + \Psi \cdot w_h, \tag{12}$$

where $w_d + w_h <= 1$, and $0 <= \Omega <= 1$. Higher measures of Ω indicate a more stable sub-swarm.

Promotion and Demotion. The stability measure controls the emergence of leaders via a modified logistic function. The probability $P(x)$ is given by:

$$P(\Omega) = \frac{p_{\max}}{1 + e^{-k_p\left(\Omega - \left(b_p - \frac{\log\left(\frac{1}{p_{\max}}\right)}{k_p}\right)\right)}}, \tag{13}$$

Here, $p_{\max}$ is the max probability, k_p controls the steepness of the curve, and b_p controls the point at which the function reaches half of its maximum value.

Leaders track the cumulative distance since their last step sampling. Once this distance meets or exceeds the sampled step length, leaders resample both step length and turning angle. Leaders are demoted after 5 samplings, promoting dynamic leadership rotation and diverse navigation patterns.

3 Experiments and Discussion

We conduct two sets of experiments to analyze the emergent random walk strategy at the collective level. Firstly, we analyze the behavior of the swarm's center of mass using the Velocity Auto-Correlation Function (VACF) of a Brownian particle, which decays as $x^{\tau/2}$ [8]. We fit the VACF data from the swarm's centroid trajectory using the above power law, estimating a decay rate parameter. As Brownian motion implies no velocity correlations, instantaneous decay should be observed. Less negative decay rates indicate more Lévy-like behavior.

In the second set (analysis of turning angles), we study the turning angle distribution of the swarm centroid. We also use the KL divergence metric, which calculates relative entropy between two distributions [15], to measure the difference between correlated and non-correlated turning angle distributions. In both experimental sections, we test different swarm sizes [25, 50, 100] and leader position restrictions [free, outer, inner]. In "free," no restrictions are applied; in

Table 1. Experimental Parameters

γ_1	γ_1	γ_3	σ	σ (leaders)	ϵ	U_{max}	U_c	ω_{max}	k_1	k_2	w_d	w_h	p_{max}	k_p	b_p
2	0.5	4	0.7	0.6	15	0.1	0.05	$\pi/2$	0.6	0.05	0.7	0.3	0.05	100	0.91

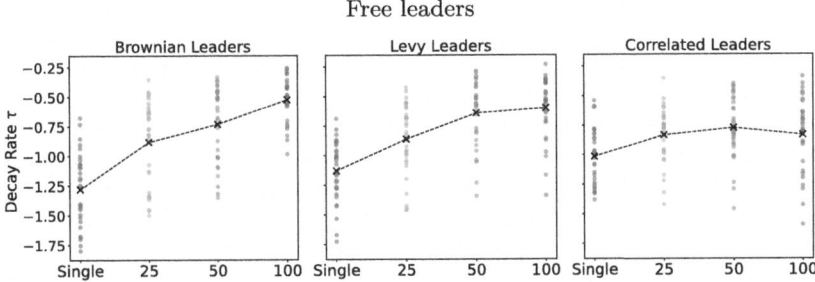

Fig. 1. VACF τ-decay rate values.

"inner," leaders emerge only if surrounded by neighbors; in "outer," leaders are only at the swarm's border. We also test different leader types: $\beta = 0$ for all, Brownian ($\alpha = 2$, $\rho = 0$), Lévy ($\alpha = 1$, $\rho = 0$), and CRW ($\alpha = 2$, $\rho = 0.7$). Each combination runs 100 individual trials of 50000 steps. Single-agent random walkers are studied as a baseline. Step lengths are clamped between [4, 14]; other parameters are in Table 1.

3.1 Analysis of Step Lengths

The main finding of this paper is depicted in Fig. 1, in which we report the extracted τ coefficients from the power-law curve we fitted on the VACF (see the beginning of Sect. 3) for experiments with free leader positions. As explained, larger values of τ indicate slower decays of the power-law, meaning velocities remain correlated for longer time-lags. As the step size distribution cannot easily be analyzed directly for a collective, we use the decay rate of the VACF as a proxy of the step length: a slower decay of VACF (a higher τ value) indicates collective movement patterns with long-range correlations, akin to Lévy walks. Interestingly, we observe an increase of the τ coefficient as a function of the swarm size (similar relation found in restricted leader positions; not shown). In particular, Brownian single agents are characterized by low values of τ (faster decay of the VACF), when emergent leaders are introduced in a group and swarm size increases the dynamics are increasingly more Lévy-like. In the Lévy leaders case, the effect is similar: the group becomes even more Lévy-like than the single walker. Finally, only for CRW leaders, we observe a less pronounced increase of Lévy-like tendency compared to the other two cases.

Table 2. KL Scores Per Setup

	Free	Inner	Outer
25	2.98	1.66	1.40
50	2.77	2.35	1.66
100	2.68	2.01	1.48

Table 3. Mean Leaders Per Setup

	Free	Inner	Outer
25	51.6%	28.8%	19.6%
50	34%	24.7%	7.7%
100	27.29%	27.23%	4.44%

3.2 Analysis of Turning Angles

When examining the turning angle distribution of the swarm centroid, we see significant differences appear in the single walkers' behavior and turning angle distribution (Brownian and CRW): a uniform distribution at $\rho = 0$ contrasts with a centered distribution at $\rho = 0.7$, with a KL divergence of 4.41. Table 2 shows the KL divergence scores for swarm experiments, comparing Brownian and CRW leaders. Notably, restricting leader emergence, especially to outer agents, makes CRW walkers' turning angle distribution less uniform, with a lower KL divergence. We discuss this effect in Sect. 3.3.

3.3 Discussion

The main findings of this paper are in the analysis of the step lengths: we have observed the emergence of Lévy Walk characteristics in swarm collective motion independently from the nature of the walk of the emergent leader. Notably, we observe an emerging dynamic across all leader position restrictions. Additionally, this effect is stronger for Brownian agents. This result is interesting because it may suggest a pathway that evolution may have chosen when selecting for group behaviours: if resources in the environment become scarce, animals that do not have good individual-level exploration strategies may have evolved to collectively move in a group, so that Lévy-like characteristics emerge, as these are known to be more efficient in environment exploration [26].

In our turning angle analysis, the model consistently produces collective motion strategies that retain CRW characteristics. The KL divergence analysis shows a trend: as leader positions are more constrained, angle distributions become uniform across swarm sizes. This is due to the leader-to-leader ratio (Table 3): with more leaders, new leaders with potentially conflicting directions have less impact which suggests a "wisdom of the crowd" effect.

4 Conclusion

In this study, we introduced a pioneering approach to obtain random walks during collective motion in swarms, through the emergence of implicit leaders. The experiments conducted across varying swarm sizes and leader position constraints have provided invaluable insights into the dynamics of collective random walks and the critical role of emergent leadership within these processes. Our findings reveal that the model exhibited a tendency to transition towards Lévy-like behaviors as swarm sizes increased, independently from other factors. This result could reveal an evolutionary motivation for animals to move in groups: to explore the environment in Lévy walk fashion, that is deemed to be efficient [27]. A second finding is the ability of the model to retain correlations in random walk directions, when introduced at the individual level. Although in general independent from other factors, the effect was less pronounced in cases where constraints on the leader positions resulted in fewer leaders overall. The influence of leader position and number of leaders at a given time is one of the avenues for further exploration.

References

1. Atkinson, R., Rhodes, C.J., Macdonald, D., Anderson, R.: Scale-free dynamics in the movement patterns of jackals. Oikos **98**(1), 134–140 (2002)
2. Bartumeus, F., da Luz, M.G.E., Viswanathan, G.M., Catalan, J.: Animal search strategies: a quantitative random-walk analysis. Ecology **86**(11), 3078–3087 (2005)
3. Berdahl, A., Torney, C.J., Ioannou, C.C., Faria, J.J., Couzin, I.D.: Emergent sensing of complex environments by mobile animal groups. Science **339**(6119), 574–576 (2013)
4. Berman, S., Kumar, V., Nagpal, R.: Design of control policies for spatially inhomogeneous robot swarms with application to commercial pollination. In: 2011 IEEE International Conference on Robotics and Automation, pp. 378–385. IEEE (2011)
5. Bernardi, S., Eftimie, R., Painter, K.J.: Leadership through influence: what mechanisms allow leaders to steer a swarm? Bull. Math. Biol. **83**(6), 69 (2021)
6. Brent, L.J., Franks, D.W., Foster, E.A., Balcomb, K.C., Cant, M.A., Croft, D.P.: Ecological knowledge, leadership, and the evolution of menopause in killer whales. Curr. Biol. **25**(6), 746–750 (2015)
7. Brown, R.: Xxvii. a brief account of microscopical observations made in the months of June, July and August 1827, on the particles contained in the pollen of plants; and on the general existence of active molecules in organic and inorganic bodies. Philos. Mag. **4**(21), 161–173 (1828)
8. Chakraborty, D.: Velocity autocorrelation function of a Brownian particle. Eur. Phys. J. B **83**, 375–380 (2011)
9. Couzin, I.D., Krause, J., Franks, N.R., Levin, S.A.: Effective leadership and decision-making in animal groups on the move. Nature **433**(7025), 513–516 (2005)
10. Dimidov, C., Oriolo, G., Trianni, V.: Random walks in swarm robotics: an experiment with Kilobots. In: Dorigo, M., et al. (eds.) ANTS 2016. LNCS, vol. 9882, pp. 185–196. Springer, Cham (2016). https://doi.org/10.1007/978-3-319-44427-7_16
11. Falcón-Cortés, A., Boyer, D., Aldana, M., Ramos-Fernández, G.: Lévy movements and a slowly decaying memory allow efficient collective learning in groups of interacting foragers. PLoS Comput. Biol. **19**(10), e1011528 (2023)
12. Ferrante, E., Turgut, A.E., Huepe, C., Stranieri, A., Pinciroli, C., Dorigo, M.: Self-organized flocking with a mobile robot swarm: a novel motion control method. Adapt. Behav. **20**(6), 460–477 (2012)
13. Karagüzel, T.A., Turgut, A.E., Eiben, A., Ferrante, E.: Collective gradient perception with a flying robot swarm. Swarm Intell. **17**(1), 117–146 (2023)
14. Kareiva, P., Shigesada, N.: Analyzing insect movement as a correlated random walk. Oecologia **56**, 234–238 (1983)
15. Kullback, S.: Information Theory and Statistics. Courier Corporation (1997)
16. Levandowsky, M., Klafter, J., White, B.: Swimming behavior and chemosensory responses in the protistan microzooplankton as a function of the hydrodynamic regime. Bull. Mar. Sci. **43**(3), 758–763 (1988)
17. Nauta, J., Van Havermaet, S., Simoens, P., Khaluf, Y.: Enhanced foraging in robot swarms using collective lévy walks. In: 24th European Conference on Artificial Intelligence (ECAI), vol. 325, pp. 171–178. IOS (2020)
18. Olson, R.S., Hintze, A., Dyer, F.C., Knoester, D.B., Adami, C.: Predator confusion is sufficient to evolve swarming behaviour. J. R. Soc. Interface **10**(85), 20130305 (2013)
19. Patlak, C.S.: Random walk with persistence and external bias. Bull. Math. Biophys. **15**, 311–338 (1953)

20. Reebs, S.G.: Can a minority of informed leaders determine the foraging movements of a fish shoal? Anim. Behav. **59**(2), 403–409 (2000)
21. Sardinha, H., Dragone, M., Vargas, P.A.: Combining Lévy walks and flocking for cooperative surveillance using aerial swarms. In: Bassiliades, N., Chalkiadakis, G., de Jonge, D. (eds.) EUMAS/AT -2020. LNCS (LNAI), vol. 12520, pp. 226–242. Springer, Cham (2020). https://doi.org/10.1007/978-3-030-66412-1_15
22. Sims, D.W., et al.: Scaling laws of marine predator search behaviour. Nature **451**(7182), 1098–1102 (2008)
23. Sutantyo, D., Levi, P., Möslinger, C., Read, M.: Collective-adaptive lévy flight for underwater multi-robot exploration. In: 2013 IEEE International Conference on Mechatronics and Automation, pp. 456–462. IEEE (2013)
24. Sutantyo, D.K., Kernbach, S., Levi, P., Nepomnyashchikh, V.A.: Multi-robot searching algorithm using lévy flight and artificial potential field. In: 2010 IEEE Safety Security and Rescue Robotics, pp. 1–6. IEEE (2010)
25. Vishwakarma, M., Di Russo, J., Probst, D., Schwarz, U.S., Das, T., Spatz, J.P.: Mechanical interactions among followers determine the emergence of leaders in migrating epithelial cell collectives. Nat. Commun. **9**(1), 3469 (2018)
26. Viswanathan, G.M., Afanasyev, V., Buldyrev, S.V., Murphy, E.J., Prince, P.A., Stanley, H.E.: Lévy flight search patterns of wandering albatrosses. Nature **381**(6581), 413–415 (1996)
27. Viswanathan, G.M., Buldyrev, S.V., Havlin, S., da Luz, M.G., Raposo, E.P., Stanley, H.E.: Optimizing the success of random searches. Nature **401**(6756), 911–914 (1999)

Decentralized Conflict Resolution for Navigation in Swarm Robotics

Sebastian Mai[1(✉)] and Sanaz Mostaghim[1,2]

[1] Chair of Computational Intelligence, Faculty of Computer Science,
Otto von Guericke University, Magdeburg, Germany
{sebastian.mai,sanaz.mostaghim}@ovgu.de
[2] Fraunhofer Institute for Transportation and Infrastructure Systems IVI,
Dresden, Germany

Abstract. The main challenge in multi-robot navigation is the resolution of navigation conflicts between agents caused by intersecting paths. The contributions of this paper are: The creation of a robot framework that allows us to study the emergent behavior of a swarm of robots in a navigation task that involves frequent conflicts, a baseline strategy for decentralized conflict resolution and a more advanced strategy called Decentralized Collective Conflict Resolution (DCCR). To resolve conflicts, the DCCR strategy uses communication between robots and collective decision-making to create planning constraints, while the baseline strategy creates implicit priorities from communication timing. To show the effectiveness of our methods, we use a swarm of up to ten TurtleBot3 robots applying the DCCR and the baseline strategies in simulated and real-world experiments. In scenarios with more than six robots, the baseline strategy is outperformed by the DCCR strategy, which has more resilience against congestion.

1 Introduction

The development of efficient and versatile algorithms for multi-robot navigation can improve the quality of life in various sectors of society. By enabling multiple robots to cooperate seamlessly, automation can bring benefits to fields such as agriculture, mobility, and logistics. When multiple robots navigate a shared environment, negative emergent effects like delays and deadlocks caused by congestion limit the scalability of robot swarms. These effects are caused by navigation conflicts that occur when the planned paths of two robots intersect. The contributions of this paper are the creation of a robotic framework which allows the evaluation of decentralized navigation behaviors with a swarm of real TurtleBot3 robots and two strategies for conflict resolution called Implicit Priorities and Decentralized Collective Conflict Resolution (DCCR). The DCCR strategy is based on our previous work [19], an algorithm called Collective Conflict Resolution (CCR), which uses collective decision-making to find priorities between robots. We implemented the behavior using the TurtleBot3 platform and the Driving Swarm [20] package for the ROS 2 middleware. Our experiments with up to ten robots indicate that both the baseline approach (Implicit Priorities) and the novel DCCR approach show linear performance with less than six robots, while our approach outperforms the baseline as the number of robots is increased and congestion limits performance.

The remainder of this paper is structured as follows: Sect. 2 describes related works. Next, we explain the navigation behavior of individual robots and the conflict resolution strategies. Section 4 describes which experiments we conducted and how we interpret the results. Finally, Sect. 5 concludes the paper and highlights relevant questions for future research.

2 Related Works

Multi-robot navigation is an emerging topic in centralized multi-robot systems [10–12] and swarm robotics [1,30]. Centralized multi-robot systems often use the MAPF problem [26,27] to formalize the problem of conflict free navigation. Many algorithms exist, that can solve the MAPF problem optimally [24] or suboptimally [16,22]. Sampling-based planners can also be used to optimize the joint plan of all robots [18]. The disadvantage of directly optimizing the high dimensional state space renders real-time implementations infeasible for large groups of robots. The plan execution of MAPF plans with real robots was conducted by Barták et al. [3,4]. In this experiment, the robots were unable to accurately follow the timing in each plan, which sometimes led to plan failures. One possibility to address this problem is to use a planning algorithm, capable of finding robust solutions [2,4]. In our paper, we take a different approach and convert the discrete plans to precedence constraints, an approach which is first presented in [11]. Other experiments with real robots include [12], where MAPF planning is successfully applied to a swarm of nano-copters, while [10] is focused on high-level planning in warehouse applications with pre-existing roadmaps. In contrast to our roadmap, other planners perform the planning on a grid-based environment, but translate the grid-plan directly into continuous plans [5]. This requires the solver to act on a different set of constraints (MAPF-with large agents, cf. [14]). While MAPF-based approaches inherently aim to solve conflicts between robots at a global level, in recent studies there are new attempts to allow a low number of conflicts and require an additional local planning to resolve those conflicts [31]. Strategies for local conflict resolution include Control Barrier Functions [6], Velocity Obstacles [28] and Buffered Voronoi cells [21].

In contrast, we aim to solve the same navigation problem in a decentralized manner, which is different in two key ways: (1) The computation is concurrently for all robots and (2) communication between robots is limited and not synchronized with communication. One approach to solve such problems in a decentralized manner is to train a machine learning model (in this case graph neural networks) with optimal plans. The model can then output a local plan for a robot in a given state without global knowledge [15,32]. Raymond et al. [23] use the concept of fairness to resolve conflicts between agents in local planning.

A different approach is used by [30] that uses position swapping between robots to solve a MAPF problem in an efficient and distributed way. In contrast to our approach, the robots must be able to swap their position (state in the roadmap graph), which is not possible in our scenario. [1] performs experiments with a swarm of kilobots, that do not rely on path planning to find a way. Instead, robots can use the structure of the swarm for navigation, for example by following another robot or going towards the nearest robot.

3 Navigation Behavior and Conflict Resolution

The behavior of the robots is defined by two components: The navigation of an individual robot, and the strategy for conflict resolution. Robots navigate using a roadmap Graph G, that allows high-level planning and conflict resolution in a discrete domain. Each state s_k in the graph G corresponds to a region within the workspace of the robots. When each robot occupies the inner area of such a region (squares, shaded in blue), collisions between robots are impossible due to the shape of each region (cf. [17]). If robot i can transition from state s_k^i to s_{k+1}^i without the possibility of collision with robots in other regions, there is an edge s_k^i, s_{k+1}^i in the roadmap graph (visualized as a blue line). The high-level planner uses the discrete graph G for planning, while the low-level planner uses the geometry of the regions in the workspace associated with each state $s \in G$ for plan execution.

Planning and Plan Execution: Upon receiving a goal position, a robot converts its current pose $(x, y, \theta) = \boldsymbol{x}(t) \in SE(2)$ and its goal pose $\boldsymbol{x}_g \in SE(2)$ into discrete states $(s_0; s_g)$ using the roadmap graph G. Within in the roadmap graph G, the high-level planner computes a discrete trajectory for robot i: $\pi^i = (s_0^i, s_1^i, ... s_k^i, ..., s_g^i)$ using the Spacetime-A* algorithm [25]. The algorithm computes the most cost-effective trajectory from start to goal, subject to a set of constraints $C_i = \{(s_k^i \neq s)|s_k^i \in G, s \in G\}$ which is created by the conflict resolution strategies. Due to the planning constraints, a trajectory may contain a state twice $s_{k_1}^i = s_{k_2}^i$, for example when the robot must wait for another robot to move out of the way. The cost of a path is set as follows: Each edge (s_k^i, s_{k+1}^i) has a cost equal to the center to center distance between the two regions (states), which for our map is equal to 0.5 m. The low-level planner executes the discrete plan $\pi_i = (s_0^i, s_1^i, ..., s_g^i)$ computed by the high-level planner. During plan execution, the robot i moves through the regions in the workspace corresponding to the states s_k^i of the discrete plan. The planner uses a visibility graph and the Rotate-Translate-Rotate vehicle model [13] to compute a continuous trajectory, which is executed using the dynamic window approach [7]. All robots while navigating simultaneously compute, publish, and execute their planned trajectory $\pi_i = (s_0^i, s_1^i, s_2^i, ... s_g^i)$. A conflict between two plans π_i, π_j occurs, when two equal states $s_{k_1}^i = s_{k_2}^j$ are present in both plans π_i, π_j at adjacent timesteps $|k_1 - k_2| \leq 1$. In conflict resolution, we only consider conflicts which occur before a cutoff point of $k_c = 6$ time-steps. All robots use a set of constraints

$C_i = \{(s_k^i \neq s) | s_k^i \in G, s \in G\}$ for planning, which allows finding conflict free plans, given the right set of constraints. Each constraint restricts the robot i to move to a specific state s at time k. The high-level planner can then find the plan that satisfies the constraints. Each robot broadcasts their plan $\pi_i = (s_0^i, s_1^i, ...)$ to the other robots.

Implicit Priorities: The most common approach to resolving inter robot conflicts in navigation is prioritized planning [29]. In this method, robots use predefined priority (e.g., their IP-address) and the robot with smaller priority uses constraints in its planning to avoid the conflict. In our method, we assume that the plan that was communicated first has the higher priority. The robot's plan is recomputed if the state, or the goal of the robot changes or if an updated plan is received from another robot, which leads to a new set of planning constraints. Given the plan π_j communicated by robot j, robot i adds the following constraints for high-level planning $C_i \cup \{s_{k-1}^i \neq s_k^j, s_k^i \neq s_k^j, s_{k+1}^i \neq s_k^j | \forall s_k^j \in \pi_j\}$. The new plan π_i, which is computed using the constraints C_i, cannot contain any conflicts with the plan π_j.

With implicit priorities, the robots can dynamically change priorities, while in some scenarios a set of fixed priorities would lead to a deadlock. The cost of this flexibility is the amount of re-planning and communication needed to come up with a conflict free plan. Because all robots do their planning concurrently, plans are usually updated multiple times before a conflict free plan is found for each robot.

Decentralized Collective Conflict Resolution: Similar to the Implicit Priorities, the navigation with Collective Conflict Resolution (CCR) [19] uses priorities to resolve navigation conflicts. The main idea of the CCR [19] algorithm is the way how priorities between robots are found using collective decision-making. To make an informed (collective) decision about the priorities, we need to model the priorities in a way that lets a robot estimate the quality of a decision. The quality estimate of all robots is then used to find a consensus on the set of priorities used for planning (collective decision-making).

A conflict occurs, by a robot entering a specific region (state s) in the environment, which is already occupied. We define the priorities to decide, which robot is allowed to enter a state s first. The edges of the roadmap graph G describe the possible state transitions. We assign priorities $B = \{b_r, s \in \mathbb{R} | \forall (r, s) \in G\}$ to each edge of the roadmap graph. In the high-level planner we make sure that robot i using edge (r_1, s) moves to state s before robot j which uses the edge (r_2, s), if $b_{r_1,s} > b_{r_2,s}$. The waiting action is always prioritized, i.e., $b_{s,s} > b_{r,s} \forall r \neq s$. In the DCCR method, we use the concept of collective decision-making [8] in two ways: (1) to find a consensus on the priorities $b_{r,s} \in B$, such that all robots in the same neighborhood use the same set of priorities B for planning. (2) to find and update such a consensus without a central server keeping track of the priorities. The definition of priorities allows us to determine whether a set of priorities $b_{r,s}$ for a single state s is beneficial for a given robot i. If the path $\pi_i = (s_0^i, ..., s_k^i, s_{k+1}^i, ...)$ contains two states s_k^i, s_{k+1}^i, the robot is quicker if the priority is aligned with the robot's plan, i.e., $b_{s_k^i, s_{k+1}^i}$ is higher than the other priorities $b_{r,s_{k+1}^i}, r \neq s_k^i$ leading to the state s_{k+1}^i.

The quality of each priority $b_{r,s}$ is computed based on a network flow, which takes alternative routes into account. A specific priority value is more important, when no alternative route with similar cost exists. We set the capacity of all edges $c(r, s)$ for a state s as follows, with the highest priority to 100, the priorities of all edges with no set priority to 50. We sort the remaining edges by their priority and assign a capacity of $1, 2, 3...$ according to the sorted priority values of the edges. Using those capacity values, we compute a maximum (s_0, s_g)-flow of minimum cost $F(B, G)$, where we can read the flow values $f(r, s)$ for each edge. We use this flow network to compute the quality of an edge is $q(r, s) = \frac{f(r,s)}{f_{\max}}$, i.e., the value of flow going through the edge normalized by the flow value of the source of the flow network F. This means the quality $q(r, s)$ is high, if all flow goes through the edge (r, s) and lower if only part of the flow goes through the edge. Computing the quality metric with the flow graph has several advantages: A robot which is not using state s will compute $q(r, s) = 0$ for all edges leading to s, i.e., the decision is not affected by robots that cannot use state s in their plan. If robots have no alternative route to the goal $q(r, s) = 1$, while robots that do have an alternative route will compute a lower quality $q(r, s) < 1$. Therefore, robots without an alternative route will get higher priority. When a robot plans to use state s_k^i in its plan π_i and there is a conflict at state s_k^i with the plan of another robot π_j, a message is sent to all other robots to trigger decision-making. Each robot receiving the message computes quality values $q(r, s_k^i)$ for all edges $(r, s_k^i) \in G$ leading to state s_k^i. As a decision function, we compute the average of the quality-values for each robot participating in the decision: $b_{r,s} = \sum \frac{q_i(r,s)}{N}$. In using this decision function, we assume a fully connected and error-free communication network.

A further novelty of the DCCR method is that the priorities can expire, to account for the dynamic configuration of the robot swarm. In contrast to [19] we use a task that does not end., a way to reconfigure the priorities is beneficial. A priority $b_{r,s}$ which is older than the priority lifetime of the robot, which is set randomly to $\mathcal{N}(t_b, \sigma_b)$ once the robot is started. The priorities $b_{r,s_k^i} \forall (r, s_k^i) \in G$ are also deleted, once a robot has executed its plan past the state s_k^i $s_k^i \notin \pi_i$, assuming that the priority is no longer relevant. If there is a conflict at state s_k^i in the plan π_i after the priorities associated to s_k^i were deleted from the memory of robot i, a new message is sent to trigger decision-making.

4 Experiments and Results

In our experiments, we want to observe the emergent behavior of the robot swarm across different scales (cf. [9]), from a single robot $N = 1$ to $N = 10$ robots. Figure 2 shows the swarm of 10 robots navigating the benchmark environment. The task of the robots is to drive back and forth between a predefined start- and goal position. Each pair of start- and goal position is shared by two robots. This

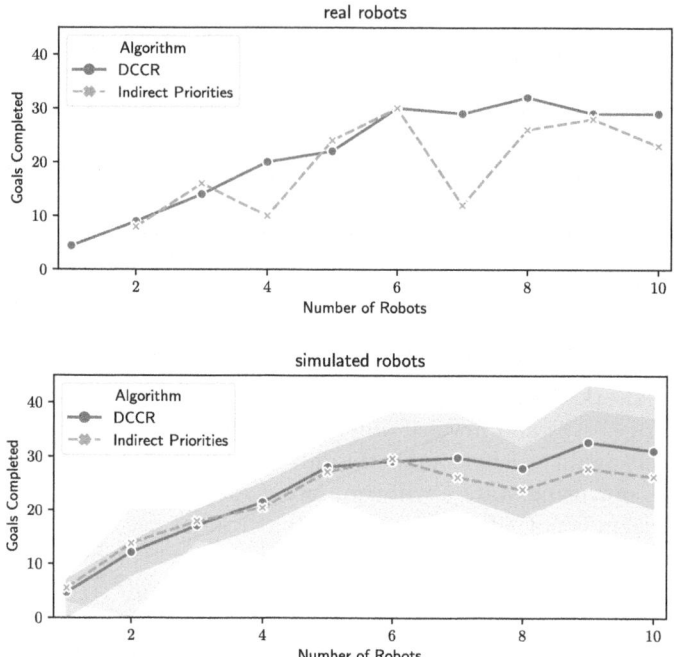

Fig. 1. Goals completed during one run by all robots. For the simulated runs, the 80%-percentile interval is shown as a colored band. The gray line indicates linear performance with 5 goals per robot.

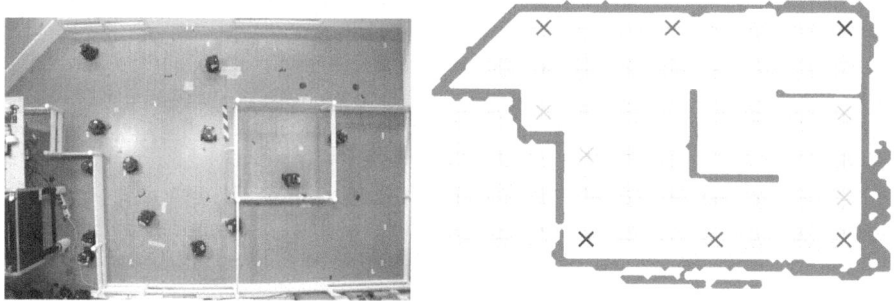

Fig. 2. Real experiments with $N = 10$ robots (left) and roadmap Graph G used for navigation (right). The working area has a size of 3×5 m. The start and goal position for each pair of robots is marked with ×.

task leads to a dynamic behavior, as navigation conflicts are frequent and plans must be updated and adapted during runtime, once robots reach a goal position. During each experiment, we gather data on the position ($\boldsymbol{x}(t)$, orientation $\theta(t)$), state $s(t)$ and goal $s_g(t)$ for each robot. In addition, we keep track of the number of goals completed. Each experiment runs for 15 min, a single robot completes 5 goals during this time. We implemented the robot behavior in Python using ROS2 and the Driving Swarm [20] framework.[1]

The results confirm that the behavior works in a swarm of real robots. In preliminary experiments without conflict resolution, deadlocks occur even with $N = 2$ robots. During one 15 min run of the experiments, a single robot completes an average of 5 goals. Figure 1 shows how the performance of the swarm scales with increasing number of robots with real robots and in simulation. We only perform one run for each swarm size N with real robots, we performed a total of 10 runs per configuration in simulation (bottom). Our results show linear performance for $N \leq 5$ robots and in some cases even super linear performance, i.e., the swarm completes more goals than predicted. The performance stagnates or decreases when using more than 5 robots. When we compare the baseline approach (implicit priorities) and DCCR, we see that while both approaches show linear performance, the DCCR approach is slightly better for $N > 6$ robots. This is surprising when we compare the results to our previous paper, where the success rate for prioritized planning is much lower than for the CCR approach (47% vs. 78%) [19].

5 Conclusion

In this paper, we analyze the emergent behavior of a swarm of up to ten real TurtleBot3 robots that navigate cooperatively. The robots replan their paths in case the plans are in conflict using two different methods: (1) Plan using Implicit Priorities, which are created using communication timing, and (2) Decentralized Collective Conflict Resolution (DCCR), which uses explicit communication and collective decision-making to generate planning constraints. The results of our experiments are that the number of goals completed per robot is constant when less than six robots operate at once. When employing the DCCR method, congestion has a lower impact on performance in comparison to the baseline approach. The conflict resolution works in our experiments, more research is required to find out how the results transfer to other scenarios. In the future, we want to use the robot framework to find out how different quality metrics and decision function affect the conflict resolution in different scenarios.

References

1. Ali, Z., Meehan, K., Hyndman, J., Dowling, T.: Optimising swarm robotic navigation: a comparative analysis of fastest path vs. nearest neighbour path projection strategies. In: 2023 31st Irish Conference on Artificial Intelligence and Cognitive Science (AICS), pp. 1–6 (2023). https://doi.org/10.1109/AICS60730.2023.10470717

---

[1] The code for the planner is available as a python package at https://github.com/ovgu-FINken/polygonal_roadmaps and for the ROS2 implementation at https://github.con/ovgu-FINken/driving_swarm_infrastructure.

2. Atzmon, D., Stern, R., Felner, A., Wagner, G., Bartak, R., Zhou, N.F.: Robust multi-agent path finding. In: Proceedings of the International Symposium on Combinatorial Search, vol. 9, no. 1, pp. 2–9 (2018). https://doi.org/10.1609/socs.v9i1.18445
3. Barták, R., Mestek, J.: OzoMorph: demonstrating colored multi-agent path finding on real robots. In: Proceedings of the AAAI Conference on Artificial Intelligence, vol. 35, pp. 15991–15993 (2021). https://doi.org/10.1609/aaai.v35i18.17990
4. Barták, R., Švancara, J., Škopková, V., Nohejl, D., Krasičenko, I.: Multi-agent path finding on real robots. AI Commun. **32**(3), 175–189 (2019). https://doi.org/10.3233/aic-190621
5. Chen, J., Li, J., Fan, C., Williams, B.: Scalable and safe multi-agent motion planning with nonlinear dynamics and bounded disturbances. arXiv (2020)
6. Chen, Y., Singletary, A., Ames, A.D.: Guaranteed obstacle avoidance for multi-robot operations with limited actuation: a control barrier function approach. IEEE Control Syst. Lett. **5**(1), 127–132 (2021). https://doi.org/10.1109/LCSYS.2020.3000748
7. Fox, D., Burgard, W., Thrun, S.: The dynamic window approach to collision avoidance. IEEE Robot. Autom. Mag. **4**(1), 23–33 (1997). https://doi.org/10.1109/100.580977
8. Hamann, H.: Swarm Robotics: A Formal Approach. Springer, Cham (2018). https://doi.org/10.1007/978-3-319-74528-2
9. Hamann, H., Reina, A.: Scalability in computing and robotics. IEEE Trans. Comput. **71**(6), 1453–1465 (2022). https://doi.org/10.1109/TC.2021.3089044
10. Holdener, E., Chemodanov, D.: EVA : An evolutionary architecture for network virtualization, pp. 1192–1197
11. Hönig, W., et al.: Summary: multi-agent path finding with kinematic constraints *. Technical report (2017)
12. Honig, W., Preiss, J.A., Kumar, T.K., Sukhatme, G.S., Ayanian, N.: Trajectory planning for quadrotor swarms. IEEE Trans. Rob. **34**(4), 856–869 (2018). https://doi.org/10.1109/TRO.2018.2853613
13. LaValle, S.M.: Planning Algorithms (2006). https://doi.org/10.1017/CBO9780511546877
14. Li, J., Felner, A., Koenig, S.: Multi-agent path finding for large agents. In: AAAI Conference on Artificial Intelligence (AAAI) (2019)
15. Li, Q., Gama, F., Ribeiro, A., Prorok, A.: Graph neural networks for decentralized multi-robot path planning (2019)
16. Ma, H., Harabor, D., Stuckey, P.J., Li, J., Koenig, S.: Searching with consistent prioritization for multi-agent path finding. In: 33rd AAAI Conference on Artificial Intelligence, AAAI 2019, 31st Innovative Applications of Artificial Intelligence Conference, IAAI 2019 and the 9th AAAI Symposium on Educational Advances in Artificial Intelligence, EAAI 2019, pp. 7643–7650 (2019)
17. Mai, S., Deubel, M., Mostaghim, S.: Multi-objective roadmap optimization for multiagent navigation. In: 2022 IEEE Congress on Evolutionary Computation (CEC), pp. 1–8 (2022). https://doi.org/10.1109/CEC55065.2022.9870300
18. Mai, S., Mostaghim, S.: Modeling pathfinding for swarm robotics. In: Dorigo, M., et al. (eds.) ANTS 2020. LNCS, vol. 12421, pp. 190–202. Springer, Cham (2020). https://doi.org/10.1007/978-3-030-60376-2_15
19. Mai, S., Mostaghim, S.: Collective decision-making for conflict resolution in multi-agent pathfinding. In: Dorigo, M., et al. (eds.) ANTS 2022. LNCS, vol. 13491, pp. 79–90. Springer, Cham (2022). https://doi.org/10.1007/978-3-031-20176-9_7

20. Mai, S., Traichel, N., Mostaghim, S.: Driving swarm: a swarm robotics framework for intelligent navigation in a self-organized world. In: 2022 IEEE International Conference on Robotics and Automation (ICRA), pp. 4958–4964. IEEE, Philadelphia (2022). https://doi.org/10.1109/ICRA46639.2022.9811852
21. Pierson, A., Schwarting, W., Karaman, S., Rus, D.: Weighted buffered voronoi cells for distributed semi-cooperative behavior. In: ICRA 2020, pp. 5611–5617 (2020)
22. Rahman, M., Alam, M., Islam, M.M., Iqbal, T., Rahman, I., Khan, M.: An adaptive agent-specific sub-optimal bounding approach for multi-agent path finding. IEEE Access **10**, 1 (2022). https://doi.org/10.1109/ACCESS.2022.3151092
23. Raymond, A., Malencia, M., Paulino-Passos, G., Prorok, A.: Agree to disagree: subjective fairness in privacy-restricted decentralised conflict resolution. Front. Robot. AI **9** (2022). https://doi.org/10.3389/frobt.2022.733876
24. Sharon, G., Stern, R., Felner, A., Sturtevant, N.: Meta-agent conflict-based search for optimal multi-agent path finding. In: Proceedings of the 5th Annual Symposium on Combinatorial Search, SoCS 2012, pp. 97–104 (2012)
25. Silver, D.: Cooperative pathfinding the problem with A * reservation table. In: Proceedings of the First Conference on Artificial Intelligence and Interactive Digital Entertainment (2005)
26. Stern, R.: Multi-agent path finding – an overview. In: Osipov, G.S., Panov, A.I., Yakovlev, K.S. (eds.) Artificial Intelligence. LNCS (LNAI), vol. 11866, pp. 96–115. Springer, Cham (2019). https://doi.org/10.1007/978-3-030-33274-7_6
27. Stern, R., et al.: Multi-agent pathfinding: definitions, variants, and benchmarks. In: AAAI Conference on Artificial Intelligence (AAAI) (2019)
28. Van Den Berg, J., Lin, M., Manocha, D.: Reciprocal velocity obstacles for real-time multi-agent collision avoidance. In: Proceedings of IEEE International Conference on Robotics and Automation, pp. 1928–1935 (2007)
29. Van Den Berg, J.P., Overmars, M.H.: Prioritized motion planning for multiple robots. In: 2005 IEEE/RSJ International Conference on Intelligent Robots and Systems, IROS, pp. 430–435 (2005). https://doi.org/10.1109/IROS.2005.1545306
30. Wang, H., Rubenstein, M.: Walk, stop, count, and swap: decentralized multi-Agent path finding with theoretical guarantees. IEEE Robot. Autom. Lett. **5**(2), 1119–1126 (2020). https://doi.org/10.1109/LRA.2020.2967317
31. Weise, J., Mai, S., Zille, H., Mostaghim, S.: On the scalable multi-objective multi-agent pathfinding problem. In: Accepted at Congress on Evolutionary Computing CEC 2020 (2020)
32. Wu, W., Bhattacharya, S., Prorok, A.: Multi-robot path deconfliction through prioritization by path prospects. In: 2020 IEEE International Conference on Robotics and Automation (ICRA), pp. 9809–9815. IEEE (2020)

On the Design of Control Mechanisms for a Site Selection Task in a Simulated Swarm of Robots

Ahmed Almansoori[1,2]([✉])[iD], Dari Trendafilov[1][iD], Muhanad Alkilabi[1][iD], and Elio Tuci[1][iD]

[1] Faculty of Computer Science, University of Namur, Namur, Belgium
ahmed.almansoori@unamur.be
[2] Faculty of Engineering, University of Kerbala, Kerbala, Iraq

Abstract. Collective decision-making refers to a decision process by a group of agents in which, once the decision is made, it cannot be attributed to any of its group members. In this study, we design decision-making mechanisms, using evolutionary methods, to allow a swarm of simulated robots to make a collective decision in a site selection scenario. That is, the robots have to reach a consensus on which site is the best among those available in the environment. The original contribution of this study is in demonstrating that the design process can be free from several assumptions, made in previous related research work, on crucial elements underpinning the individual and group-level response.

1 Introduction

Swarm robotics is a field of research which draws inspiration from collective behaviour observed in nature [8]. Despite the relative simplicity of individuals, these systems exhibit strikingly complex and coordinated group actions that emerge from decentralised local interactions [6]. In a swarm of robots, simple rules followed by many individuals can produce intricate group behaviours. The challenge for roboticists is to design the individual mechanisms that underpin the group-level response. This process not only guides the development of efficient adaptable robots, but also feeds back insights, providing a deeper understanding of collective behaviour in natural swarms [10]. To address this design problem, researchers have applied various approaches, such as hand-coded mechanisms and automatic design methods, such as those based on the use of evolutionary computation techniques and artificial neural networks, generally referred to as evolutionary swarm robotics [21,22]. Hand-crafted mechanisms produce behavioural strategies that can be easily described in operational terms, even if they tend to lack flexibility in unforeseen situations [5]. On the other hand, the evolutionary approach in swarm robotics creates mechanisms that support relatively robust and adaptable behavioural strategies. Unfortunately, these strategies are hard to interpret in operational terms [13,15,19,23].

A few years ago, we launched a research project aiming to evaluate the robustness, adaptability and scalability of mechanisms designed using the evolutionary swarm robotics approach for swarms of robots engaged in collective decision-making scenarios [25]. In particular, we focused on perceptual discrimination tasks, where the robots have to collectively decide which element is the most represented in the arena. We demonstrated that neuro-controllers synthesised using evolutionary computation techniques underpin group responses that are equally accurate and also more robust and more scalable with respect to the group size than group responses generated with hand-coded mechanisms in perceptual discrimination tasks [1–3]. The objective of this study is to further contribute to the analysis of neural mechanisms for swarms of robots engaged in collective decision-making by extending the work to the site-selection scenario.

The swarm robotics literature targeting site selection scenarios studies tasks in which the group has to collectively decide which site is the best among those available in the environment [11,16,17]. In [26], a swarm of robots converge to consensus in a binary site-selection task using the voter model. The latter refers to a process which makes a robot switch to the opinion of a randomly chosen robot among those that are spatially close. In subsequent studies [24,27], the authors looked at alternative opinion formation mechanisms, such as the majority rule, which makes a robot switch to the most frequent opinion among the n spatially close robots. In these studies, robots alternate between exploring options and disseminating their current opinion for a time proportional to the option's quality. This modulation of the dissemination time is the element that allows the group to achieve consensus on the best quality option. These studies have inspired further research [7,18], which investigated the effect different opinion formation strategies have on the decision dynamics in a binary site selection scenario. [20] explores the relationship between maximum communication distance and the accuracy of collective decision-making processes in dynamic environments with three sites of different quality, using the voter model. Moreover, individual and social information is combined through behaviour based on the cross-inhibition pattern, where conflicting information between two communicating robots causes the robots to reset their opinions and poll other robots' opinions.

To the best of our knowledge, the large majority of research on site selection in swarm robotics is based on hand-coded mechanisms. Contrary to prior work, in this study, we employ the evolutionary swarm robotics methods with the intent to question a series of important assumptions that characterise most previous studies. In particular, we have developed a setup in which there is no built-in correlation between environmental features and robot behaviour, such as the positive correlation between option quality and dissemination time used in [7,18,24,26,27] and in several other studies on site selection. Moreover, in our work, there is no built-in assumption concerning how the robots exploit perceptual cues and communication signals to generate their opinion, as it is with the voter or majority model. In our setup, evolution builds the individual mechanisms to integrate perceptual and communication signals, as well as the relationships between the elements that generate the positive/negative modulations that break the initial equilibrium and push the group toward consensus on the best option. As shown in Sect. 3, this approach generates effective, robust and scalable group strategies. One final consideration is related to the task. As discussed in Sect. 2, in spite of the simplifications dictated by the necessity to limit the complexity of this initial study, the task not only possesses all the relevant characteristics of classic site selection scenarios, but also features some properties that make it particularly challenging. For example, outside the social context, single robots do not have the means to systematically make correct decisions on the best quality site. This makes it particularly hard for the swarm designer to predict and potentially hand-code the actions underpinning virtuous collective responses. In view of the promising results that we obtain in terms of accuracy and robustness of the decision-making strategies, we will discuss the potential further developments of this research in Sect. 4.

2 Methods

The site-selection scenario investigated in this study is implemented using a single square arena of 2 m × 2 m in which 21 simulated robots pseudo-randomly move for the entire duration of the evaluation while avoiding collisions with other robots and arena walls. Alike similar studies, the robots alternate between two phases of the task: i) dissemination phase, in which they are allowed to communicate their respective opinions on the best quality site to spatially close robots, but they cannot sample the sites' quality; and ii) exploration phase, in which the robots perceive the cue signalling the respective sites' quality, but they cannot communicate. As in other studies on site selection with simulated agents (e.g., see [18]), our experimental setup is simplified with respect to those elements that can be omitted while preserving the distinctive challenges related to the development of a consensus to the best quality option through a process based on a collective decision. In our scenario, depicted in Fig. 1a, the sites are not distinctive locations which the robots join and leave based on the task phase. The robots never leave the square arena, therefore, the entire swarm operates

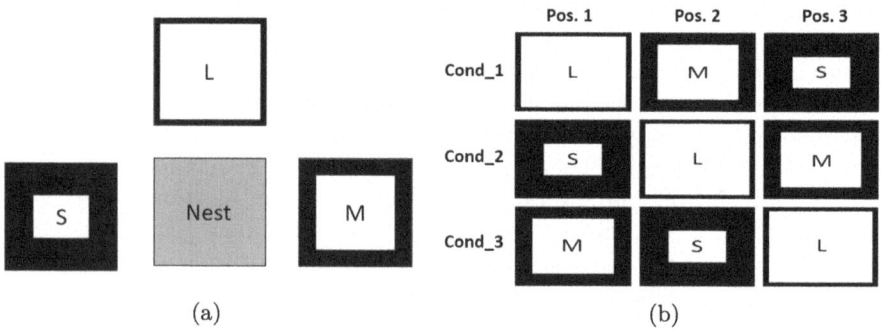

Fig. 1. a) Graphical representation of the site selection task. b) The position of sites in the three experimental conditions.

in the same space during evaluation. The back-and-forth movement between the nest and the sites is implemented through modifications of the robots' perceptual system, for which either communication is allowed (this consists of the robot being in a dissemination phase and metaphorically placed in the nest) or the site quality sampling is allowed (this consists of the robot being in an exploration phase and metaphorically placed in the site corresponding to its current opinion). This way of instantiating the nest/sites is meant to simplify the scenario with respect to the following: i) it eliminates the requirement to develop control mechanisms underpinning the back-and-forth navigation between nest and sites; ii) it reduces the evaluation time by eliminating the nest-sites transfer time. This is particularly helpful for reducing the design time of decision mechanisms which tends to be quite large with the use of evolutionary computation techniques.

The scenario features three different sites (or options) which differ from each other in the proportion of white floor. In site L, 80% of the floor is white and 20% is black; in site M, 50% of the floor is white and 50% is black; in site S, 20% of the floor is white and 80% is black. The higher the proportion of white the better the site's quality. The nest is a location in which the colour of the floor is grey. The positions of the sites in the metaphorical space change based on the experimental condition: in Cond_1, site L is in position 1, site M is in position 2, and site S in position 3; in Cond_2, site S is in position 1, site L is in position 2, and site M in position 3; in Cond_3, site M is in position 1, site S is in position 2, and site L in position 3 (see Fig. 1b). Evaluation takes place in all three conditions. The robots have to decide which position hosts the best quality site on the current evaluation or trial. At the beginning of each trial, the robots' opinions are uniformly distributed over the three sites. For each robot the duration of both the exploration and the dissemination phase are sampled independently from an exponential distribution with 10 s mean. Each robot starts the trial in the exploration phase by sampling the quality of the site corresponding to its current opinion. At the end of its exploration phase, each robot switches to the dissemination phase in which it has no access to the cue signalling the quality of the site, but is allowed to disseminate its opinion through communication to spatially close robots. At the end of its dissemination phase, each robot returns to the exploration phase that takes place in the site corresponding to the robot's current opinion. The latter is derived from the most frequently expressed opinion during the latest dissemination phase. The metaphorical back-and-forth movement between the nest and the sites of the robots lasts 800 s—this is the trial time within which the group has to reach a consensus. Note that the independently sampled exploration and dissemination phases result in a variety of durations for different robots, which eventually brings the group into an "asynchronous" state where different robots operate in different phases simultaneously. A trial is considered successful whenever all robots hold the same opinion concerning the position of the best quality site for at least 10 s consecutively.

We simulate an e-puck2 robot [14], a platform frequently used in swarm robotic experiments. Our simulated e-puck2 is equipped with eight infrared sensors located around the robot's body, a floor sensor placed on the bottom of the chassis, and a communication system implemented with a range and bearing board. During exploration, the floor sensor detects the colour of the floor underneath the robot and returns a single value, 0 for black and 1 for white. During dissemination, the floor sensor constantly reads 0.5, indicating that the robot is on a grey floor in the nest. The robot opinion is a three-bit signal which is: $\{1, 0, 0\}$ if the best quality site is considered located in position 1 (Pos. 1); $\{0, 1, 0\}$ if the best quality site is considered located in position 2 (Pos. 2); $\{0, 0, 1\}$ if the best quality site is considered located in position 3 (Pos. 3). For the entire duration of its dissemination phase, each robot emits a three-bit signal corresponding to its opinion. Every robot in the dissemination phase receives, for the entire duration of this phase, a signal from the spatially closest robot among those in the dissemination phase and located at less than 50 cm distance. If there is no disseminating robot within the communication range (<50 cm), the communication signal of a robot receiver is set to $\{0, 0, 0\}$. When a robot is in the exploration phase, the communication signal is constantly set to $\{0.5, 0.5, 0.5\}$. To compensate for the simulation-reality gap, uniform noise is added to all sensor readings, motor outputs and robot position (see [12] for a similar approach). We invite the reader to consider that this task is genuinely collective, since there is nothing that allows a single robot to unambiguously associate the quality of sites (i.e., L, M, and S) to specific positions (i.e., Pos. 1, Pos. 2, Pos. 3) as requested by the nature of the decision problem. As shown in Sect. 3, the consensus is achieved through a complex and difficult to disentangle dynamic process, in which perceptual experience and social interactions are both essential to bring forth virtuous collective dynamics leading to the emergence of an agreement on the best quality option.

The robot controller is made of two modules. The walk module generates the pseudo-random walk and obstacle avoidance behaviour. The decision module generates the robot's opinion. The walk module makes the robot move according to an isotropic random walk consisting of straight motion for 5 s at a speed of 20 cm/s and rotation with turning angles sampled from a wrapped Cauchy distribution [9]. The decision module is a three-layer dynamic neural network [4], with a fully recurrent hidden layer made of four neurons. The output layer, made of three neurons, generates the robot opinion by normalising the activation of the output neurons with a softmax function. A tournament selection evolutionary algorithm using linear ranking is employed to set the network parameters. At the beginning of each randomly seeded evaluation trial, the decision module is reset (i.e., robots have no memories from previous evaluations) and cloned on each of the 21 robots forming a homogeneous swarm. The robots are randomly positioned in the arena with a randomly chosen orientation in $[0, 2\pi]$. Each robot performs a pseudo-random walk as illustrated above for the entire duration of the trial (i.e., 800 s, $T = 8000$ simulation cycles). The fitness of a genotype is defined as the average group evaluation score after it has been assessed once in each condition. The group evaluation score is computed by counting how many robots hold the correct opinion during the second half of each trial.

3 Results

We performed ten differently seeded evolutionary runs, each one lasting 10,000 generations. During evolution, we evaluated each genotype once in each condition, for a total of three trials. We remind the reader that the objective of the robots is to collectively decide in which position (among the three available) is located the site with the largest portion of the white floor. Consensus is achieved whenever all robots share the same "correct" opinion for an interval of at least 10 s. At the end of the evolution, we evaluated the best genotypes (the elites) from the last 100 generations of each run in 150 trials (i.e., 50 trials per condition) with the fitness function used during evolution. Multiple genotypes from different evolutionary runs managed to produce groups capable of highly effective performance in these 150 evaluation trials. We selected one of these groups (hereafter, referred to as the "best" group) for further post-evaluation tests in order to learn more about the strategy used by the robots to repeatedly achieve consensus on the best option in different conditions and to test the robustness and scalability of the solution. In the remainder of this section, we illustrate the results of these post-evaluation tests. Note that, although only a single group is discussed here, similar strategies with equally effective performance have been observed in other groups selected from the elites of different evolutionary runs.

Figure 2a shows the distribution of the number of robots of the best group with the correct opinion in Cond_1 (white), Cond_2 (grey), and Cond_3 (black) measured at regular time intervals of 40 s over 50 differently seeded trials. This graph shows that the frequency of the correct opinion progressively increases in each condition, with a similar progression in Cond_2 and Cond_3. In Cond_1, the group systematically goes through an initial stage in which the correct opinion almost disappears from the group for progressively increasing its frequency up to consensus that systematically emerges around simulation cycle 400. In Cond_2 and Cond_3 the persistent variability that can be observed, even after half-trial

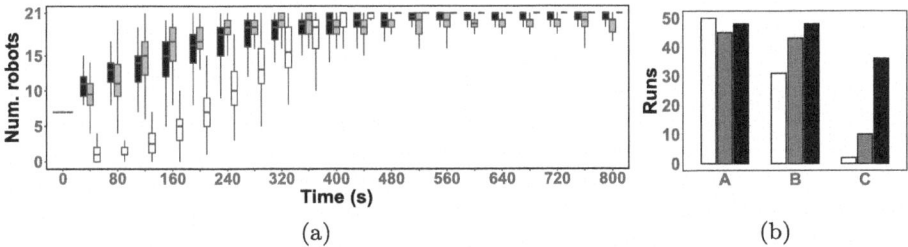

Fig. 2. a) Distribution of robots with correct opinion in Cond_1 (white), Cond_2 (grey), and Cond_3 (black) measured at regular time intervals of 40 s over 50 trials. b) Swarm consensus accuracy over 50 trials in three difficulty levels: A – white covers 80% in site L, 50% in site M, and 20% in site S; B – white covers 80% in site L, 60% in site M, and 40% in site S; C – white covers 80% in site L, 70% in site M, and 60% in site S.

time, is mainly due to a small number of robots (maximum three) that keep changing their opinions until the end of the trial.

To test the robustness of the evolved decision-making mechanisms, we analysed the accuracy (i.e., the number of successful trials out of 50) in scenarios in which the difference in quality between the sites is progressively reduced. In particular, we considered three cases: scenario A (same as during evolution), in which the white covers 80% of the floor in site L, 50% in site M, and 20% in site S; scenario B (potentially more difficult than A), in which the white covers 80% of the floor in site L, 60% in site M, and 40% in site S; and scenario C (potentially more difficult than B), in which the white covers 80% of the floor in site L, 70% in site M, and 60% in site S. Scenario C has the smallest difference between sites' quality and is potentially the most challenging among the three studied scenarios. The results of the robustness test, shown in Fig. 2b, indicate that: i) the swarm accuracy is above 80% in all conditions in scenario A; ii) the progressive reduction of the difference in quality between the sites tends to generate a drop in performance. However, the magnitude of this performance drop varies among the conditions, with Cond_1 being the most affected (see Fig. 2b, white) and Cond_3 the least affected (see Fig. 2b, black).

4 Conclusion

We demonstrated that accurate and robust collective decisions in a site selection scenario can be made by a swarm of robots without explicitly imposing processes that control: i) how perceptual cues are integrated to sample the environment; ii) how perception and communication interact to generate individual opinions; iii) how environmental features modulate behavioural responses eventually leading to consensus on the best quality option. The results of the quantitative evaluations suggest that our approach is largely accurate and surprisingly effective to cope with environmental variability. Future work will concentrate on: i) a more comprehensive analysis of the operational principles underlying opinion formation in single robots; ii) a progressive elimination of the simplifications that currently make it hard to port solutions to physical robots; iii) changing the site selection task by looking at more complex scenarios in which the cues signalling the site quality are distributed in a non-homogeneous way in the environment; iv) developing integrated neuro-controllers that support the robots in the opinion formation process as well as in the exploration of the environment.

Acknowledgements. This project has received funding from the CERUNA doctoral fellowship by the University of Namur and the European Union's Horizon 2020 research and innovation programme under the Marie Skłodowska-Curie grant agreement No 101034383. Computational resources have been provided by the Consortium des Équipements de Calcul Intensif (CÉCI), funded by the Fonds de la Recherche Scientifique de Belgique (F.R.S.-FNRS) under Grant No. 2.5020.11 and by the Walloon Region.

References

1. Almansoori, A., Alkilabi, M., Colin, J.N., Tuci, E.: On the evolution of mechanisms for collective decision making in a swarm of robots. In: Schneider, J.J., Weyland, M.S., Flumini, D., Füchslin, R.M. (eds.) WIVACE 2021. CCIS, vol. 1722, pp. 109–120. Springer, Cham (2021). https://doi.org/10.1007/978-3-031-23929-8_11
2. Almansoori, A., Alkilabi, M., Tuci, E.: On the evolution of mechanisms for three-option collective decision-making in a swarm of simulated robots. In: Proceedings of the Genetic and Evolutionary Computation Conference, pp. 4–12 (2023)
3. Almansoori, A., Alkilabi, M., Tuci, E.: On the evolution of adaptable and scalable mechanisms for collective decision-making in a swarm of robots. Swarm Intell. **20**(1), 1–21 (2024)
4. Beer, R.D.: A dynamical systems perspective on agent-environment interaction. Artif. Intell. **72**, 173–215 (1995)
5. Brambilla, M., Ferrante, E., Birattari, M., Dorigo, M.: Swarm robotics: a review from the swarm engineering perspective. Swarm Intell. **7**(1), 1–41 (2013)
6. Camazine, S., Deneubourg, J.L., Franks, N.R., Sneyd, J., Theraulaz, G., Bonabeau, E.: Self-organization in Biological Systems. Princeton University Press, Princeton (2001)
7. De Masi, G., Prasetyo, J., Zakir, R., Mankovskii, N., Ferrante, E., Tuci, E.: Robot swarm democracy: the importance of informed individuals against zealots. Swarm Intell. **15**, 315–338 (2021)
8. Dorigo, M., Şahin, E.: Guest editorial. Special issue: swarm robotics. Auton. Rob. **17**(2–3), 111–113 (2004)
9. Kato, S., Jones, M.C.: An extended family of circular distributions related to wrapped Cauchy distributions via Brownian motion. Bernoulli **19**(1), 154–171 (2013). https://doi.org/10.3150/11-BEJ397
10. Kennedy, J., Eberhart, R.C., Shi, Y.: Swarm Intelligence. Evolutionary Computation Series. Morgan Kaufmann (2001)
11. Lee, C., Lawry, J., Winfield, A.F.: Negative updating applied to the best-of-n problem with noisy qualities. Swarm Intell. **15**(1), 111–143 (2021)
12. Ligot, A., Birattari, M.: Simulation-only experiments to mimic the effects of the reality gap in the automatic design of robot swarms. Swarm Intell. **14**(1), 1–24 (2020)
13. Mendiburu, F.J., Ramos, D.G., Morais, M.R., Lima, A.M., Birattari, M.: Automode-mate: automatic off-line design of spatially-organizing behaviors for robot swarms. Swarm Evol. Comput. **74**, 101118 (2022)
14. Mondada, F., et al.: The e-puck, a robot designed for education in engineering. In: Proceedings of the 9th Conference on Autonomous Robot Systems and Competitions, vol. 1, pp. 59–65. IPCB: Instituto Politécnico de Castelo Branco (2009)
15. Nelson, A.L., Barlow, G.J., Doitsidis, L.: Fitness functions in evolutionary robotics: a survey and analysis. Robot. Auton. Syst. **57**(4), 345–370 (2009)
16. Parker, C.A.C., Zhang, H.: Biologically inspired collective comparisons by robotic swarms. Int. J. Robot. Res. **30**(5), 524–535 (2011)
17. Parker, C.A., Zhang, H.: Cooperative decision-making in decentralized multiple-robot systems: the best-of-n problem. IEEE/ASME Trans. Mechatron. **14**(2), 240–251 (2009)
18. Prasetyo, J., De Masi, G., Ferrante, E.: Collective decision making in dynamic environments. Swarm Intell. **13**(3), 217–243 (2019)

19. Salman, M., Garzón Ramos, D., Birattari, M.: Automatic design of stigmergy-based behaviours for robot swarms. Commun. Eng. **3**(1), 30 (2024)
20. Talamali, M.S., Saha, A., Marshall, J.A., Reina, A.: When less is more: robot swarms adapt better to changes with constrained communication. Sci. Robot. **6**(56), eabf1416 (2021)
21. Trianni, V., Nolfi, S.: Engineering the evolution of self-organizing behaviors in swarm robotics: a case study. Artif. Life **17**(3), 183–202 (2011)
22. Trianni, V., Tuci, E., Ampatzis, C., Dorigo, M.: Evolutionary swarm robotics: a theoretical and methodological itinerary from individual neurocontrollers to collective behaviors. In: Vargas, P.A., Di Paolo, E.A., Harvey, I., Husbands, P. (eds.) The Horizons of Evolutionary Robotics, pp. 153–178. MIT Press (2014)
23. Tuci, E., Rabérin, A.: On the design of generalist strategies for swarms of simulated robots engaged in a task-allocation scenario. Swarm Intell. **9**, 267–290 (2015)
24. Valentini, G., Ferrante, E., Hamann, H., Dorigo, M.: Collective decision with 100 kilobots: speed versus accuracy in binary discrimination problems. Auton. Agent. Multi-Agent Syst. **30**(3), 553–580 (2016)
25. Valentini, G., Ferrante, E., Dorigo, M.: The best-of-n problem in robot swarms: formalization, state of the art, and novel perspectives. Front. Robot. AI **4**, 9 (2017). https://doi.org/10.3389/frobt.2017.00009
26. Valentini, G., Hamann, H., Dorigo, M.: Self-organized collective decision making: the weighted voter model. In: Proceedings of the 13th International Conference on Autonomous Agents and Multiagent Systems (AAMAS), pp. 45–52 (2014)
27. Valentini, G., Hamann, H., Dorigo, M.: Efficient decision-making in a self-organizing robot swarm: on the speed versus accuracy trade-off. In: Proceedings of the 2015 International Conference on Autonomous Agents and Multiagent Systems, pp. 1305–1314 (2015)

Development of a Pheromone-Based Aggregation Method for Swarm Robots

Atakan Botasun[1(✉)], Mehmet Şahin[1], Ali Emre Turgut[1,2], and Erol Şahin[1,3]

[1] Center for Robotics and Artificial Intelligence (ROMER),
Middle East Technical University, Ankara, Turkey
atakanbotasun@gmail.com
[2] Department of Mechanical Engineering, Middle East Technical University, Ankara, Turkey
[3] Department of Computer Engineering, Middle East Technical University, Ankara, Turkey

Abstract. Although the use of "pheromones" in coordinating swarm robotic tasks such as foraging, coverage, and exploration is common, its use in aggregation still needs further exploration. In this paper, we introduce a pheromone-based aggregation (PBA) algorithm that extends BEECLUST, which uses an environmental cue, such as temperature, to regulate the aggregation of agents. We use pheromones for guidance, laid dynamically by the agents in the form of trails to aid the aggregation process by expanding the region of attraction. Through systematic simulations of the Kobot mobile platform, we show that PBA achieves a higher performance, measured by the Normalized Aggregation Size, than the BEECLUST algorithm in environments where the relative size of the cue within the environment is small and the robot density is low.

Swarm aggregation is defined as a phenomenon of individuals gathering together where environmental cues, such as light or temperature, are used as guidance, observed in honeybees [3], amoeba [6], and German cockroaches [10]. The BEECLUST algorithm proposed by Schmickl et al. [8] models honeybee aggregation using temperature cues but necessitates a relatively large "cue region" and high agent density in its basic state.

Pheromones are common in social insects as a stigmergic coordination mechanism, where individuals modify the environment, which in turn modulates the behaviors of all individuals within the swarm [4]. Ants use pheromones to communicate and optimize the path to food sources, spread in the environment [2]. German cockroaches use gut pheromones to coordinate aggregation behavior [10].

Pheromone-inspired stigmergic communication is compelling in swarm robotics due to its low reliance on individual robot capabilities and high scalability. The use of pheromones to increase the aggregation performance of the BEECLUST algorithm was first proposed in ΦClust [5]. ΦClust uses attractive pheromones deposited during aggregation, which diffuse into the workspace and form a gradient. Tang et al. emulate pheromones through RFID-based codes to extend pheromonal trails from the cue into the workspace [9].

In BEECLUST [8], robots move forward. Encountering other robots triggers waiting based on the cue intensity at their location, while obstacles are avoided with random turns. Figure 1 illustrates a finite-state diagram of BEECLUST. The actions taken within this algorithm can be described as the following:

Move Forward: Move forward until encountering an object.
Avoid: Turn randomly to an angle $\theta \in [-90°, 90°]$.
Measure Cue: Compute $w(I) = I^2/(I^2 + I_c)$, where I_c is a constant.
Wait: Wait $w(I)$ seconds, then randomly turn to an angle $\theta \in [-90°, 90°]$.

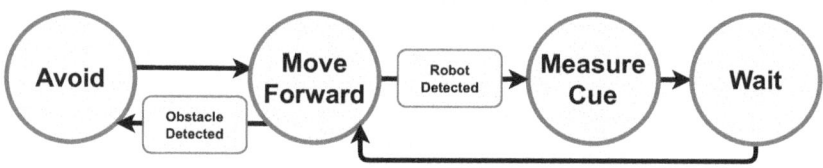

Fig. 1. Finite-state diagram of BEECLUST.

In ΦClust [5], pheromone laying and gradient following are introduced. As pheromones are laid during any encounter, an observed side effect occurs when multiple robots simultaneously track pheromones, where robots repeatedly encounter and break away from each other within a short period, effectively aggregating within strong pheromones and releasing more in the process. This sometimes results in pheromones taking over as the cue or expanding it wider. Two different actions exist within ΦClust's finite-state diagram, seen in Fig. 2 compared to BEECLUST:

Release Pheromone and Wait: Wait $w(I)$ seconds, and release pheromone at a constant rate J. Then randomly turn to an angle $\theta \in [-90°, 90°]$.

Follow Pheromone Gradient: Track pheromones towards stronger values.

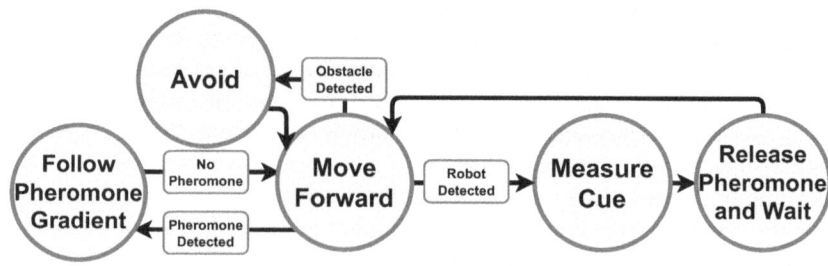

Fig. 2. Finite-state diagram of ΦClust.

This paper proposes an aggregation method named PBA that builds upon BEECLUST and uses pheromones for improved performance. We utilize pheromones for navigation but ensure aggregation primarily occurs near the intended location based on the cue by laying pheromones as trails instead of in place like ΦClust. Furthermore, our method can utilize pheromonal navigation without requiring additional information like odometry or external position measurement.

1 Methods

Our method builds on BEECLUST by adding pheromones. Robots track pheromones with local detection of at least 3×3 grids and inject pheromones behind them after aggregation to create trails leading away from the cue (Fig. 4). If pheromones are unavailable, robots will exhibit BEECLUST. An attention threshold prevents them from getting stuck in local maxima. We use a threshold for cue detection for convenience due to our circular cue setup. Figure 3 shows the state diagram. The actions within the algorithm are described as follows:

Move Forward: Move forward unless encountering an object or detecting a pheromone trail. Exponentially weaken pheromone injections at a rate ρ_J. Additionally, ignore pheromones if injecting pheromones above threshold J^*.

Measure Cue: Apply thresholding to decide if the sensor reading points at a cue. If cue readings are above the threshold, compute waiting time $w(I)$ and initial injection strength $J(I)$.

Wait and Reset Injection: Wait $w(I)$ seconds, and stop pheromone injection. Then randomly turn by angle $\theta \in [-90°, 90°]$, and start pheromone injection at strength $J(I)$.

Avoid: Turn randomly at an angle $\theta \in [-90°, 90°]$. Exponentially weaken pheromone injections at a rate ρ_J.

Follow Pheromone Gradient: Track pheromones towards stronger pheromone values. Set an attention threshold period upon beginning pheromone tracking. If this attention threshold period is exceeded, stop pheromone tracking. If no pheromones are detected, stop pheromone tracking.

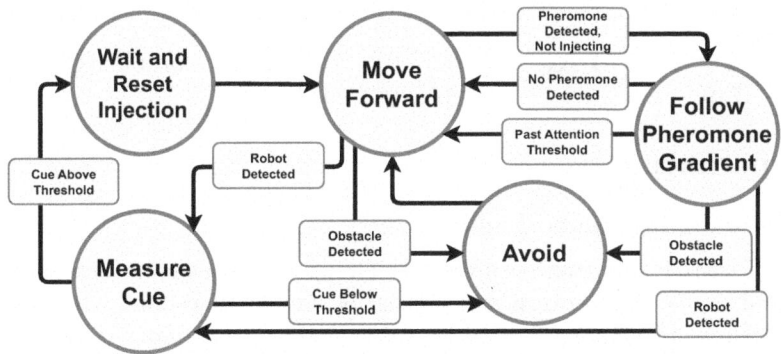

Fig. 3. Finite-state diagram of PBA.

Setting Injection Strength. The strength $J(I)$ at which a robot will inject pheromones after aggregation depends on the cue intensity I detected during aggregation to make the pheromone trails have quality-sensitive properties. Given the maximal intensity I_{max} of cue readings and maximal strength J_{max} of pheromone injections possible by a robot, a mapping is done accordingly.

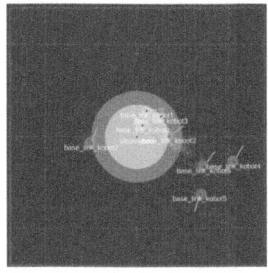
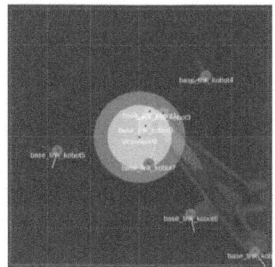
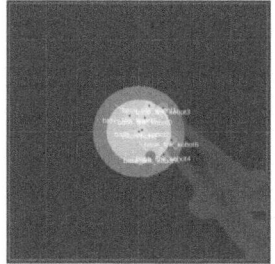

Fig. 4. Pheromone trail injections. Bright purple regions indicate stronger pheromones. Deep blue regions indicate weaker pheromones. As robots move away from the cue, their injection strength decays. (Color figure online)

$$J(I) = J_{max} \cdot (1 + m_J \cdot (I/I_{max} - 1)), m_J \in [0, 1] \quad (1)$$

Injection strength J will exponentially decay over time to form gradients, as seen in Fig. 4 and eventually pheromone injections cease entirely.

Gradient Ascent with Scharr Operator. Pheromone tracking is done by convolving pheromone readings $\mathbf{P}$ with the Scharr operator [7] to produce a desired heading. The Scharr operator generates image gradients for optical flow applications and has high derivative accuracy, making it suitable for tracking. Difference values along $\mathbf{P}$ will be found as $\mathbf{G}_x$ and $\mathbf{G}_y$. The angle $\mathbf{D}_{2,2}$ between $\mathbf{G}_x$ and $\mathbf{G}_y$ at the center cell will yield the direction in which pheromones are stronger.

$$\mathbf{G}_x = \begin{bmatrix} 47 & 0 & -47 \\ 162 & 0 & -162 \\ 47 & 0 & -47 \end{bmatrix} * \mathbf{P}, \mathbf{G}_y = \begin{bmatrix} 47 & 162 & 47 \\ 0 & 0 & 0 \\ -47 & -162 & -47 \end{bmatrix} * \mathbf{P}, \mathbf{D} = \arctan 2(\mathbf{G}_y, \mathbf{G}_x) \quad (2)$$

2 Experiments

2.1 Virtual Pheromone Fields

We have created a custom utility that simulates pheromone activity as an augmented reality (AR) overlay for actual or simulated robots, which receives pheromone injection requests and transmits detected pheromone data to robots. Virtual pheromones are represented in a grid $\mathcal{G}(t)$ with the following properties:

Injection. Robots mark the environment by injecting pheromones within a circular region in the environment. Distance-based masking is used for approximations. Figure 5 (Left) describes how injection occurs on the pheromone grid.

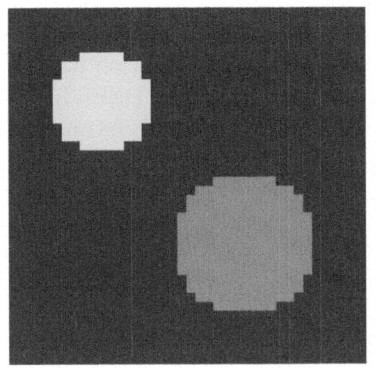
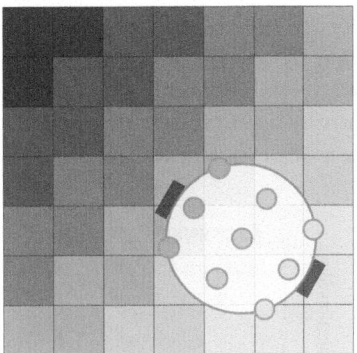

Fig. 5. Left: Pheromone injection at two different locations, strengths, and radii. **Right:** Pheromone detection from the grid. Small circles on the robot represent virtual sensors. For clarity, the grid is represented with coarse cells.

Diffusion. Pheromones within the environment spread out over a 2D plane, emulated by Gaussian blurring. The diffusion rate is controlled by the standard deviation term σ. The pheromone amount in the grid is restored by Eq. 5 due to discretization.

$$K = \exp(-0.5\sigma^{-2})/(\sqrt{2\pi\sigma^2}) \tag{3}$$

$$\bar{\mathcal{G}}(t + \Delta t) = \begin{bmatrix} K \\ 0 \\ K \end{bmatrix} * ([K\ 0\ K] * \mathcal{G}(t)) \tag{4}$$

$$\mathcal{G}(t + \Delta t) = \bar{\mathcal{G}}(t + \Delta t) \cdot \frac{||\mathcal{G}(t)||}{||\bar{\mathcal{G}}(t + \Delta t)||} \tag{5}$$

Evaporation. Existing pheromones within the environment decay over time, emulated by exponential decay at a rate ρ, applied to the entire pheromone grid.

$$\mathcal{G}(t + \Delta t) = (1 - \rho) \cdot \mathcal{G}(t) \tag{6}$$

Detection. Robots detect pheromones near them through virtual sensors. The cells on the pheromone grid corresponding to virtual sensor positions yield strength readings. The pheromone grid's cell dimensions must be scaled down appropriately to generate sensible pheromone readouts. Figure 5 (Right) describes how detection is carried out on the pheromone grid.

2.2 Kinematic Simulations

We have conducted our experiments within the Kobot kinematic simulator[12], with puck-shaped robots modeled after the Kobot-W, as seen in Fig. 4. Kobot-W [1] is a wheeled swarm robot platform equipped with floor sensors, IMUs, time-of-flight sensors, infrared sensors, and cameras, along with communication modules. In an actual test, a Kobot only lacks stigmergic feedback, which is compensated by an AR interface. The simulator provides sensing and actuation to ROS nodes, which would otherwise come from Kobot units. PBA and BEECLUST Monte-Carlo trials are done in three different square-shaped workspaces and two sizes of robot populations.

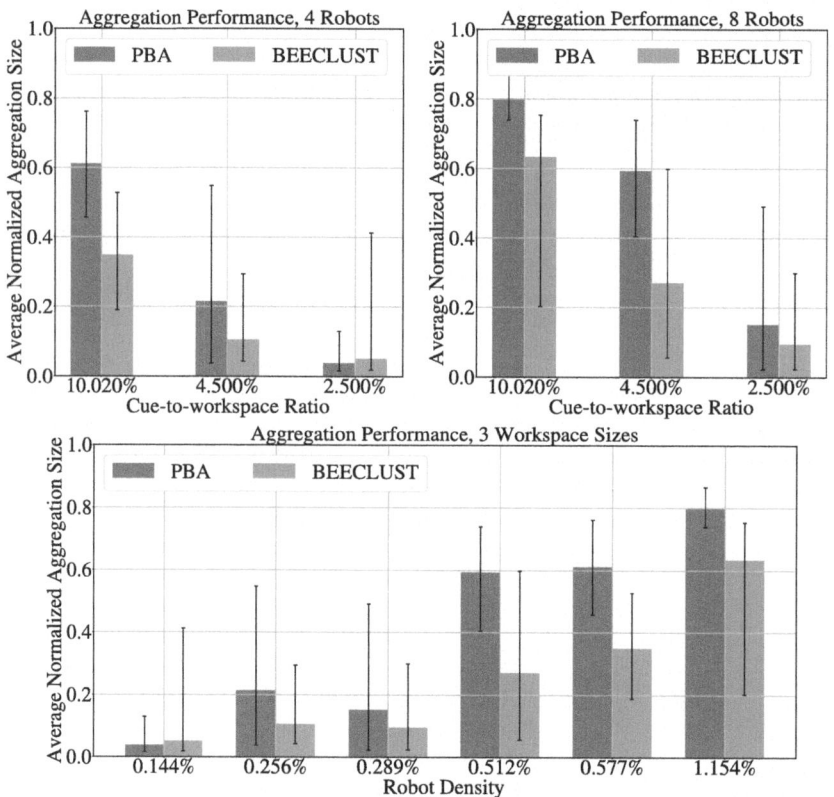

Fig. 6. Average NAS results with variations for PBA and BEECLUST for differing cue-to-workspace ratios and robot densities. Experiments are done 50 times.

We evaluated performance using the mean value of Normalized Aggregation Size (NAS), the proportion of robots within designated aggregation areas at a

---

[1] Sample experiments can be found in https://tinyurl.com/5d6wk4jx.
[2] Source code available in https://tinyurl.com/r5vamjhx.

given time. The cue is placed as two concentric zones of 0.35 m and 0.5 m at the center of the workspace, where the outer one has increasing intensities towards the center, and the inner one has uniform and maximal intensity.

We will make our discussions based on *robot density*, the total area covered by robot footprints divided by the workspace size, and *cue-to-workspace ratio*, the area covered by the cue divided by the total workspace size. We provide the mean NAS value and its variation for different experimental settings in Fig. 6.

According to the results in Fig. 6, *robot density* has a similar positive effect on PBA as on BEECLUST. Higher robot density increases encounter chance within the cue area. It increases the chance of a robot exploiting available pheromone trails rather than roaming the free workspace, making the performance gain more pronounced for PBA. For lower robot densities, the encounter chance reduction is adverse for both methods. However, PBA tends to compensate by increasing subsequent encounters through pheromone trail exploitation. *Cue-to-workspace ratio* is also another criterion for both methods. The total aggregation duration highly depends on the chance of encounters within intense cue areas. Therefore, a higher cue-to-workspace ratio has a positive effect on both algorithms. For environments of lower cue-to-workspace ratios, PBA can alleviate the decrease in encounter chances by leading robots toward the cue area. However, this is only possible if the robot density is high enough to sustain and fully exploit the pheromone trail feedback to use this advantage. As seen in the 4 robot scenario with 2.5% cue-to-workspace ratio in Fig. 6, PBA and BEECLUST both suffer from a severe lack of encounter chance in a similar manner.

3 Conclusion

This paper introduces a pheromone-based aggregation method, PBA, that extends BEECLUST, a seminal aggregation method. We leverage pheromones without other forms of positional guidance, such as odometry. We outperform BEECLUST for smaller exploitable cue areas and lower robot densities. Contrary to ΦClust, we discourage robots from aggregating near pheromones and strictly adhere to the cue for aggregation.

For future work, we plan to gather more experimental data from simulated or real Kobot units and implement other pheromone-based aggregation methods in our framework for contrast. We intend to improve pheromone following feedback mechanisms and simplify their tuning, or to explore parameters other than cue intensity for pheromone injection strength to enable a broader range of robot densities and workspace sizes.

Acknowledgements. This paper is partially supported by METU ADEP-302-2024-11468.

References

1. Bilaloğlu, C.: Development of an extensible heterogeneous swarm robot platform. Master's thesis, Middle East Technical University (2022)
2. Goss, S., Aron, S., Deneubourg, J.L., Pasteels, J.M.: Self-organized shortcuts in the argentine ant. Naturwissenschaften **76**, 579–581 (1989). https://doi.org/10.1007/bf00462870
3. Grodzicki, P., Caputa, M.: Social versus individual behaviour: a comparative approach to thermal behaviour of the honeybee (APIs mellifera l.) and the American cockroach (periplaneta americana l.). J. Insect Physiol. **51**, 315–322 (2005). https://doi.org/10.1016/j.jinsphys.2005.01.001
4. Karlson, P., Lüscher, M.: 'Pheromones': a new term for a class of biologically active substances. Nature **183**, 55–56 (1959). https://doi.org/10.1038/183055a0
5. Na, S., et al.: Bioinspired artificial pheromone system for swarm robotics applications. Adapt. Behav. **29**, 395–415 (2020). https://doi.org/10.1177/1059712320918936
6. Rappel, W.J., Nicol, A., Sarkissian, A., Levine, H., Loomis, W.F.: Self-organized vortex state in two-dimensional dictyostelium dynamics. Phys. Rev. Lett. **83**, 1247-1250 (1999).https://doi.org/10.1103/PhysRevLett.83.1247, https://arxiv.org/abs/patt-sol/9811001
7. Scharr, H.: Optimale Operatoren in der Digitalen Bildverarbeitung. Ph.D. thesis, Heidelberg University (2000). https://doi.org/10.11588/heidok.00000962
8. Schmickl, T., et al.: Get in touch: cooperative decision making based on robot-to-robot collisions. Auton. Agents Multi-Agent Syst. **18**, 133–155 (2009). https://doi.org/10.1007/s1045800890585
9. Tang, Q., Ding, L., Li, J., Zhang, Y., Yu, F.: A stigmergy-based aggregation method for swarm robotic system. In: 2017 IEEE Symposium Series on Computational Intelligence (SSCI), pp. 1–6 (2017). https://doi.org/10.1109/SSCI.2017.8285372
10. Wada-Katsumata, A., Zurek, L., Nalyanya, G., Roelofs, W.L., Zhang, A., Schal, C.: Gut bacteria mediate aggregation in the German cockroach. PNAS **112**, 15678–15683 (2015). https://doi.org/10.1073/pnas.1504031112

Extended Abstracts

Ant-Search Algorithm for Distributed Knowledge Graphs

Oleksandr Chepizhko(✉), Péter Forgács, and Melanie Schranz

Lakeside Labs GmbH, Klagenfurt, Austria
{chepizhko,peter.forgacs,schranz}@lakeside-labs.com

The introduction of edge computing as a local processing capacity has compelled the cloud to become more flexible. This shift is driven by the array of benefits that edge computing brings to future processing tasks, including enhanced security, reliability, reduced latency and energy consumption. As edge applications proliferate across distributed devices and infrastructures with autonomous requirements, the volume and heterogeneity of data and users dynamically increase.

Conventional cloud-based database perspectives prove to be inadequate for the dynamic storage needs introduced by the edge, which requires decentralized data operations. To address this, we propose a framework for a Distributed Knowledge Graph (DKG) spanning the edge-fog-cloud continuum, dynamically mapping data across different layers. The DKG offers real-time insights into distributed data while facilitating scalable, energy-efficient, and dependable operations. Central to its functionality is its adaptability in search operations, minimizing costs and data transfers. It is represented using the Resource Description Framework (RDF) format that stores the data as a directed graph. RDF represents data in a form of triplets: subject-predicate-object [1].

This study focuses on integrating the well-established Ant Colony Optimization algorithm (ACO) [2] into the highly adaptive DKG within the edge-fog-cloud continuum. We detail the DKG's engineering and network topology modeling to address current edge challenges, see Fig. 1(a). We assume three layers of nodes (devices) connected onto a network: powerful cloud nodes in the core, intermediate fog nodes attached to them and edge devices in the outer layer. Each of them possess a partition of the DKG with a subset of triples.

We adapted the ant algorithm to the edge continuum and evaluated its performance in resource utilization scenarios. Our implementation, based on ACO for peer-to-peer networks [3], differs in application domain, RDF query properties, and network structure. Each query generates a forward ant to explore the network using pheromone trails until it finds nodes with matching triples. It then sends a backward ant to the origin node to lay the pheromone trails, which evaporate at a rate p.

We show that the efficiency of the algorithm depends on the network configuration. Hit rate, defined as the ratio of the matching nodes found during a query to the total number of matching nodes available in the network, serves as the measure of efficiency of our algorithm. Stable pheromone trails emerge for low evaporation coefficient values, $p = 9 \times 10^{-5}, 9 \times 10^{-3}$, while such trails vanish immediately if the evaporation coefficient is high, $p = 9 \times 10^{-1}$, leaving no hints to the ants and this leading to random walk. For low evaporation rates we observe bi-modal distribution with some of the query finding almost all the matching nodes while others finding only a limited number, see Fig. 1(b). We can speculate that due to the structure of the network the queries starting from core nodes have advantage over the others, which allow them to explore the network in a more efficient manner. For the high evaporation rate the hit rates are low, since there is no supporting pheromone structure to guide the search agents to the matching nodes.

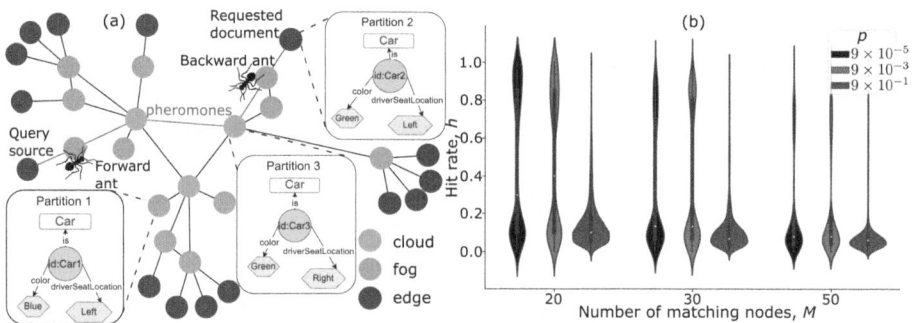

Fig. 1. (a) DKG in edge-cloud-continuum scheme. Nodes belong to either cloud, fog, or edge. Each node stores a partition of DKG. Only 3 of them are shown as example. For a subject-empty query, $Q_s = \langle *, \text{color}, \text{Blue} \rangle$, the result will be $\langle \text{Car 1} \wedge \text{Car 3} \rangle$. (b) Hit-rate distribution. Evaporation coefficient p goes from small to large values. Total number of nodes in the network $N = 128$.

Our future work will focus on a detailed distribution of queries and assessing the algorithm's robustness against network structure perturbations. Additionally, the pheromone trails from the current study will be reused in a new data movement engine. These established routes will indicate where specific data is needed and guide its movement through the DKG.

Acknowledgments. This work was performed in the course of the EU-project GLACIATION (HORIZON-CL4-2022-DATA-01-02).

References

1. Abdelaziz, I., Harbi, R., Khayyat, Z., Kalnis, P.: A survey and experimental comparison of distributed SPARQL engines for very large RDF data. Proc. VLDB Endowment **10**(13), 2049–2060 (2017)

2. Dorigo, M., Stützle, T.: The ant colony optimization metaheuristic: algorithms, applications, and advances. In: Glover, F., Kochenberger, G.A. (eds.) Handbook of Metaheuristics. International Series in Operations Research and Management Science, vol. 57, pp. 250–285. Springer, Boston (2003). https://doi.org/10.1007/0-306-48056-5_9
3. Michlmayr, E.: Ant Algorithms for Self-Organization in Social Networks. Ph.D. thesis, Vienna University of Technology, Vienna, Austria (2007)

LARS: Light Augmented Reality System for Swarm

Mohsen Raoufi[1,2](✉)[iD], Pawel Romanczuk[1,2][iD], and Heiko Hamann[1,3][iD]

[1] Science of Intelligence, Research Cluster of Excellence, Berlin, Germany
mohsenraoufi@icloud.com
[2] Department of Biology, Humboldt University of Berlin, Berlin, Germany
[3] Department of Computer and Information Science, University of Konstanz, Konstanz, Germany

Extended reality (XR) technology has found its applications in various systems, including multi-robot systems [1]. Augmented Reality and Mixed Reality tools are becoming increasingly influential in educational technology and robotics. They enrich learning experiences and expand research methodology. Multi-robot systems, as a complex collective system, have great potential for educational purposes, showcasing collective behaviors. In such systems, XR tools set up a dynamic virtual environment observable by humans as well as a medium to interact with multi-robot systems. Making non-tangible and complex concepts in multi-robot systems easier to comprehend by human users will help in understanding, analyzing, and conveying how these systems do what they do.

We present the Light Augmented Reality System (LARS) as an open-source online[1] and cost-effective tool. LARS leverages light-projected visual scenes for indirect robot-robot and human-robot interaction through the real environment. It operates in real-time and is compatible with a range of robotic platforms, from miniature to middle-sized robots. LARS can support researchers in conducting experiments with increased freedom, reliability, and reproducibility. This XR tool makes it possible to enrich the environment with full control by adding complex and dynamic objects while keeping the properties of robots as realistic as they are. The system promotes stigmergy as a natural method of indirect communication between robots. However, it also keeps the possibility of directly transmitting messages to robots using central information. Furthermore, such interactive systems set the scene to investigate the principles across various disciplines, e.g., biology and social sciences, as we used in our previous studies [2].

Key features of LARS include light-based interaction, marker-free cross-platform tracking, real-time performance, scalability, and ease of setup. The system uses light as a medium for (indirect) communication. Indirect communication via the environment, Stigmergy, is the key to making LARS independent of a specific platform, compared to the previous robot-specific tools [3]. Light is easily observable by both humans and robots and allows for human-robot interaction in the *real* world. This is in contrast to other related works that simulate a virtual environment for the interaction of the system [3]. Marker-free tracking enhances the versatility of the system and reduces the time and effort needed

---

[1] https://github.com/mohsen-raoufi/LARS

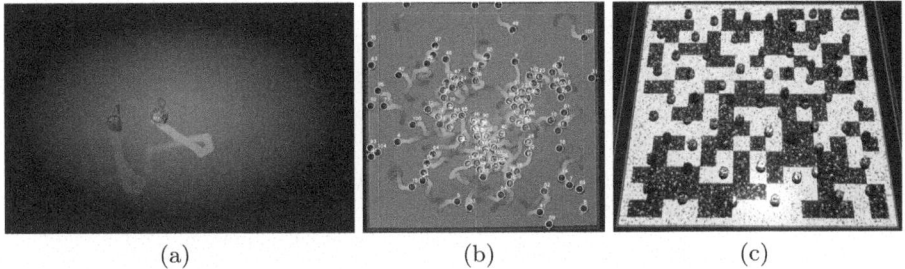

Fig. 1. Example scenarios with (a) two Thymio robots with different ring colors, (b) 109 Kilobots cleaning up the environment, and (c) 63 Kilobots making a collective decision on a tiled environment with dynamic noise.

to integrate new platforms. LARS operates in real-time with approximately 38 frames per second on a standard desktop PC. We implemented the software in C++ language and used the OpenCV library due to their real-time performance and being free and open-source. At the core of the detection and tracking code we employed and optimized the tracker developed for ARK [3], which was originally developed specifically for Kilobots [4]. Given the prevalent circular shape of swarm robot platforms, LARS harnesses this regularity and detects various robotic platforms using a simple Hough method for circle detection.

LARS is a test-bed system and a tool to conduct multi-robot experiments with; a recording, tracking, and logging system to save experiment data; and a medium through which the reality is extended by augmenting virtual objects. Without altering the limitation of a robotic system, LARS enriches the environment, for example, by adding user-defined noise. The augmented environment is a step toward bridging the gap between simulation and reality. We show examples in Fig. 1, where (a) Thymio robots explore the environment, leaving an elusive pheromone trail on the environment, (b) Kilobots diffuse in the environment to clean it up, and (c) Kilobots make a collective decision in a noisy environment.

Acknowledgements. Funded by the DFG (German Research F.) under Germany's Exc. Strat. - EXC 2002/1 "Science of Intelligence" - proj. num. 390523135.

References

1. Makhataeva, Z., Varol, H.A.: Augmented reality for robotics: a review. Robotics **9**(2), 21 (2020)
2. Raoufi, M., Romanczuk, P., Hamann, H.: Individuality in swarm robots with the case study of Kilobots: noise, bug, or feature? In: ALIFE 2023: Proceedings of the Artificial Life Conference. MIT Press (2023)
3. Reina, A., Cope, A.J., Nikolaidis, E., Marshall, J.A., Sabo, C.: ARK: Augmented reality for Kilobots. IEEE Robot. Autom. Lett. **2**(3) (2017)
4. Rubenstein, M., Ahler, C., Nagpal, R.: Kilobot: a low cost scalable robot system for collective behaviors. In: IEEE International Conference on Robotics and Automation (ICRA) (2012)

Moving Depot (MOD): An Efficient Depot Motion Strategy for Multi-Robot Foraging

Pratik Ingle[1](✉), Ananya Gandhi[2], and Sujit Baliyarasimhuni[2](✉)

[1] Department of Computer Science, IT University of Copenhagen, Copenhagen, Denmark
prin@itu.dk
[2] Department of Electrical and Computer Engineering, IISER Bhopal, Bhopal, India
{ananya20,sujit}@iiserb.ac.in

Foraging behaviours in insect communities, like ants and bees, have inspired multi-robot foraging algorithms [1] aimed at maximizing resource collection at a designated depot through coordinated efforts. Most swarm robotics research focuses on stationary depot strategies (SDS), where robots collect resources from various patches and deposit them at a fixed depot to enhance overall foraging efficiency. Common foraging strategies include stigmergy-based foraging [5] and response threshold approaches [2,3]. While SDS is effective in obstacle-free conditions, its performance decreases in dynamic environments with obstacles or changing patch quality. Multiple-Place Foraging Algorithm (MPFA) considers depot motion [4], but fails to address depot mobility and optimal repositioning under obstacle settings. Inspired by the adaptability of insect colonies to uncertain environments, we propose a novel approach for dynamic depot placement in robotic swarms. Our proposed Moving Depot (MOD) algorithm employs potential-based path planning and the weighted centroid method to dynamically adjust depot positions based on resource quality, minimizing relocation costs. This model adapts to real-time changes in the environment, optimizing foraging efficiency. We evaluate the MOD algorithm through agent-based simulations, comparing its performance with the MPFA algorithm across varying agent numbers and resource qualities in the presence of obstacles.

The MOD algorithm's overall behavior is shown as a finite state machine in Fig. 1. In MOD, depot and foraging agents do not have a priori information about the environment. Foraging agents decision-making involves three phases – (i) Exploration: Agents perform random walks until a patch p_i is discovered, then the agents enter into next phase (ii) Foraging: Agent obtain patch information such as location, quality, and size, and collect resources that will be deposited in the depot (iii) Decision-Making: returned agents either enter into phase (i) or (ii) with a transition probability of γ. If the agent enters phase (ii), then it selects the next patch to be visited either through uniform decision-making (where all m patches have the same probability), or heuristic decision-making (probability is directly proportional to the distance between the patch and the depot's current position $D*$).

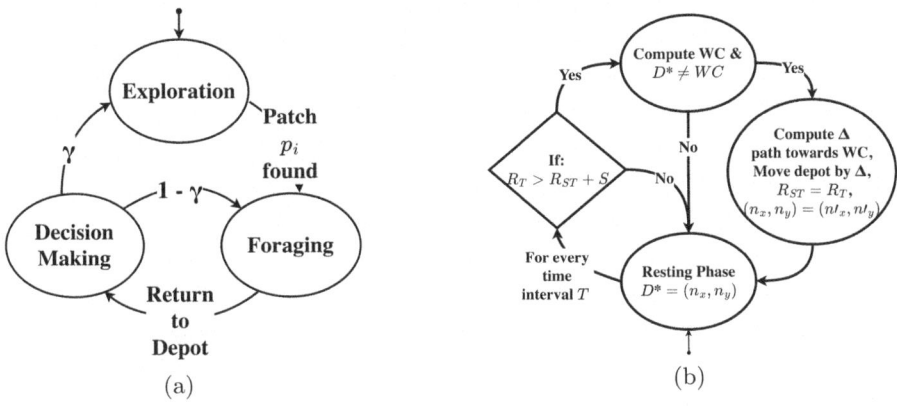

Fig. 1. Finite state machine (a) individual robot behaviour (b) depot behaviour

The MOD algorithm strategically changes the depot's position based on total resources collected (R_T), movement cost (C), time elapsed (T), and surplus (S). Initially, the depot is randomly placed at $D = (n_x, n_y)$ with $R_T = 0$. The depot remains stationary until $R_T \geq S$ at each interval T. If this condition is met, the depot calculates the weighted centroid (WC) of m discovered patches as

$$WC = \left(\frac{\sum_{i=1}^{m} q_i x_i}{\sum_{i=1}^{m} q_i}, \frac{\sum_{i=1}^{m} q_i y_i}{\sum_{i=1}^{m} q_i} \right), \tag{1}$$

and uses a potential field path planner to move towards WC by a distance Δ.

We conducted agent-based simulations to evaluate the MOD algorithm against SDS and MPFA, considering various patch qualities, obstacles, and agent populations. The simulation arena was 100×100 m, with random placement of depots and patches. Each simulation ran for 10,000 steps and was repeated 10 times. For varying patch qualities, MOD prioritized higher-quality patches, enhancing foraging efficiency by more than 50%. Comparing MOD to MPFA with a single depot and 20 agents, MOD collected nearly 30% more high-quality resources, showcasing its long-term efficiency despite movement restrictions. Overall, MOD algorithm showed higher performance across different scenarios, demonstrating its adaptability and efficiency.

References

1. Brambilla, M., Ferrante, E., Birattari, M., Dorigo, M.: Swarm robotics: a review from the swarm engineering perspective. Swarm Intell. **7**(1), 1–41 (2013)
2. Castello, E., Yamamoto, T., Libera, F.D., Liu, W., Winfield, A.F., Nakamura, Y., Ishiguro, H.: Adaptive foraging for simulated and real robotic swarms: the dynamical response threshold approach. Swarm Intell. **10**, 1–31 (2016)
3. Liemhetcharat, S., Yan, R., Tee, K.P.: Continuous foraging and information gathering in a multi-agent team. In: AAMAS, pp. 1325–1333 (2015)

4. Lu, Q., Hecker, J.P., Moses, M.E.: Multiple-place swarm foraging with dynamic depots. Auton. Robot. **42**(4), 909–926 (2018). https://doi.org/10.1007/s10514-017-9693-2
5. Talamali, M.S., Bose, T., Haire, M., Xu, X., Marshall, J.A., Reina, A.: Sophisticated collective foraging with minimalist agents: a swarm robotics test. Swarm Intell. **14**(1), 25–56 (2020)

Statistical Study of Worker Activity Relying on Location in Ant Colonies

Masashi Shiraishi(✉) and Hiraku Nishimori

Meiji Institute for Advanced Study of Mathematical Sciences, Meiji University, Tokyo, Japan
shiraishi.mu@gmail.com

Ants are eusocial insects that perform coordinated and self-organized tasks like nest building, foraging, and brood care. These collective behaviors have been studied in various disciplines, including ecology and mathematical biology. The fixed response threshold (FRT) model proposed by Bonabeau et al. [1] is based on Robinson's theoretical work [3]. Recently, many researchers have collected extensive behavioral data. Mersch et al. found that the tasks and locations of the ants within the nest change over time. In contrast, Yamanaka et al. [4] verified the FRT model using radio frequency identity (RFID) technology.

We observed two colonies of *Camponotus japonicus*, grown from single-mated queens in May 2022 at the Laboratory of the National Institute for Advanced Industrial Science and Technology, Tsukuba, Japan. Experiments were conducted for colonies A and B from December 10, 2023, to February 22, 2024. The colonies A and B comprise $N = 153$ and 169 workers, respectively. We set up the experiment systems and installed sensors, as shown in Fig. 1(a), and followed the protocol in a previous study [4]; the RFID system and the postprocess of the measured data to obtain the n-th timestamp of the worker with ID i as $t_i^{(n)}$ shown in Fig. 1(b). We define N_i as the count of passing events of each worker, the time difference $\Delta t_i^{(n)}$, and the frequency distribution:

$$f(\tau_k) = \frac{1}{\mathcal{N}} \sum_{j=1}^{N} f_j(\tau_k) = \sum_{j=1}^{N} \sum_{j=i}^{N_i} 1_{\tau_k, \Delta\tau}[t_i^{(j+1)} - t_i^{(j)}], \quad (1)$$

where $1 \leq k \leq M$, $\mathcal{N}$ is the normalization constant, $\Delta t_i^{(n)} = t_i^{(n+1)} - t_i^{(n)}$, and the $f_i(\tau)$ is for each worker varied widely due to intrinsic differences. We focused on the average distribution to identify differences in worker activity among locations in the nest. Using this averaged frequency distribution $f(\tau_k)$, we obtain the probability distribution:

$$P(\tau > \tau_k) = 1 - \sum_{n=1}^{k} f(\tau_n)\Delta\tau, \quad (2)$$

calculated for each sensor data and compared with the results.

We found that the distributions of N_i and $\Delta t_i^{(n)}$ varied with sensor location in the ant colony. The N_i rank distributions in Figs. 2 (a) and (b) show that the

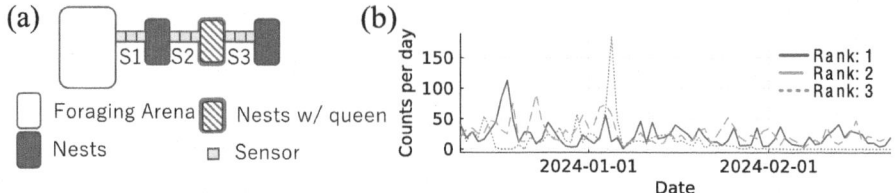

Fig. 1. a) Schematic figures of experiment setups. (b) Examples of the time series of activity of individual workers in Colony A.

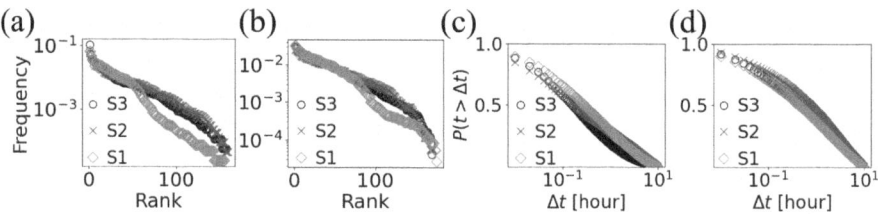

Fig. 2. (a), (b), Rank distribution. (c), (d) The time difference distribution. (a) and (c) are figures for Colony A, and (b) and (d) are figures for Colony B.

S1 sensor exhibits two-step distributions, while S2 and S3 have broad activity distributions. The specialized foraging activity in S1 contrasts with the broad activity distribution within the nest chamber. For $\Delta t_i^{(n)}$ distributions in Figs. 2 (c) and (d), S1 shows a structure weighted towards shorter time scales, indicating foraging activity, while other sensors show longer time scale activities. Fujioka et al. [2] noted that the 24-hour care of nurses results in distributions of time difference in fat tails. In summary, we observed significant differences in statistical distribution data between S1 and the other sensors. Due to limited data from only two colonies, further measurements are needed to confirm these findings' universality. Furthermore, colony size N might influence these distributions, as larger colonies could have different activity demands.

References

1. Bonabeau, E., Theraulaz, G., Deneubourg, J.L.: Phase diagram of a model of self-organizing hierarchies. Phys. A **217**(3–4), 373–392 (1995). https://doi.org/10.1016/0378-4371(95)00064-e
2. Fujioka, H., Abe, M.S., Fuchikawa, T., Tsuji, K., Shimada, M., Okada, Y.: Ant circadian activity associated with brood care type. Biol. Let. **13**(2), 20160743 (2017). https://doi.org/10.1098/rsbl.2016.0743
3. Robinson, G.E.: Regulation of honey bee age polyethism by juvenile hormone. Behav. Ecol. Sociobiol. **20**(5), 329–338 (1987). https://doi.org/10.1007/bf00300679
4. Yamanaka, O., Shiraishi, M., Awazu, A., Nishimori, H.: Verification of mathematical models of response threshold through statistical characterisation of the foraging activity in ant societies. Sci. Rep. **9**(1), 8845 (2019). https://doi.org/10.1038/s41598-019-45367-w

Swarm in the City: Inspirations from Urban Street Networks for Swarm Robot Aggregation

Dalia S. Ibrahim[1,2] and Andrew Vardy[3(✉)]

[1] Department of Computer Science, Memorial University of Newfoundland, St. John's, Canada
dsibrahim@mun.ca
[2] Department of Computer Systems, Faculty of Computer and Information Science, Ain-Shams University, Cairo, Egypt
[3] Department of Computer Science, Department of Electrical and Computer Engineering, Memorial University of Newfoundland, St. John's, Canada
av@mun.ca

In this work, we design a scalar field inspired by urban street layouts [5] to reduce interference among simple robots. These robots do not have localization or communication capabilities. They rely only on local sensing and the provided scalar field. We address the aggregation task where all robots seek to reach a predefined location. Our usage of a scalar field is inspired by the use of pheromones which help guide social insects in many of their behaviours, including their impressive nest-building activities [4]. We are inspired by the grid layouts of cities like Manhattan and Chicago [2] as they allow for the division of the city into easily organized blocks and predictable routes (see Fig. 1). We designed a street layout which is projected onto the operating environment of our robots (see Fig. 1). The blue area represents a desired location that the robots will visit frequently, such as an aggregation area or collection station. The diagonal line represents a highway which intersects all streets, and ends at the desired location. We used a sampling-based path planning algorithm, Rapidly-exploring Random Trees (RRT*) [3] to draw the street layout. Streets are then interpolated to avoid sharp curves. To connect the streets with the highway, we use Bézier curves as ramps.

Our simulation was implemented in C++ using CWaggle[1] an open-source simulator for swarm robotics. We define the Road-Following control algorithm for aggregation task using a state machine. The goal of the aggregation task is for all robots to gather at a pre-determined location (blue area). We studied the performance of the Road-Following controller for varying numbers of robots. As a benchmark we compare against a controller known as Random-Walk which uses a Brownian motion-inspired approach to explore the environment and eventually reach the desired location. Figure 1 presents a comparison between Road-Following and Random-Walk controllers for 50 robots. It shows that earlier in the simulation run, Random-Walk exhibits faster aggregation due to fewer collisions. However, after $t = 10,000$, Road-Following provides improved flow of the robots' movements, leading to faster aggregation.

---

[1] https://github.com/davechurchill/cwaggle.

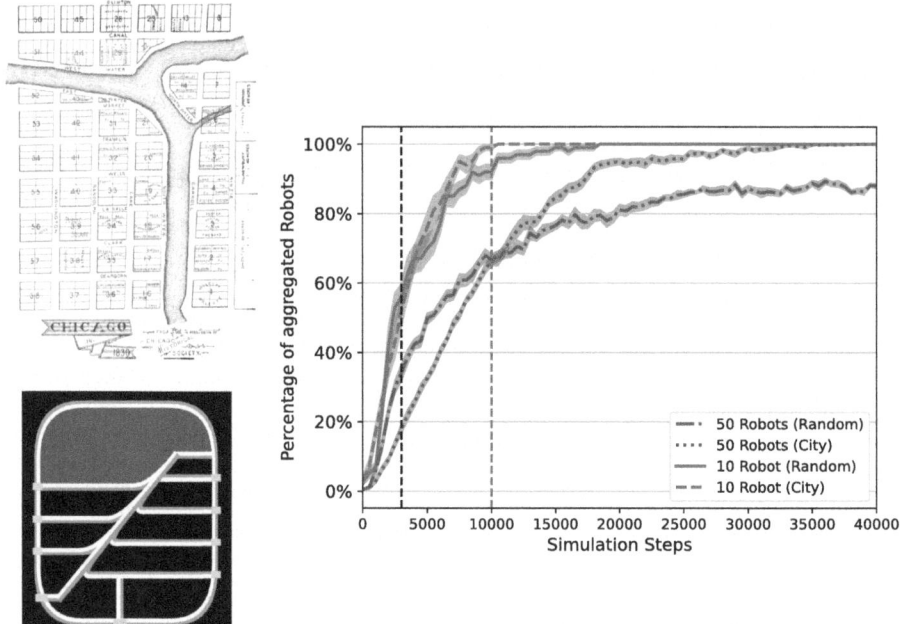

Fig. 1. (top left) Chicago in 1830 [1]; (bottom left) Scalar field design; (right) Comparison between Road-Following and Random-Walk.

We deployed the Road-Following on Zumo-based robots[2] using simple proximity and color sensors operating on a 75-inch diagonal LCD screen. A video showing 1, 2, 3 and 4 robots aggregating is available at https://youtu.be/L4XW3Rz2fgA.

References

1. Heyen, M.: Understanding the Chicago grid: how to never be lost in Chicago (2012). https://terribuseman.wordpress.com/tag/map/
2. Maxemchuk, N.: Routing in the Manhattan street network. IEEE Trans. Commun. **35**(5), 503–512 (1987)
3. Noreen, I., Khan, A., Habib, Z.: Optimal path planning using RRT* based approaches: a survey and future directions. Int. J. Adv. Comput. Sci. Appl. **7**(11) (2016)
4. Perna, A., Theraulaz, G.: When social behaviour is moulded in clay: on growth and form of social insect nests. J. Exp. Biol. **220**(1), 83–91 (2017)
5. Smets, M.: Foundations of Urban Design. Actar Publishers, New York (2022)

---

[2] https://www.pololu.com/docs/0J63.

The Two-Bridge Ant Experiment as an Interactive NetLogo Library Model

Martina Umlauft(✉) and Melanie Schranz

Lakeside Labs GmbH, Klagenfurt, Austria
{umlauft,schranz}@lakeside-labs.com

Research shows that working with models is beneficial to teaching and understanding complex systems and swarm intelligence [3, 4, 6, 7]. Since there is often not enough budget to use swarms of robots in an educational setting, simulation using a beginner-friendly platform, where users can explore system behavior by interacting with the model is an alternative approach that can be used effectively.

NetLogo [5] is a well-known agent-based simulation platform which is often used as a tool in education. It comes with a library with a large number of models from a variety of scientific fields, and even an Ants model of simulated foraging. However, it was still missing a model of the Double-Bridge Experiment [2] that inspired the creation of technical ant algorithms [1]. We will submit such a model (see Fig. 1) for introduction into the NetLogo Models Library[1].

In the Double-Bridge Experiment [2], the nest of an ant colony in a laboratory was connected to a food source via two bridges of different lengths. Despite individual ants being unable to detect the lengths of the paths when first encountering the fork in the path, collectively, the ants were able to find the shorter path quickly by using pheromones when travelling the paths. This experiment is simulated in our NetLogo model; the ants move away from the nest at a constant speed, exiting the nest at the selected interval. The `look-for-food` procedure with which ants find the path showcases the new `who-are-not` keyword that has been introduced in NetLogo 6.4.0.

When encountering the fork in the path, an ant will choose which path to use with a probability proportional to the pheromone values at it detects on each path. Here, we demonstrate how to program roulette-wheel selection. With a certain probability, an ant will become an explorer, though, ignoring the pheromone values and choosing its path randomly. Our model ants remember all the patches of the path they used moving forward in a "pathlist". When ants reach the food source, they pick up food, turn around and walk back towards the nest along the remembered path. On the way back to the nest, the ants drop a pheromone value of `1000/length pathlist` on each patch.

Unlike most library models, we use a `startup` procedure to avoid having to re-draw the paths when the model is restarted with the `setup` button. Our NetLogo model can be downloaded here: http://mumlauft.blogspot.com/2024/04/ants-double-bridge-experiment-model-for.html.

---

[1] https://ccl.northwestern.edu/netlogo/models/index.cgi [Online; last access: 26-03-2024]

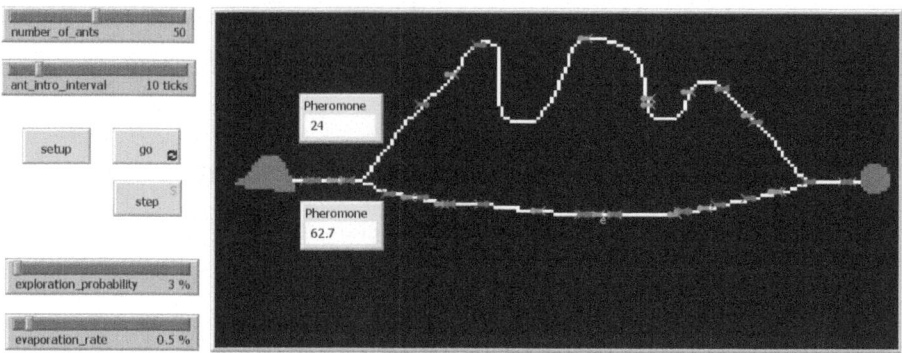

Fig. 1. The Double-Bridge Model User Interface: the user sets the_number_of_ants and the interval of introduction into the model before resetting and starting the simulation with the setup and go buttons. The exploration_probability (probability that an ant will ignore pheromone information and choose its path randomly) and the evaporation_rate of the pheromones can be changed via the respective sliders while the model is running. Ants looking for food are colored orange, with exploring ants labelled with an "e". Ants carrying food back to the nest (and laying down pheromones) are colored blue. Two monitors show the pheromone values at the junction. An additional graphic plot (not shown) allows to track and compare the pheromone values of both branches of the path over time.

Acknowledgements.. This work was performed in the course of project SwarmIn supported by FFG under contract number 894072.

References

1. Dorigo, M., Maniezzo, V., Colorni, A.: Ant system: optimization by a colony of cooperating agents. IEEE Trans. Syst. Man Cybern. Part B **26**(1), 29–41 (1996)
2. Goss, S.A., Denebourg, J.L., Pasteels, J.: Self-organized shortcuts in the argentine ant. Naturwissenschaften **76**(12), 579–581 (1989)
3. Hmelo-Silver, C.E., Azevedo, R.: Understanding complex systems: some core challenges. J. Learn. Sci. **15**(1), 53–61 (2006)
4. Papert, S.: What is logo? Who needs it. Logo philosophy and implementation, pp. 4–16 (1999)
5. Wilensky, U.: Netlogo (1999). http://ccl.northwestern.edu/netlogo/
6. Wilensky, U., Jacobson, M.J.: Complex systems and the learning sciences. In: The Cambridge Handbook of the Learning Sciences. 2nd edn., pp. 319–338. Cambridge University Press, Cambridge (2014)
7. Yoon, S.A., Goh, S.E., Park, M.: Teaching and learning about complex systems in k-12 science education: a review of empirical studies 1995–2015. Rev. Educ. Res. **88**(2), 285–325 (2018)

Author Index

A
Agrawal, Swadhin 127
Alkilabi, Muhanad 224
Almansoori, Ahmed 224

B
Babič, Jan 29
Baliyarasimhuni, Sujit P. 127, 248
Belhaddad, Emir Khaled 197
Beltrame, Giovanni 181, 197
Bessone, Nicolas 168
Botasun, Atakan 233
Braga, Rafael Gomes 181

C
Chepizhko, Oleksandr 243
Chiu, Darren 57

D
Dah-Achinanon, Ulrich 197
De Neve, Wesley 84
Diggelen, Fuda van 206
Dorigo, Marco 42, 98, 112

F
Ferrante, Eliseo 98, 155, 206
Forgács, Péter 243

G
Gandhi, Ananya 248
Groß, Roderich 3
Gupta, Himank 98

H
Haghighat, Bahar 57
Hamann, Heiko 127, 246
Harandi, Negin 84
Hauert, Sabine 16, 71

I
Ibrahim, Dalia S. 253
Ingle, Pratik 248

J
Jhawar, Jitesh 127

K
Karagüzel, Tugay Alperen 155, 206

L
Lajoie, Pierre-Yves 42
Lee, Suet 16
Li, Liang 127

M
Mai, Sebastian 215
McConville, Alex 71
Miyauchi, Genki 3
Moroncelli, Angelo 42
Mostaghim, Sanaz 215

N
Nagpal, Radhika 57
Natalizio, Enrico 98
Nishimori, Hiraku 251

O
Oddi, Fabio 141

P
Pacheco, Alexandre 42, 98

R
Ramshanker, Sneha 57
Raoufi, Mohsen 246
Reina, Andreagiovanni 42, 112, 127, 141
Ricard, Guillaume 197
Richardson, Tom 71

Rincon, Andres Garcia   155, 206
Romanczuk, Pawel   246

S
Şahin, Erol   233
Şahin, Mehmet   233
Sajko, Gal   29
Salahshour, Mohammad   112
Salina, Lucio   71
Schranz, Melanie   243, 255
Shiraishi, Masashi   251
Siemensma, Thiemen   57
St-Onge, David   181
Stoy, Kasper   168
Strobel, Volker   42, 98

T
Talamali, Mohamed S.   3
Trendafilov, Dari   224

Trianni, Vito   141
Tuci, Elio   224
Turgut, Ali Emre   233
Tzoumas, Georgios   71

U
Umlauft, Martina   255

V
van Diggelen, Fuda   155
Van Messem, Arnout   84
Vankerschaver, Joris   84
Varadharajan, Vivek Shankar   181
Vardy, Andrew   253

Z
Zahadat, Payam   168
Zakir, Raina   112

SPRINGER NATURE

GPSR Compliance

The European Union's (EU) General Product Safety Regulation (GPSR) is a set of rules that requires consumer products to be safe and our obligations to ensure this.

If you have any concerns about our products, you can contact us on ProductSafety@springernature.com

In case Publisher is established outside the EU, the EU authorized representative is:

Springer Nature Customer Service Center GmbH
Europaplatz 3
69115 Heidelberg, Germany

The manufacturer's authorised representative in the EU is Springer Nature Customer Service Centre GmbH, Europaplatz 3, 69115 Heidelberg, Germany. If you have any concerns regarding our products, please contact ProductSafety@springernature.com

Printed and bound by CPI Group (UK) Ltd, Croydon, CR0 4YY
26/03/2026
02078973-0004